# Hardware Design Verification

# Prentice Hall Modern Semiconductor Design Series

Farzad Nekoogar
*Timing Verification of Application-Specific Integrated Circuits (ASICs)*

Samir Palnitkar
*Design Verification with* **e**

David Pellerin and Scott Thibault
*Practical FPGA Programming in C*

David Pellerin and Douglas Taylor
*VHDL Made Easy!*

Christopher T. Robertson
*Printed Circuit Board Designer's Reference: Basics*

Chris Rowen
*Engineering the Complex SOC*

Frank Scarpino
*VHDL and AHDL Digital System Implementation*

Wayne Wolf
*FPGA-Based System Design*

Wayne Wolf
*Modern VLSI Design: System-on-Chip Design, Third Edition*

Kiat-Seng Yeo, Samir S. Rofail, and Wang-Ling Goh
*CMOS/BiCMOS ULSI: Low Voltage, Low Power*

Brian Young
*Digital Signal Integrity: Modeling and Simulation with Interconnects and Packages*

Bob Zeidman
*Verilog Designer's Library*

# Hardware Design Verification

## Simulation and Formal Method-Based Approaches

**William K. Lam**
**Sun Microsystems**

**PRENTICE HALL**
**Professional Technical Reference**
Upper Saddle River, NJ • Boston • Indianapolis • San Francisco
New York • Toronto • Montreal • London • Munich • Paris • Madrid
Capetown • Sydney • Tokyo • Singapore • Mexico City

The author and publisher have taken care in the preparation of this book, but make no expressed or implied warranty of any kind and assume no responsibility for errors or omissions. No liability is assumed for incidental or consequential damages in connection with or arising out of the use of the information or programs contained herein.

The publisher offers excellent discounts on this book when ordered in quantity for bulk purchases or special sales, which may include electronic versions and/or custom covers and content particular to your business, training goals, marketing focus, and branding interests. For more information, please contact:

U. S. Corporate and Government Sales
(800) 382-3419
corpsales@pearsontechgroup.com

For sales outside the U. S., please contact:

International Sales
international@pearsoned.com

Visit us on the Web: www.phptr.com

*Library of Congress Cataloging-in-Publication Data*

Hardware design verification : simulation and formal method-based approaches / William K. Lam.
    p.  cm.
  Includes bibliographical references and indexes.
  ISBN 0-13-143347-4 (hardcover : alk. paper)
  1. Integrated circuits—Verification.  I. Lam, William K. C.  II. Title.

TK7874.58 .L36 2005
621.39/2 22

                                                                    2004026386

ISBN 0-13-143347-4
Text printed in the United States on recycled paper at Courier in Westford, Massachusetts.
First printing, February 2005

*To Grace, James, Serene, and Rachel*

# Contents

# Preface

Two groups of people are essential to a successful design project: a design team and a verification team. Designers usually have formal training from schools. Many colleges have comprehensive curricula on logic design, ranging from introduction to digital design to advanced computer architecture. On the contrary, most verification engineers learn their trade on the job; few were educated as verification engineers in academia, although many schools are beginning to teach verification curriculum. In fact, a majority of verification engineers started out as designers and gradually transitioned to design verification. Unlike design techniques and methodologies, a broad range of verification knowledge is loosely organized and informally acquired through hands-on experience. Furthermore, the horizon of verification has been expanding at a rapid pace: The verification landscape is evolving every six months with new techniques, standards, and tools. Nevertheless, there are principles and techniques that have survived the test of time and prove to be cornerstones of verification.

This book collects and organizes a wide range of digital design verification techniques and methodologies commonly used in industry and presents them in a systematic fashion. The focus of the book is on digital logic design and verification. It does not cover verification of circuits with mixed-signal or radio frequency components. A goal of the book is to pass the vast amount of verification knowledge to college students and engineers so that they are better prepared for the workforce, and to speed up their learning pace. It is tempting to write a ten-minute verification engineering book that lists detailed practical tips that can be used immediately on the job. These quick-fix tips often become obsolete over a short period of time. On the other hand, presenting the principles only will not be immensely useful for practicing engineers. Thus, I decided to strike a balance between the two and present verification principles as well as common practices. It is my belief that only by understanding the principles can one truly grasp the essence of the practices and be creative about using them.

## To the Audience

One targeted audience for this book includes undergraduate students at the junior or senior level or first-level graduate students. I will assume that these readers have a sound understanding of a hardware descriptive language, preferably Verilog, because Verilog is

used for illustration purposes in most parts of the book. In addition, a rudimentary knowledge of logic design is beneficial. This book is an introduction to design verification.
Through its study, students will learn the main ideas, tools, and methodologies used in
simulation-based verification, and the principles behind formal verification. The materials presented are industry tested and widely used. At the end of each chapter, problems
are presented as a means of refreshing the knowledge covered in the chapter. For those
who want to explore certain topics in more depth, please refer to the citations listed in the
bibliography.

The other targeted audience includes verification professionals who may have some verification experience but would like to get a systematic overview of the different areas in verification and an understanding of the basic principles behind formal verification. The
prerequisites for this audience are similar to those for the first group: basic design knowledge and a hardware descriptive language. The first part of the book provides a comprehensive treatment of simulation-based verification methodology and serves as a refresher or an
introduction to verification professionals. For many practicing engineers, formal verification tools appear to be a form of black magic that requires a doctorate in mathematics to
comprehend. This book explains these tools and the working principles of formal verification. Before formal verification is discussed in depth, Chapter 7 reviews the basics of mathematics and computer algorithms required to understand verification algorithms fully.

As with every technology, there is no substitution for hands-on experience. I encourage
you to get acquainted with verification tools by running some of the examples and problems in the book or designing a project that makes use of the tools described. Free CAD
tools such as Verilog simulator, test bench development aid, and waveform viewer are
available at www.verilog.net/free.html.

## To the Instructor
This book is comprised of two parts. The first part addresses the conventional verification
strategy—simulation-based verification—whereas the second part addresses the aspects
of formal verification that are well established in academia and tested in industry. The two
parts are self-contained and thus can be taught independently as deemed fit.

The first part describes many verification tools. Because the specifics of these tools differ
from vendor to vendor and change from time to time, only pseudo common commands
are used in this book. To reinforce learning these tools, I recommend that industrial tools
be part of a verification laboratory (for instance, a simulator, a waveform viewer, a coverage tool, a bug tracking system, and a revision control system). Similarly, to solidify the
knowledge of formal verification, commercial formal verification tools such as an HDL
linter, a model checker, and an equivalence checker should be used during lab sessions.

This book has a companion instructor's manual that contains the solutions to all the problems at the end of each chapter. The instructor's manual is available on request to the publisher.

## Organization of the Book

Again, this book is intended as an introduction to design verification. I will assume that you have an understanding of basic Verilog constructs. Even though the book is written using Verilog as the hardware description language, I made an effort to present ideas independent of Verilog. When the use of Verilog is unavoidable, the simplest Verilog constructs are used to allow readers unfamiliar with the language to grasp the main ideas.

As previously stated, this book consists of two parts. The first part is devoted to simulation-based verification and the second part discusses formal verification. Simulation-based verification is by far the most widely used methodology and is a necessary requirement for all verification engineers. Formal verification is a relatively new technology and it complements simulation-based verification. I believe that to utilize a technology best, one must first be equipped with an in-depth understanding of the internal working principles of that technology. As a result, instead of just studying a verification tool's operations at the user level—a topic better suited for user manuals—this book spends much time studying the fundamental principles of simulation and formal technology.

The first part—simulation technology—consists of Chapters 2 through 6, and these chapters are ordered similar to the usual sequence of operations encountered during a simulation verification process. We start with Chapter 2, checking for static errors. These are errors that can be detected without input vectors and must be eliminated before extensive simulation begins. In Chapter 3, we study the basic architectures of simulators. Event-driven and cycle-based simulation algorithms are presented first, followed by simulator operations and applications for which we discuss cosimulation, design profiling, common simulator options, and the user interface.

Before one can begin a simulation, one must construct a test bench to host the design. Chapter 4 discusses test bench design, initialization, stimuli generation, clock network, error injection, result assessment, and test configuration. After a design is free of static errors and is embedded in a test bench, it is ready to be simulated. But how should the design be simulated? Chapter 5 addresses the issues of what to simulate and how to measure the quality of the simulation. We will look at test plan design, generation of tests for the items in the test plan, output response assessment, assertion (particularly, SystemVerilog assertions), and verification coverage.

After a circuit is simulated and the bugs are discovered, the next step is to debug problems found during simulation. Chapter 6 presents widely used debugging techniques, including case reduction, check pointing, error tracing, trace dumping, and forward and backward debugging. In addition, Chapter 6 examines the four basic views of design: source code, schematic, waveform, and finite-state machine. We then examine the scenario after the bugs are fixed. The tools and methodology discussed include the revision control system, regression mechanism, and tape-out criteria.

The second part of the book—formal verification—consists of Chapters 7 through 9. Several chapters in the first part of the book can be studied in conjunction with the second. For example, Chapters 2, 4, 5, and 6 are also applicable to formal verification.

The key to understanding formal verification lies in an understanding of the theory behind it. Chapter 7 provides the basic mathematical background for the later chapters. The materials cover Boolean functions and representations, symmetric Boolean functions, finite-state machines and the equivalence algorithm, and graphic algorithms such as depth-first search and breadth-first search, and strongly connected components.

Chapter 8 presents a survey of decision diagrams, with an emphasis on binary decision diagrams. We then review SAT (satisfiability) as an alternative to decision diagrams. The chapter concludes with a look at applications of decision diagrams and SAT in equivalence checking and symbolic simulation.

Chapter 9 presents an in-depth study of symbolic model checking. First it presents automata and computational tree logic as a means of modeling temporal behavior with fairness constraints. It then discusses algorithms for checking a model against a temporal specification. Based on the model-checking algorithms, efficient symbolic model-checking algorithms are presented in which graphical operations are accomplished through Boolean function computations. Next, equivalence checking is revisited for general circuits for which one-to-one state correspondence does not exist. To conclude, algorithms for better managing symbolic computation are studied.

## Errata

A book of this size is bound to have errors and areas for improvement. Please report errors and send comments to the author at williamlamemail@gmail.com.

**Simulation part**                    **Formal verification part**

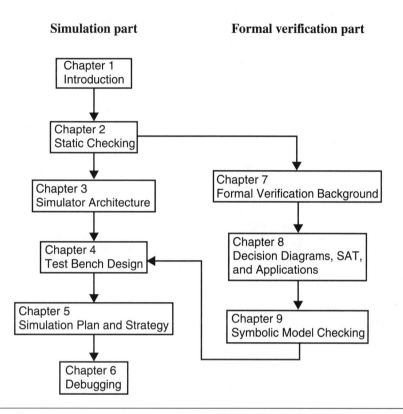

**Figure P.1**    An overview of the chapters

# Acknowledgments

First and foremost I thank my wife, Serene, for her patience, understanding, and encouragement. Her love and support sustained me through this long and lonely journey. I also thank my parents, Grace and James Lam, for their continuing guidance and support.

I am grateful to the many reviewers who spent their valuable time and effort reading the manuscript and pointing out errors and improvements. In particular, I thank Rajeev Alur, K. C. Chen, Thomas Dillinger, Manoj Gandhi, Yu-Chin Hsu, Sunil Joshi, Shrenik Mehta, Vigyan Singhal, Ed Liu, Paul Tobin, and John Zook for their insightful comments, suggestions, and encouraging words. The Web site for Verilog freeware was suggested by Thomas Dillinger. Furthermore, I appreciate the enthusiasm and encouragement of Bernard Goodwin, editor at Prentice Hall, and the professionalism of the editing staff at Prentice Hall.

# About the Author

William K. Lam has held senior manager and senior staff engineer positions at Sun Microsystems. During his years at Sun and Hewlett-Packard, he has worked on design verification, low power design and analysis, logic synthesis, and ATPG. In addition, he designed ICs and PC boards for HP test and measurement instruments. William received the 2002 Sun Microsystems Chairman's Award for Innovation; the 1994 D. J. Sakrison Award for distinguished Ph.D. thesis from the EECS department at the University of California, Berkeley; and a John Hertz foundation doctoral fellowship during graduate study.

He has two books and an extensive collection of publications and U.S. patents to his credit.

# An Invitation to Design Verification

## Chapter Highlights

- What is design verification?

- The basic verification principle

- Verification methodology

- Simulation-based verification versus formal verification

- Limitations of formal verification

- A quick overview of Verilog scheduling and execution semantics

From the makeup of a project team, we can see the importance of design verification. A typical project team usually has an equal number of design engineers and verification engineers. Sometimes verification engineers outnumber design engineers by as many as two to one. This stems from the fact that to verify a design, one must first understand the specifications as well as the design and, more important, devise a different design approach from the specifications. It should be emphasized that the approach of the verification engineer is different from that of the design engineer. If the verification engineer follows the same design style as the design engineer, both would commit the same errors and not much would be verified.

From the project development cycle, we can understand the difficulty of design verification. Statistical data show that around 70% of the project development cycle is devoted to design verification. A design engineer usually constructs the design based on cases representative of the specifications. However, a verification engineer must verify the design under all cases, and there are an infinite number of cases (or so it seems). Even with a heavy investment of resources in verification, it is not unusual for a reasonably complex chip to go through multiple tape-outs before it can be released for revenue.

The impact of thorough design verification cannot be overstated. A faulty chip not only drains budget through respin costs, it also delays time-to-market, impacts revenue, shrinks market share, and drags the company into playing the catch-up game. Therefore, until people can design perfect chips, or chip fabrication becomes inexpensive and has a fast turnaround time, design verification is here to stay.

## 1.1    What Is Design Verification?

A design process transforms a set of specifications into an implementation of the specifications. At the specification level, the specifications state the functionality that the design executes but does not indicate how it executes. An implementation of the specifications spells out the details of how the functionality is provided. Both a specification and an implementation are a form of description of functionality, but they have different levels of concreteness or abstraction. A description of a higher level of abstraction has fewer details; thus, a specification has a higher level of abstraction than an implementation. In an abstraction spectrum of design, we see a decreasing order of abstraction: functional specification, algorithmic description, register–transfer level (RTL), gate netlist, transistor netlist, and layout (Figure 1.1). Along this spectrum a description at any level can give rise to many forms of a description at a lower level. For instance, an infinite number of circuits at the gate level implements the same RTL description. As we move down the ladder, a less abstract description adds more details while preserving the descriptions at higher levels. The process of turning a more abstract description into a more concrete description is called *refinement*. Therefore, a design process refines a set of specifications and produces various levels of concrete implementations.

Design verification is the reverse process of design. Design verification starts with an implementation and confirms that the implementation meets its specifications. Thus, at every step of design, there is a corresponding verification step. For example, a design step that turns a functional specification into an algorithmic implementation requires a verification step to ensure that the algorithm performs the functionality in the specification. Similarly, a physical design that produces a layout from a gate netlist has to be verified to ensure that the layout corresponds to the gate netlist. In general, design verification encompasses many areas, such as functional verification, timing verification, layout verification, and electrical verification, just to name a few. In this book we study only functional verification and refer to it as *design verification*. Figure 1.2 shows the relationship between the design process and the verification process.

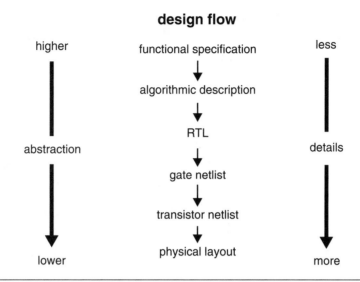

**Figure 1.1**   A ladder of design abstraction

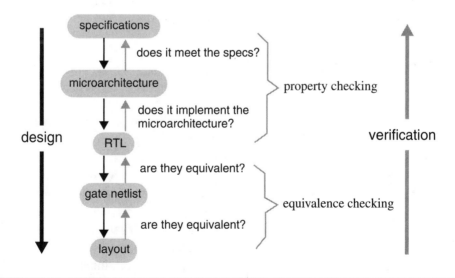

**Figure 1.2**   The relationship between design and verification

On a finer scope, design verification can be further classified into two types. The first type verifies that two versions of design are functionally equivalent. This type of verification is called *equivalence checking*. One common scenario of equivalence checking is comparing two versions of circuits at the same abstraction level. For instance, compare the gate netlist of a prescan circuit with its postscan version to ensure that the two are equivalent under normal operating mode.

However, the two versions of the design differ with regard to abstraction level. For example, one version of the design is at the level of specification and the other version is at the gate netlist level. When the two versions differ substantially with regard to level of abstraction, they may not be functionally equivalent, because the lower level implementation may contain more details that are allowed, but are unspecified, at the higher level. For example, an implementation may contain timing constraints that are not part of the original specification. In this situation, instead of verifying the functional equivalence of the two versions, we verify instead that the implementation satisfies the specifications. Note that equivalence checking is two-way verification, but this is a one-way verification because a specification may not satisfy an unspecified feature in the implementation. This type of verification is known as *implementation verification, property checking*, or *model checking*. Based on the terminology of property checking, the specifications are properties that the implementation must satisfy. Based on the terminology of model checking, the implementation or design is a model of the circuit and the specifications are properties. Hence, model checking means checking the model against the properties.

## 1.2  The Basic Verification Principle

There are two types of design error. The first type of error exists not in the specifications but in the implementations, and it is introduced during the implementation process. An example is human error in interpreting design functionality. To prevent this type of error, we can use a software program to synthesize an implementation directly from the specifications. Although this approach eliminates most human errors, errors can still result from bugs in the software program, or usage errors of the software program may be encountered. Furthermore, this synthesis approach is rather limited in practice for two reasons. First, many specifications are in the form of casual conversational language, such as English, as opposed to a form of precise mathematical language, such as Verilog or C++. We know that automatic synthesis from a loose language is infeasible. In fact, as of this writing, there is no high-level formal language that specifies both functional and timing requirements. A reason for this is that a high-level functional requirement does not lend itself to timing requirements, which are more intuitive at the implementation level. Therefore, timing requirements such as delay and power, when combined with high-level functional specifications, are so overtly inaccurate that people relegate timing specifications to levels of lower abstraction. Second, even if the specifications are written in a precise mathematical language, few synthesis software programs can produce implementations that meet all requirements. Usually, the software program synthesizes from a set of functional specifications but fails to meet timing requirements.

Another method—the more widely used method—to uncover errors of this type is through redundancy. That is, the same specifications are implemented two or more times using different approaches, and the results of the approaches are compared. In theory, the more times and the more different ways the specifications are implemented, the higher the confidence produced by the verification. In practice, more than two approaches is rarely used, because more errors can be introduced in each alternative verification, and costs and time can be insurmountable.

The design process can be regarded as a path that transforms a set of specifications into an implementation. The basic principle behind verification consists of two steps. During the first step, there is a transformation from specifications to an implementation. Let us call this step *verification transformation*. During the second step, the result from the verification is compared with the result from the design to detect any errors. This is illustrated in Figure 1.3 (A). Oftentimes, the result from a verification transformation takes place in the head of a verification engineer, and takes the form of the properties deduced from the specifications. For instance, the expected result for a simulation input vector is calculated by a verification engineer based on the specifications and is an alternative implementation.

Obviously, if verification engineers go through the exact same procedures as the design engineers, both the design and verification engineers are likely to arrive at the same conclusions, avoiding and committing the same errors. Therefore, the more different the design and verification paths, the higher confidence the verification produces. One way to achieve high confidence is for verification engineers to transform specifications into an implementation model in a language different from the design language. This language is called *verification language*, as a counterpart to design language. Examples of verification languages include Vera, C/C++, and *e*. A possible verification strategy is to use C/C++ for the verification model and Verilog/VHSIC Hardware Description Language (VHDL) for the design model.

During the second step of verification, two forms of implementation are compared. This is achieved by expressing the two forms of implementation in a common intermediate form so that equivalency can be checked efficiently. Sometimes, a comparison mechanism can be sophisticated—for example, comparing two networks with arrival packets that may be out of order. In this case, a common form is to sort the arrival packets in a predefined way. Another example of a comparison mechanism is determining the equivalence between a transistor-level circuit and an RTL implementation. A common intermediate form in this case is a binary decision diagram.

Here we see that the classic simulation-based verification paradigm fits the verification principle. A simulation-based verification paradigm consists of four components: the circuit, test patterns, reference output, and a comparison mechanism. The circuit is simulated on the test patterns and the result is compared with the reference output. The implementation result from the design path is the circuit, and the implementation results from the verification path are the test patterns and the reference output. The reason for considering the test patterns and the reference output as implementation results from the verification path is that, during the process of determining the reference output from the test patterns, the verification engineer transforms the test patterns based on the specifications into the reference output, and this process is an implementation process. Finally, the comparison mechanism samples the simulation results and determines their equality with the reference output. The principle behind simulation-based verification is illustrated in Figure 1.3 (C).

Verification through redundancy is a double-edged sword. On the one hand, it uncovers inconsistencies between the two approaches. On the other hand, it can also introduce incompatible differences between the two approaches and often verification errors. For example, using a C/C++ model to verify against a Verilog design may force the verification engineer to resolve fundamental differences between the two languages that otherwise could be avoided. Because the two languages are different, there are areas where one language models accurately whereas the other cannot. A case in point is modeling timing and parallelism in the C/C++ model, which is deficient. Because design codes are susceptible to errors, verification code is equally prone to errors. Therefore, verification engineers have to debug both design errors as well as verification errors. Thus, if used carelessly, redundancy strategy can end up making engineers debug more errors than those that exist in the design—design errors plus verification errors—resulting in large verification overhead costs.

As discussed earlier, the first type of error is introduced during an implementation process. The second type of error exists in the specifications. It can be unspecified functionality, conflicting requirements, and unrealized features. The only way to detect the type of error is through redundancy, because specification is already at the top of the abstraction hierarchy and thus there is no reference model against which to check. Holding a design review meeting and having a team of engineers go over the design architecture is a form of verification through redundancy at work. Besides checking with redundancy directly, examining the requirements in the application environment in which the design will

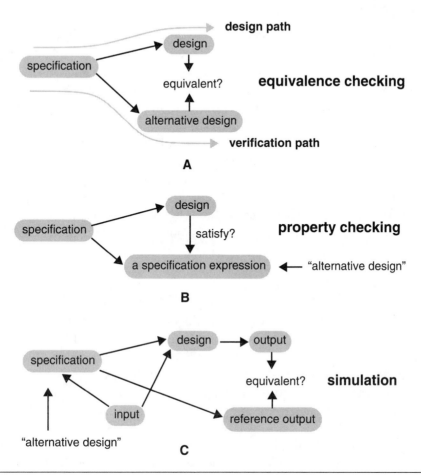

**Figure 1.3**   The basic principle of design verification. (A) The basic methodology of verification by redundancy. (B) A variant of the basic methodology adapted in model checking. (C) Simulation methodology cast in the form of verification by redundancy.

reside when it has become a product also detects bugs during specification, because the environment dictates how the design should behave and thus serves as a complementary form of design specification. Therefore, verifying the design requirements against the environment is another form of verification through redundancy. Furthermore, some of these types of errors will eventually be uncovered as the design takes a more concrete form. For example, at a later stage of implementation, conflicting requirements will surface as consistencies, and features will emerge as unrealizable given the available technologies and affordable resources.

## 1.3    Verification Methodology

A sound verification methodology starts with a test plan that details the specific functionality to verify so that the specifications are satisfied. A test plan consists of features, operations, corner cases, transactions, and so forth, such that the completion of verifying them
constitutes verifying the specifications. To track progress against a test plan, a "scoreboarding" scheme is used. During scoreboarding, the items in the test plan are checked off
when they are completely verified. In reality, it is infeasible to verify a set of specifications
completely, in the sense that all bugs are uncovered. Thus, a measure of verification quality is desirable. Some commonly used coverage metrics are functional coverage and code
coverage. Functional coverage approximates the percentage of functionality verified,
whereas code coverage measures the percentage of code simulated. Neither functional
coverage nor code coverage directly corresponds to how thoroughly bugs are eliminated.
To compensate this shortcoming, bug rate and simulation cycles are widely used to gauge
bug absence.

Besides a test plan, a verification methodology has to decide what language or languages
should be used during verification. Verilog or VHDL are common design languages. Verification code often demands a language of its own, because the code is usually of a higher
level and does not have to conform to strict coding style guidelines like design code. For
example, arithmetic operators occur often in verification code but seldom in design code.
On the contrary, design code usually has to be able to be synthesized, but verification code
does not. Hence, an ideal verification language resembles a software language more than a
hardware language. Popular verification languages include Vera, e, C/C++, and Java.

To study verification methodologies further, let's group verification methodologies into two
categories: simulation-based and formal method-based verification. The distinguishing
factor between the two categories is the existence or absence of vectors. Simulation-based
verification methodologies rely on vectors, whereas formal method-based verification
methodologies do not. Another way to distinguish simulation-based and formal method
based verification is that simulation is input oriented (for example, the designer supplies
input tests) and formal method verification is output oriented (for example, the designer
supplies the output properties to be verified). There is a hybrid category called *semiformal
verification* that takes in input vectors and verifies formally around the neighborhood of
the vectors. Because the semiformal methodology is a combination of simulation-based
and formal technology, it is not described separately here.

### 1.3.1    Simulation-Based Verification

The most commonly used verification approach is simulation-based verification. As mentioned earlier, simulation-based verification is a form of verification by redundancy. The

variant or the alternative design manifests in the reference output. During simulation-based verification, the design is placed under a test bench, input stimuli are applied to the test bench, and output from the design is compared with reference output. A test bench consists of code that supports operations of the design, and sometimes generates input stimuli and compares the output with the reference output as well. The input stimuli can be generated prior to simulation and can be read into the design from a database during simulation, or it can be generated during a simulation run. Similarly, the reference output can be either generated in advance or on the fly. In the latter case, a reference model is simulated in lock step with the design, and results from both models are compared.

Before a design is simulated, it runs through a linter program that checks static errors or potential errors and coding style guideline violations. A linter takes the design as input and finds design errors and coding style violations. It is also used to glean easy-to-find bugs. A linter does not require input vectors; hence, it checks errors that can be found independent of input stimuli. Errors that require input vectors to be stimulated will escape linting. Errors are static if they can be uncovered without input stimuli. Examples of static errors include a bus without a driver, or when the width of a port in a module instantiation does not match the port in the module definition. Results from a linter can be just alerts to potential errors. A potential error, for example, is a dangling input of a gate, which may or may not be what the designer intended. A project has its own coding style guidelines enforced to minimize design errors, improve simulation performance, or for other purposes. A linter checks for violations of these guidelines.

Next, input vectors of the items in the test plan are generated. Input vectors targeting specific functionality and features are called *directed tests*. Because directed tests are biased toward the areas in the input space where the designers are aware, bugs often happen in areas where designers are unaware; therefore, to steer away from these biased regions and to explore other areas, pseudorandom tests are used in conjunction with directed tests. To produce pseudorandom tests, a software program takes in seeds, and creates tests and expected outputs. These pseudorandomly generated tests can be vectors in the neighborhood of the directed tests. So, if directed tests are points in input space, random tests expand around these points.

After the tests are created, simulators are chosen to carry out simulation. A simulator can be an event-driven or cycle-based software or hardware simulator. An event simulator evaluates a gate or a block of statements whenever an input of the gate or the variables to which the block is sensitive change values. A change in value is called an *event*. A cycle-based simulator partitions a circuit according to clock domains and evaluates the subcircuit in a clock domain once at each triggering edge of the clock. Therefore, event count affects the speed a simulator runs. A circuit with low event counts runs faster on event-driven simulators, whereas a circuit with high event counts runs faster on cycle-based simulators. In practice, most circuits have enough events that cycle-based simulators out-perform their event-driven counterparts. However, cycle-based simulators have their own shortcomings. For a circuit to be simulated in a cycle-based simulator, clock domains in the circuit must be well defined. For example, an asynchronous circuit does not have a clear clock domain definition because no clock is involved. Therefore, it cannot be simulated in a cycle-based simulator.

A hardware simulator or emulator models the circuit using hardware components such as processor arrays and field programmable gate arrays (FPGAs). First, the components in a hardware simulator are configured to model the design. In a processor array hardware simulator, the design is compiled into instructions of the processors so that executing the processors is tantamount to simulating the design. In an FPGA-based hardware acceleration, the FPGAs are programmed to mimic the gates in the design. In this way, the results from running the hardware are simulation results of the design. A hardware simulator can be either event driven or cycle based, just like a software simulator. Thus, each type of simulator has its own coding style guidelines, and these guidelines are more strict than those of software simulators. A design can be run on a simulator only if it meets all coding requirements of the simulator. For instance, statements with delays are not permitted on a hardware simulator. Again, checking coding style guidelines is done through a linter.

The quality of simulating a test on a design is measured by the coverage the test provides. The coverage measures how much the design is stimulated and verified. A coverage tool reports code or functional coverage. Code coverage is a percentage of code that has been exercised by the test. It can be the percentage of statements executed or the percentage of branches taken. Functional coverage is a percentage of the exercised functionality. Using a coverage metric, the designer can see the parts of a design that have not been exercised, and can create tests for those parts. On the other hand, the user could trim tests that duplicate covered parts of a circuit.

When an unexpected output is observed, the root cause has to be determined. To determine the root cause, waveforms of circuit nodes are dumped from a simulation and are viewed through a waveform viewer. The waveform viewer displays node and variable values over time, and allows the user to trace the driver or loads of a node to determine the root cause of the anomaly.

When a bug is found, it has to be communicated to the designer and fixed. This is usually done by logging the bug into a bug tracking system, which sends a notification to the owner of the design. When the bug is logged into the system, its progress can be tracked. In a bug tracking system, the bug goes through several stages: from opened to verified, fixed, and closed. It enters the opened stage when it is filed. When the designer confirms that it is a bug, he moves the bug to the verified stage. After the bug is eradicated, it goes to the fixed stage. Finally, if everything works with the fix, the bug is resolved during the closed stage. A bug tracking system allows the project manager to prioritize bugs and estimate project progress better.

Design codes with newly added features and bug fixes must be made available to the team. Furthermore, when multiple users are accessing the same data, data loss may result (for example, two users trying to write to the same file). Therefore, design codes are maintained using revision control software that arbitrates file access to multiple users and provides a mechanism for making visible the latest design code to all.

The typical flow of simulation-based verification is summarized in Figure 1.4. The components inside the dashed enclosure represent the components specific to the simulation-based methodology. With the formal verification method, these components are replaced by those found in the formal verification counterparts.

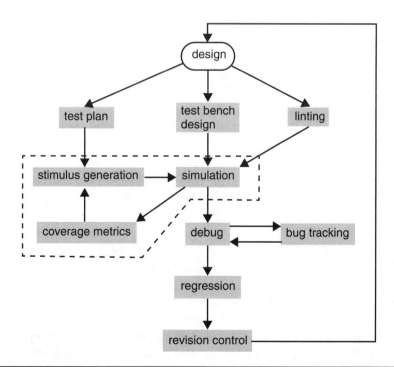

**Figure 1.4**   Flow of simulation-based verification

## 1.3.2    Formal Method-Based Verification

The formal method-based verification methodology differs from the simulation-based methodology in that it does not require the generation of test vectors; otherwise, it is similar. Formal verification can be classified further into two categories: equivalence checking and property verification.

Equivalence checking determines whether two implementations are functionally equivalent. Checking equivalence is infeasible using a simulation-based methodology because there are practically an infinite number of vectors in the input space. During formal verification, the decision from a equivalence checker is clear-cut. However, industrial equivalence checkers have not yet reached the stage of turnkey solution and often require the user to identify equivalent nodes in the two circuits to limit the input search space for the checkers. These user-identified equivalent nodes are called *cut points*. At times, a checker can infer equivalent nodes from the node names.

Two basic approaches are behind equivalence checking. The first approach searches the input space in a systematic way for an input vector or vectors that would distinguish the two circuits. This approach, called SAT (satisfiability), is akin to automatic test pattern generation algorithms. The other approach converts the two circuits into canonical representations, and compare them. A canonical representation has the characteristic property that two logical functions are equivalent if and only if their respective representations are isomorphic—in other words, the two representations are identical except for naming. A reduced ordered binary decision diagram is a canonical representation. Binary decision diagrams of two equivalent functions are graphically isomorphic.

Equivalence checking is most widely used in the following circumstances:

1. Compare circuits before and after scan insertion to make sure that adding scan chains does not alter the core functionality of the circuit.
2. Ensure the integrity of a layout versus its RTL version. This is accomplished by first extracting from the layout a transistor netlist and comparing the transistor netlist with the RTL version.
3. Prove that changes in an engineering change order (ECO) check-in are restricted to the scope intended. This is done by identifying the changed regions in the circuits.

If an equivalence check fails, the checker generates an input sequence of vectors that, when simulated, demonstrates the differences between the two circuits. From the waveforms, the user debugs the differences. A failure can result from a true error or an unintended boundary condition. An unintended boundary condition arises when the scan circuitry is not disabled. Then a checker will generate a sequence of vectors that runs the postscan circuit in test mode while running the prescan circuit in normal operating mode. Surely these two circuits are not equivalent under this input sequence. Therefore, when comparing a prescan circuit and a postscan circuit, the postscan circuit must be configured in normal operating mode.

The other type of formal verification is property checking. Property checking takes in a design and a property which is a partial specification of the design, and proves or disproves that the design has the property. A property is essentially a duplicate description of the design, and it acts to confirm the design through redundancy. A program that checks a property is also called a *model checker*, referring the design as a computational model of a real circuit. Model checking cast in the light of the basic verification principle is visualized in Figure 1.3 (B).

The idea behind property checking is to search the entire state space for points that fail the property. If a such point is found, the property fails and the point is a counterexample. Next, a waveform derived from the counterexample is generated, and the user debugs the failure. Otherwise, the property is satisfied. What makes property checking a success in industry are symbolic traversal algorithms that enumerate the state space implicitly. That is, it visits not one, but a group of points at a time, and thus is highly efficient.

Even though symbolic computation has taken a major stride in conquering design complexity, at the time of this writing only the part of the design relevant to the property should be given to a property verifier, because almost all tools are incapable of automatically carving out the relevant part of the design and thus almost certainly will run into memory and runtime problems if the entire design is processed.

Furthermore, the power and efficiency of a property verifier is highly sensitive to the property being verified. Some properties can be decided readily whereas others will never finish or may not even be accepted by the tool. For example, some property verifiers will not accept properties that are unbound in time (for example, an acknowledgment signal will eventually be active). A bound property is specified within a fixed time interval (for example, an acknowledgment signal will be active within ten cycles).

A failure can result from a true design bug, a bug in the property or unintended input, or state configurations. Because a property is an alternative way of expressing the design, it is as equally prone to errors as the design. A failure can be caused by unintended input or state configurations. If the circuit under verification is carved out from the design, the input waveforms to the circuit must be configured to be identical to them when the circuit is embedded in the design. Failure to do so sends unexpected inputs (for example, inputs that would not occur when the circuit is a part of the design) to the circuit, and the circuit may fail over these unexpected inputs because the circuit has not been designed to handle them. To remove unintended configurations from interfering with property verification, the inputs or states are constrained to produce only the permissible input waveforms. Underconstraining means that input constraining has not eliminated all unexpected inputs, and this can trigger false failures. Overconstraining, which eliminates some legal inputs, is much more dangerous because it causes false successes and the verifier will not alert the user.

Some practical issues with property verifiers include long iteration times in determining the correct constraining parameters, and debugging failures in the properties and in the design. Because only a portion of the design is given to the property verifier, the environment surrounding that portion has to be modeled correctly. Practical experience shows that a large percentage of time (around 70 percent) in verification is spent getting the correct constraints. Second, debugging a property can be difficult when the property is written in a language other than that of the design (for instance, the properties are in SystemVerilog whereas the design is in Verilog). This is because many property verifiers internally translate the properties into finite-state machines in Verilog, and verify them against the design. Therefore, the properties are shown as the generated machine code, which is extremely hard to relate to signals in the design or is hard to interpret its meaning. Finally, debugging failures in a property has the same difficulty as any other debugging.

As an alternative to a state space search in verifying a property, the theorem-proving approach uses deductive methods. During theorem proving, a property is specified as a mathematical proposition and the design, also expressed as mathematical entities, is treated as a number of axioms. The objective is to determine whether the proposition can be deduced from the axioms. If it can, the property is proved; otherwise, the property fails. A theorem prover is less automatic than a model checker and is more of an assistance to the user. The user drives the tool to arrive at the proposition by assembling relevant information and by setting up intermediate goals (in other words, lemmas), whereas the tool attempts to achieve the intermediate goals based on the input data.

Effective use of a theorem prover requires a solid understanding of the internal operations of the tool and a familiarity with the mathematical proof process. Although less automatic, efficient usage of a theorem prover can handle much larger designs than model checkers and requires less memory. Furthermore, a theorem prover accepts more complex properties. For example, a theorem prover can allow the properties written in higher order logic (HOL) whereas almost all model checkers only accept first-order logic and computation tree logic or their variants. HOL is more expressive than first-order logic and enables a concise description of complex properties.

A typical flow of formal verification is shown in Figure 1.5. Remember that the components in this diagram replace the components inside the dashed box of Figure 1.4. Although

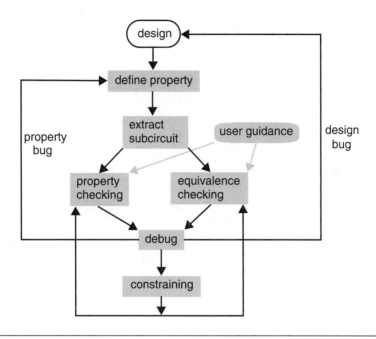

**Figure 1.5** A typical flow of formal verification

formal verification does not require test vectors, it does require a test bench that configures the circuit in the correct operational mode, such as input constraining.

## 1.4 Simulation-Based Verification versus Formal Verification

The most prominent distinction between simulation-based verification and formal verification is that the former requires input vectors and the latter does not. The mind-set in simulation-based verification is first to generate input vectors and then to derive reference outputs. The thinking process is reversed in the formal verification process. The user starts out by stating what output behavior is desirable and then lets the formal checker prove or disprove it. Users do not concern themselves with input stimuli at all. In a way, the simulation-based methodology is input driven and the formal methodology is output driven. It is often more straightforward to think in the input-driven way, and this tendency is reflected in the perceived difficulty in using a formal checker.

Another selling point for formal verification is completeness, in the sense that it does not miss any point in the input space—a problem from which simulation-based verification suffers. However, this strength of formal verification sometimes leads to the misconception that once a design is verified formally, the design is 100% free of bugs. Let's compare simulation-based verification with formal verification and determine whether formal verification is perceived correctly.

Simulating a vector can be conceptually viewed as verifying a point in the input space. With this view, simulation-based verification can be seen as verification through input space sampling. Unless all points are sampled, there exists a possibility that an error escapes verification. As opposed to working at the point level, formal verification works at the property level. Given a property, formal verification exhaustively searches all possible input and state conditions for failures. If viewed from the perspective of output, simulation-based verification checks one output point at a time; formal verification checks a group of output points at a time (a group of output points make up a property). Figure 1.6 illustrates this comparative view of simulation-based verification and formal verification. With this perspective, the formal verification methodology differs from the simulation-based methodology by verifying *groups* of points in the design space instead of *points*. Therefore, to verify completely that a design meets its specifications using formal methods, it must be further proved that the set of properties formally verified collectively constitutes the specifications. The fact that formal verification checks a group of points at a time makes formal verification software less intuitive and thus harder to use.

A major disadvantage of formal verification software is its extensive use of memory and (sometimes) long runtime before a verification decision is reached. When memory capacity is exceeded, tools often shed little light on what went wrong, or give little guidance to fix

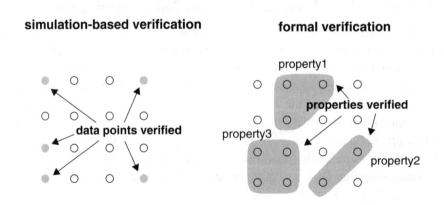

**Figure 1.6**   An output space perspective of simulation-based verification versus formal verification

the problem. As a result, formal verification software, as of this writing, is applicable only to circuits of moderate size, such as blocks or modules.

## 1.5  Limitations of Formal Verification

Formal verification, although offering exhaustive checking on properties, by no means guarantees complete functional correctness of design. Many factors can betray the confidence offered by formal verification. These factors include, but are in no way limited to, the following:

1. Errors in specification. An implementation verification requires two versions of a design. Therefore, a bug in the specifications voids the reference model, may escape implementation verification, and may trickle down to implementations. A typical type of specification error is missing specifications.
2. Incomplete functional coverage of specifications. Formal methods verify properties that, of course, can be specifications. In practice, however, it is rare that all specifications are given to a formal checker on the complete design, because of memory and runtime limitations. A partial specification is checked against the relevant portion of the design. Therefore, it leaves the door open to errors resulting from not checking all properties mandated by the specifications and from situations when, for example, a property succeeds on a subcircuit but does not succeed on the whole circuit.
3. User errors. Example user errors are incorrect representation of specifications and overconstraining the design.
4. Formal verification software bugs. Bugs in formal verification software can miss design errors and thus give false confirmation. It is well-known that there is no guarantee for a bug-free software program.

## 1.6  A Quick Overview of Verilog Scheduling and Execution Semantics

On the assumption that you have a good understanding of Verilog, we will not review the general syntax of Verilog. However, we will review the more advanced features of the language—namely, scheduling and execution semantics—they appear in various places throughout this book. If you are well versed in Verilog, you can skip this section.

### 1.6.1  Concurrent Processes

Verilog is a parallel hardware descriptive language. It differs from a software language such as C/C++ in that it models concurrency of hardware. Components in a circuit always execute in parallel, whereas a software program executes one statement at time—in a serial fashion. To model hardware concurrency, Verilog uses two types of data structures: continuous assignment and procedural block. A continuous assignment executes whenever a variable in the right side of the assignment changes. As the name implies, a continuous assignment continuously watches changes of variables on the right side and updates the left-side variables whenever changes occur. This behavior is consistent with that of a gate, which propagates inputs to outputs whenever inputs change. An example of a continuous assignment is shown here. The value of $v$ is updated whenever the value of $x$, $y$, or $z$ changes:

```
assign v = x + y + z;
```

A procedural block consists of statements called *procedural statements*. Procedural statements are executed sequentially, like statements in a software program, when the block is triggered. A procedure block has a sensitivity list containing a list of signals, along with the signals' change polarity. If a signal on a sensitivity changes according to its prescribed polarity, the procedural block is triggered. On the other hand, if a signal changes that is not on the sensitivity list but is present in the right-hand side of some procedural statements of the block, the procedural statements will not be updated. An example of a procedural block is shown here:

```
always @(posedge c or d) begin
    v = c + d + e;
    w = m - n;
end
```

The signals on the sensitivity list are $c$ and $d$. The polarity for $c$ is positive edge, and no polarity is specified for $d$, which by default means either positive or negative edge. This block is executed whenever a positive event occurs at $c$. A positive edge is one of the following: a change from 0 to 1, X, or Z; or from X or Z to 1. The block is also triggered when an event occurs at $d$, positive or negative edge. A negative edge is one of the following: a change from 1 to either 0, X, or Z; or from X or Z to 0. Changes on variables $e$, $m$, or $n$ will not cause the block to execute because these three variables are not on the sensitivity list.

In a way, a procedural block is a generalization of a continuous assignment: Any variable change on the right side of a continuous assignment triggers evaluation, whereas in a procedural block only changes in the variables on the sensitivity list can trigger evaluation, and the changes must conform to the polarity specification. Any continuous assignment can be rewritten as a procedural block, but not vice versa. The following procedural block and continuous assignment are equivalent:

```
// continuous assignment
assign v = x + y + z;

// equivalent procedural block
always @(x or y or z)
   v = x + y + z;
```

However, there is no equivalent continuous assignment for the following procedural block, because rising or falling changes are not distinguished in continuous assignment:

```
always @(posedge x)
   y = x;
```

### 1.6.2   Nondeterminism

When multiple processes execute simultaneously, a natural question to ask is how these processes are scheduled. When multiple processes are triggered at the same time, the order in which the processes are executed is not specified by Institute of Electrical and Electronics Engineers (IEEE) standards. Thus, it is arbitrary and it varies from simulator to simulator. This phenomenon is called *nondeterminism*. Let's look at two common cases of nondeterminism that are caused by the same root cause but that manifest in different ways.

The first case arises when multiple statements execute in zero time, and there are several legitimate orders in which these statements can execute. Furthermore, the order in which they execute affects the results. Therefore, a different order of execution gives different but correct results. Execution in zero time means that the statements are evaluated without advancing simulation time. Consider the following statements:

```
always @(d)
   q = d;

assign q = ~d;
```

These two processes, one procedural block and one continuous assignment, are scheduled to execute at the same time when variable d changes. Because the order in which they execute is not specified by the IEEE standard, two possible orders are possible. If the always block is evaluated first, variable q is assigned the new value of d by the always block. Then the continuous assignment is executed. It assigns the complement of the new value of d to variable q. If the continuous assignment is evaluated first, q gets the complement of the new value of d. Then the procedural assignment assigns the new value (uncomplemented) to q. Therefore, the two orders produce opposite results. This is a manifestation of nondeterminism.

Another well-known example of nondeterminism of the first type occurs often in RTL code. The following D-flip–flop (DFF) module uses a blocking assignment and causes nondeterminism when the DFFs are instantiated in a chain:

```
module DFF(clk, q, d);
input clk, d;
output q;
reg q;

always @(posedge clk)
    q = d; // source of the problem

endmodule

module DFF_chain;

DFF dff1(clk, q1, d1); // DFF1
DFF dff2(clk, q2, q1); // DFF2

endmodule
```

When the positive edge of clk arrives, either DFF instance can be evaluated first. If dff1 executes first, q1 is updated with the value of d1. Then dff2 executes and makes q2 the value of q1. Therefore, in this order, q2 gets the latest value of d1. On the other hand, if dff2 is evaluated first, q2 gets the value of q1 before q1 is updated with the value of d1. Either order of execution is correct, and thus both values of q2 are correct. Therefore, q2 may differ from simulator to simulator.

A remedy to this race problem is to change the blocking assignment to a nonblocking assignment. Scheduling of nonblocking assignments samples the values of the variables on the right-hand side at the moment the nonblocking assignment is encountered, and assigns the result to the left-side variable at the end of the current simulation time (for example, after all blocking assignments are evaluated). A more in-depth discussion of scheduling order is presented in the section "Scheduling Semantics." If we change the blocking assignment to a nonblocking assignment, the output of dff1 always gets the updated value of d1, whereas that of dff2 always gets the preupdated value of q1, regardless of the order in which they are evaluated.

Nonblocking assignment does not eliminate all nondeterminism or race problems. The following simple code contains a race problem even though both assignments are nonblocking. The reason is that both assignments to variable x occur at the end of the current simulation time and the order that x gets assigned is unspecified in IEEE standards. The two orders produce different values of x and both values of x are correct:

```
always @(posedge clk)
begin
    x <= 1'b0;
    x <= 1'b1;
end
```

The second case of nondeterminism arises from interleaving procedural statements in blocks executed at the same time. When two procedural blocks are scheduled at the same time, there is no guarantee that all statements in a block finish before the statements in the other block begin. In fact, the statements from the two blocks can execute in any interleaving order. Consider the following two blocks:

```
always @(posedge clk) // always block1
begin
    x = 1'b0;
    y = x;
end

always @(posedge clk) // always block2
begin
    x = 1'b1;
end
```

Both always blocks are triggered when a positive edge of clk arrives. One interleaving order is

```
x = 1'b0;
y = x;
x = 1'b1;
```

In this case, y gets 0.

Another interleaving order is

```
x = 1'b0;
x = 1'b1;
y = x;
```

In this case, y gets 1.

There is no simple fix for this nondeterminism. The designer has to reexamine the functionality the code is supposed to implement and, more often than not, nondeterminism is not inherent in the functionality but is introduced during the implementation process.

### 1.6.3   Scheduling Semantics

Having discussed nondeterminism, it is natural for us to examine the scheduling semantics in Verilog (that is, the order in which events are scheduled to execute). Events at simulation time are stratified into five layers of events in the order of processing:

1. Active
2. Inactive
3. Nonblocking assign update
4. Monitor
5. Future events

Active events at the same simulation time are processed in an arbitrary order. An example of an active event is a blocking assignment. The processing of all the active events is called a *simulation cycle*.

Inactive events are processed only after all active events have been processed. An example of an inactive event is an explicit zero-delay assignment, such as #0 x = y, which occurs at the current simulation time but is processed after all active events at the current simulation time have been processed.

A nonblocking assignment executes in two steps. First, it samples the values of the right-side variables. Then it updates the values to the left-side variables. The sampling step of a nonblocking assignment is an active event and thus is executed at the moment the non-blocking statement is encountered. A nonblocking assign update event is the updating step of a nonblocking assignment and is executed only after both active and inactive events at the current simulation time have been processed.

Monitor events are the execution of system tasks $monitor and $strobe, which are executed as the last events at the current simulation time to capture steady values of variables at the current simulation time. Finally, events that are to occur in the future are the future events.

## 1.7 Summary

In this chapter we defined design verification as the process of ensuring that a design meets its specifications—the reverse of the design process. We discussed the general verification principle through redundancy, and simulation-based and formal verification methodologies, and then illustrated typical flows of these two methodologies. We then contrasted simulation and formal verification, and emphasized their input- and output-oriented nature. We viewed simulation-based verification as operating on a point basis whereas formal verification operates on a subspace basis. Next, we listed some limitations of formal verification—in particular, the inability to detect errors in the specification, incomplete functional coverage caused by verifying subcircuits, user errors, and software bugs in the verification tool. As a refresher of Verilog, we reviewed scheduling and execution semantics, during which we discussed the concurrent nature of the Verilog language and its inherent nondeterminism.

## CHAPTER 2

# Coding for Verification

## Chapter Highlights:

- Functional correctness

- Timing correctness

- Simulation performance

- Portability and maintainability

- "Synthesizability," "debugability," and general tool compatibility

- Cycle-based simulation

- Hardware simulation/emulation

- Two-state and four-state simulation

- Design and use of a linter

T he best way to reduce bugs in a design is to minimize the opportunities that bugs can be introduced—in other words, design with verification in mind. Once this objective is achieved, the next step is to maximize the simulation speed of the design. These two objectives can be accomplished right from the beginning with the cooperation of Hardware Descriptive Language (HDL) designers by introducing coding style rules to which designers must adhere. These rules restrict the kinds of constructs a designer can write and make the code more regular in structure and easy to understand. As an example, consider the following code:

```
wire [8:0] in;
wire [7:0] out;
assign out = in;
```

This segment of code is perfectly legal. Because signal `out` is 1 bit shorter than `in`, the most significant bit (MSB) of `in` will be truncated, which may be the designer's intent. However, it is also very likely that an error was introduced in the declaration of signal `in`, when the designer meant `[7:0]` for the 8-bit signal. To eliminate the ambiguity in this assignment, a rule can be established to force designers to make their intent explicit by requiring that the right-side variable and the left-side variable have equal width.

One may argue that coding guidelines diminish the expressiveness of the language and thus may create inefficient code. In practice, the benefits of minimizing bug-introducing opportunities far outweigh the potential lost code efficiency.

There are generally several coding guideline categories, each emphasizing a particular domain. Typical categories include

- Functional correctness
- Timing correctness
- Simulation performance
- Portability and maintainability
- "Synthesizability," "debugability," and general tool compatibility

In addition, there are other special applications that require their own coding guidelines:

- Cycle-based simulation
- Hardware simulation/emulation
- Two-state simulation

Functional correctness rules attempt to eliminate the hidden side effects of some constructs by requiring the designer to state explicitly the coding intent. Requiring a matching width for all operands is an example of a coding guideline for functional correctness. Rules for timing correctness examine code that may cause race problems, glitches in clock trees, and other timing problems. Portability and maintainability rules enforce project code partition and structure, naming convention, comments, file organization, and so on. Rules for simulation performance flag slow code and recommend styles that yield faster simulation. Finally, rules for tool compatibility ensure the code is acceptable by other tools used in the project (for example, synthesizability for synthesis tools). The three special categories listed earlier apply to particular simulators.

Checking for coding guidelines is static—namely, no input vectors are required and violations can be determined by examining the code alone, even for the timing correctness category. The tool used to enforce coding guidelines is usually called a *linter*. Traditionally, a linter is used to check syntactical errors and warnings (for example, left and right operands have mismatched lengths). As people realized that many design errors can be checked statically, the role of the linter has expanded beyond simple syntactical checks. Examples of expanded linting are combinational loops, and timing problems on a clock tree. For simplicity's sake I refer to a linter as a tool that checks for all the errors and warnings discussed in this chapter, although I must point out that the term *linter* has become slightly abused.

Errors and warnings in coding guidelines form a spectrum of difficulty. Some design constructs may be inferred with certainty as errors, whereas others may be variants that deviate from a predetermined guideline. The former is one extreme of the spectrum and the latter is the other extreme. In between are errors and warnings with varying degrees of severity. Therefore, a linter usually issues levels of severity along with errors and warnings.

## 2.1  Functional Correctness

Some constructs in a design can be analyzed just by examining the code locally without knowing how it is connected to the rest of the circuit, whereas others require a more global scope. For example, the assignment in the sample in the previous section reveals a potential error. But an error of a loop consisting of combinational gates cannot be detected just by looking at a gate in the loop; the error can only be discovered by examining how the gates are connected (in other words, the structure of the circuits must be checked). Therefore, we classify checks into two categories: syntactical and structural. The first category can be analyzed locally and the second must be done globally.

### 2.1.1  Syntactical Checks

Because design codes are examined not just for errors but also for potential errors or undesirable coding styles, which may vary from project to project, instead of having a set of fixed rules, each project has its own set of design guidelines. It is not possible nor is it fruitful to enumerate all possible rules. Instead, let's study some common rules that all projects should enforce.

**Operands of unequal width.** All operands of an operator must be of equal width. Although most language such as Verilog automatically pad the shorter operands with zeros to make them of equal length, it's a good practice to have the designer make this explicit to avoid careless mistakes. An example of equal and unequal operand widths is shown here. Two operators are presented. The first one is a bitwise AND and the second is the assignment. The first assign is correct because the operands X, Y, and Z are all declared to be 32 bits wide. The second assign is also correct, because the operands X[10:2], Y[8:0], and Z[11:3] are all 9 bits wide, where X[10:2] means the part of bus X from bit 2 to bit 10. The third assign violates our equal-width guideline, because Z is 32 bits wide but the two operands on the right side are 9 bits and 8 bits wide respectively. Finally, in the last assignment, zeros are padded to an operand to make the operands equal width, where {} is a concatenation operator that combines the 1-bit constant 1'b0 with the 8-bit signal Y[8:1] to form a 9-bit signal:

```
reg [31:0] X;
reg [31:0] Y;
reg [31:0] Z;

Z = X & Y; //all operands have equal width
Z[11:3] = X[10:2] & Y[8:0]; // all operands have equal width
Z = X[9:1] & Y[8:1]; // error: unequal operand width
Z[8:0] = X[9:1] & {1'b0,Y[8:1]}; // pad with zeros for equal width
```

While operands should have equal width, the width of the variable holding the operation result must be able to accommodate the result. In a 64-bit adder, the width of the sum is 65 bits, with the MSB holding carry-out. Similarly, in a 64-bit multiplier, the width of the product is 128 bits.

**Implicitly embedded sequential state.** Implicitly embedded sequential states result from a specification that has memory but does not explicitly create states. One such case is incompletely specified conditional statements. Conditional statements such as if–then–else and case statements need to have all cases specified. Failure to do so may result in unexpected circuit behavior. Consider the following description:

```
case(X) // a 2-bit variable
2'b00: Q = 1'b1;
2'b11: Q = 1'b0;
endcase
```

In this case statement, the value of the 2-bit variable X is compared first with value 2'b00 (a 2-bit value of 00). If it matches, variable Q is assigned a value of 1. Next, X is compared with value 2'b11. If it matches, Q takes on value 0. Here, only two of four possible values are specified. Thus, if X takes on value 01 or 10, Q retains the old value—that is, this code segment has memory. Thus the code, although it has the appearance of a multiplexor, is of a sequential nature because of the missing two cases (`01` and `10`). This implied sequential behavior should be made explicit if it is intended to be so; otherwise, the implied sequential behavior should be removed by completing the cases, as shown here:

```
case(X)  // a 2-bit variable
2'b00:  Q = 1'b1;
2'b11:  Q = 1'b0;
default:  Q = 1'b0;
endcase
```

Similar situations apply to if–then–else statements. This implied sequential behavior is also known as the *inferred latch phenomenon*, because a latch or sequential element will be needed to preserve the value of Q in the absence of the default statement. To make this inferred latch phenomenon more concrete, let us construct a circuit implementing the previous description. An implementation is shown in Figure 2.1, where the flip-flop (FF) is required to preserve the state. Here we assume the case statement resides inside an `always` block with clock `clk`.

To avoid inferred latches, one can either fully specify all possible values of the conditioning variable or use a directive to instruct the synthesis tool not to generate inferred latches and to treat the unspecified values as don't-cares. This directive is a special comment that the synthesis tool understands and is often termed *full_case*, meaning that the case statement should be treated as if it were fully specified; if the conditioning variable takes on the unspecified cases, the output variable can take on any value (in other words, don't-cares for the synthesis tool). Note that because this directive appears as a comment, it has no effect on the simulator. If a full-case directive were used in Figure 2.1, then it would mean that when X is either `01` or `10`, Q can be any value. In other words, with these unspecified

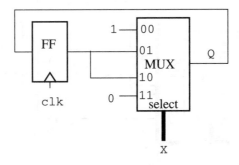

**Figure 2.1**  An FF is required to preserve state in an incomplete case statement.

cases taking on don't-care values implicitly, the conditional statement is now fully specified—it has all (full) cases. When a synthesis tool sees a *full_case* directive, it can assign any value to the unspecified cases, as opposed to retaining the current value, and thus no latches are required to retain the current value. When a don't-care is encountered, a synthesis tool uses it for optimization (for example, producing a smaller circuit). The following example illustrates the use of don't-cares in synthesis.

Consider the following code segment and two versions of its implementation, one using the unspecified cases for the purpose of minimizing gate count and the other not:

```
always @(posedge clock) begin
    case (S):
        3'b000: Q = a;
        3'b011: Q = b;
        3'b101: Q = a;
        3'b111: Q = b;
    endcase

    assign F = ( (S == 3'b100) | (S == 3'b001) ) ? (a | b) : Q;
```

An implementation is shown in Figure 2.2, where the FF and the multiplexor implement the incomplete case statement and other gates, the assign statement.

When we place a *full_case* directive next to the case statement:

```
    case (S): // synthesis full case
```

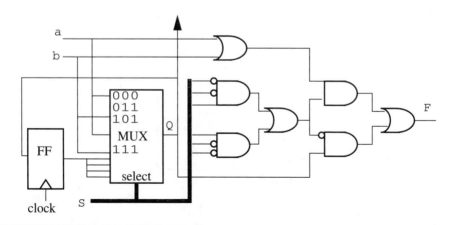

**Figure 2.2**  A circuit with an inferred FF implementing the HDL specification with incomplete cases

Then the synthesis tool recognizes the don't-care values of Q in the unspecified cases and uses them to optimize gate count. One optimization technique is to choose Q to be (a | b) when S is either 001 or 100. So, Q' and F', the resulting Q and F, become

```
case (S): // synthesis full case
    3'b000: Q' = a;
    3'b011: Q' = b;
    3'b101: Q' = a;
    3'b111: Q' = b;
    3'b001: Q' = a | b;
    3'b100: Q' = a | b;
    3'b010: 0;
    3'b110: 0;
endcase
```

Note that Q' can be written as (S==3'b100) | (S==3'b001))?(a|b):Q, which is just F. Therefore, assign

```
F = Q';
```

With this choice of don't-care, the new description yields the much simpler circuit shown in Figure 2.3.

Although using a *full_case* directive has the advantage of giving a synthesis tool more freedom for optimization, it has the side effect of creating two different versions from the same design description. One is the simulation version, for which the simulator assigns the current value to the output if the unspecified case is encountered. The second is the

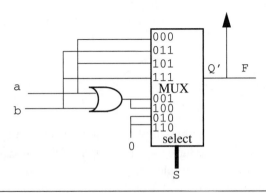

**Figure 2.3**  An optimized circuit making use of unspecified cases

synthesized circuit, for which the output takes on whatever value the synthesis tool deems optimal. The danger of having a simulation version that is different from the synthesis version is obvious: You are not verifying what you are designing.

Another common source of creating an implied sequential state is a variable read in a block that is not on the block's sensitivity list, and none of the variables on the sensitivity list is a clock. (If one variable on the sensitivity list is a clock, the block models a latch, and is less likely a user error.) This type of description can result from a user who wants to model a combinational block but inadvertently leaves out a variable on the sensitivity list. To see why an incomplete sensitivity list gives rises to a sequential circuit, consider the following code. A change in X does not cause variable Y to be updated until there is a transition on Z. Meaning, before the transition of Z, Y still holds the old value of X; therefore, the block exhibits memory and thus is sequential:

```
always @(Z)
   Y = X & Z;
```

A warning needs to be issued if such a situation occurs. However, if a warning is issued for every such occurrence, there would be many warnings for intended FFs and other sequential elements. To avoid flooding of warnings, intended sequential elements should always be instantiated from library cells for which such check should be skipped.

**Overlapping conditional cases.** Another anomaly of case statements results from the order the condition variable compares against the case items. In the following case statement, condition variable S compares its value against the case items from top to bottom— namely, first 1?? and then ??1, where ? is a wild card that matches either 1 or 0. When a match is found, the corresponding statement is executed and the remaining case items are skipped. Therefore, when S is 101, S matches 1??, and therefore Q is assigned 0 even though 101 also matches the second case item ??1:

```
Casex (S)
   3'b1??: Q = 1'b0;
   3'b??1: Q = 1'b1;
endcase
```

This first matched/first assigned nature implies a priority encoder in the case statement (in other words, the earlier case items have higher priorities). An N-bit encoder takes in $2^N$ bits, of which only 1 bit is active (for example, 1), and it produces a binary code for the bit (for example, 101 is the output if the fifth bit is 1). A priority encoder can accept more than one active bit, but it only produces the code for the active bit with the highest priority.

If this case statement is given to a synthesis tool, a priority encoder will be produced. A portion of the circuit is shown in Figure 2.4. (This case statement is incomplete; therefore, a sequential element is needed to model completely the specification. For clarity, we only show the portion involving the priority encoder.) As dictated by the description, if the MSB of S, S[2], is 1, Q is assigned 0 regardless of the value of its least significant bit (LSB). In this case, the priority encoder gives 1, which selects 0 in the multiplexor. Having a priority encoder for a case statement may not be what the designer had in mind. If S[0] is 1 and S[2] is 0, then the priority encoder gives 0, which makes the multiplexor produce a 1. Note that if we can guarantee that only one case item will match the case variable, then a simple encoder can be used.

Therefore, if the designer is certain that variable S can never have a value straddling the two ranges of the case items, then she can instruct a synthesis tool not to use a priority encoder. Having only one match also means that comparisons with case items can be done in parallel. To relay this information to a synthesis tool, the designer can place a parallel-case directive (for example, //synthesis parallel case) next to the case statement.

Although the parallel-case directive rescues designers from having an implied priority encoder, there are side effects. First, the directive affects only synthesis tools, not simulators. Thus, two models exist: A simulator sees a prioritized case statement and a synthesis tool interprets a parallel case statement. Again, the model being verified is not what is being designed—a dangerous double standard. What would happen if a *parallel-case pragma* is specified anyway, even though the case items overlap? The synthesized circuit may vary from one synthesis tool to another. An intelligent synthesis tool will use a priority

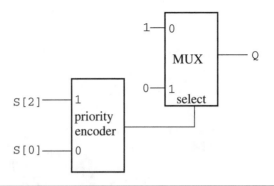

**Figure 2.4**   A priority encoder is included to implement a case statement.

encoder to resolve the conflict; others may just follow the *pragma* to use an encoder. Therefore, to ensure that the synthesis model and simulation model are consistent, case items should always be mutually exclusive.

**Connection rules.** There are ways to connect and access components, but explicit ones are preferred, thus reducing the chances of inadvertent mistakes. In Verilog there are two ways to connect ports during module instantiation. The implicit way connects instantiated ports with declared ports in a one-to-one correspondence in order. For example, in Figure 2.5, module `Gate` declares ports in order `Q1`, `Q2`, and `Q3`, and instance `M1` declares ports in order `P1`, `P2`, and `P3`. Thus, connection by order connects `P1` with `Q1`, `P2` with `Q2`, and `P3` with `Q3`.

Explicit connection connects instantiated ports and declared ports by name. Continuing the previous example, instance `M2` has the same port connection as `M1`, even though its port order is different:

```
Gate M2 (.Q2(P2), .Q3(P3), .Q1(P1)); // explicit port connection
```

Instance `FF` in Figure 2.5 uses explicit connection, and the block diagram shows the actual connections. Explicit connection is less error prone and is preferred over connection by order.

Furthermore, no expression is allowed during port connection. For example, instead of having

```
Gate M2 (.Q2(P2&R), .Q3(P3), .Q1(P1));
```

```
module Top (P1);
  Gate M1(P1, P2, P3); // instantiation of Gate
  FF M2 (.D(P3), .Q(P2), .CLK(clock));
endmodule


module Gate (Q1, Q2, Q3); // Gate definition
  input Q1;
  input Q2;
  output Q3;
  assign Q3 = Q1 & Q2 ;
endmodule
```

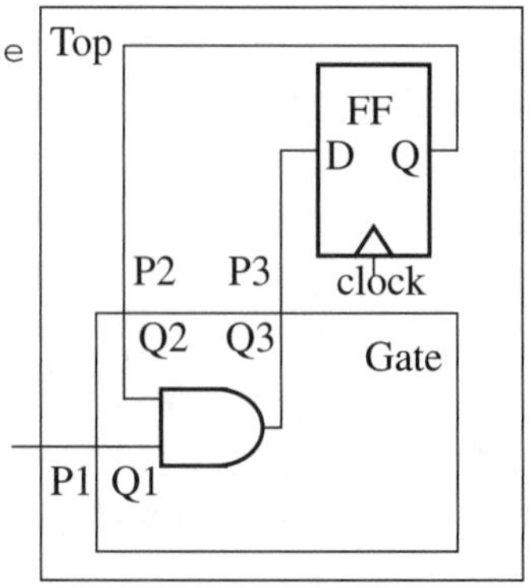

**Figure 2.5**  Port connection by order

where input port Q2 is fed with express P2&R. One should create a variable that computes the expression and then gets connected to port Q2, as in

```
assign V = P2 & R;
Gate M2 (.Q2(V), .Q3(P3), .Q1(P1));
```

The rationale is to separate logic computation and port connection for a better structured design.

Besides going through a port, the so-called hierarchical path is another way to access a signal. In a design there are usually several levels of hierarchies (or levels of module instantiation). For example, in a central processing unit (CPU) design, the top module is a CPU that contains several modules, such as a floating point unit (FPU),and a memory management unit (MMU). In turn, the FPU may contain several modules, such as an Arithmetic Logic Unit (ALU) and others. One can access net N inside the ALU from anywhere in the design through a hierarchical path, without going through the sequence of ports from the CPU to the FPU to the ALU. To write the value of N to another net M, one simply uses

```
assign M = CPU.FPU.ALU.N; // access by hierarchical path
```

In a project, two types of HDL code exist. One type belongs to the design (the code that constitutes the chip) and the other belongs to the test bench (the code that forms the testing structure). It's strongly recommended that hierarchical accesses exist only in the test bench. In other words, the design should contain no hierarchical paths. The rationale is that access in hardware is always through ports; therefore, the design description should reflect this to be consistent. However, a test bench is used only for testing purpose. No real hardware will be fabricated from it. Therefore, it is reasonable to use hierarchical access.

Finally, it's a good practice to require that the top-level module have only module instantiations. That is, no logic circuitry. The reason for this is that having logic at the top level is indicative of poor design partitioning—namely, the logic not properly assigned to any submodules. Furthermore, the partition of modules at the RTL should correspond to the physical partition. This simplifies the equivalence checks between the RTL and the physical design.

**Preferred design constructs.** Some legal constructs are discouraged for several good reasons. Because it is not possible to enumerate all of them (some are project specific), let's take a look at the common ones. Loop constructs such as FOR/WHILE/REPEAT are discouraged for design code (test bench code has more relaxed rules). The reason for this is that the loop structure is more of a software entity than a hardware entity. Using them in the RTL creates a less direct correspondence between the RTL and the real hardware, making verification more complex. Sometimes loop constructs are used to create regular structures. An example is a memory array, which consists of a large, repetitive regular structure. A macro should be used instead, because directly coding a large, repetitive structure is error prone, whereas automatic macro generation gives the user an easy way to inspect for correctness. Macros can be written in Perl, m4, or C and, when called during preprocessing, will write out a Verilog description of the code.

When coding at the RTL, lower level constructs such as gate and transistor primitives should be avoided, some of which are AND, NOR, NOT, PMOS, NMOS, CMOS, TRAN, TRIREG, PULLUP, and PULLDOWN. The first three are logic gates, the next five are transistors, and the last two are resistive PULLUPs and PULLDOWNs. Furthermore, Verilog has a syntax for signal strength that is used to resolve conflict when a net is driven to opposite values at the same time. The signal with the greater strength wins. Strength constructs belong to transistor-level design and should be avoided in RTL code.

In addition, some constructs are mainly for test benches, for which no hardware counterpart is produced and thus have a coding style that can be more software oriented than hardware oriented. Examples of such Verilog constructs are force/release, fork/join, event, and wait.

Finally, user-defined primitives (UDPs), especially sequential ones, should be avoided as much as possible, because their functionality is not immediately obvious and they often hide unexpected side effects. The workaround is to replace UDPs with small circuits made of standard library cells.

### 2.1.2  Structural Checks

Structural errors and warnings result from connections between components. To detect them, code from multiple designers may be required for examination, and they may be much less obvious to individual designers.

**Loop structure.** A loop structure is a loop made of circuit components with outputs that feed to the inputs of the next component in the loop. An example is shown in Figure 2.6. A loop structure is detrimental if, at a moment in time, the signal can travel along the loop uninhibited—that is, not blocked by an FF or a latch. A case in point is a loop made of only combinational gates. Another situation is a loop made of latches that can all become transparent at the same time. The latch loop in Figure 2.6 can become transparent if clock1 is high and clock2 is low.

A combinational loop can be detected by following combinational gates from input to output. When you arrive at the same net without going through a sequential device, you

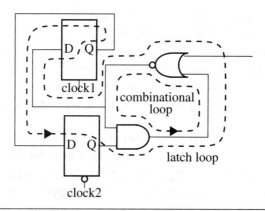

**Figure 2.6** A circuit with a combinational loop and a latch loop

have found such a loop. A latch loop is more difficult to discover because, besides finding a loop of latches and combinational gates, you need to determine whether all latches in the loop can become transparent at the same time, and this decision is computationally expensive. In practice, the designer assists the checker by telling her about the phases of latch clocks. Even with hints, the checker may still make pessimistic assumptions and may have the designer validate or invalidate the latch loops.

**Bus operation.** A bus is a communication channel among devices that have the capability to relinquish the bus by driving its outputs to high impedance Z. Thus, all gates connected to a bus must be tristate devices. In theory, a bus should be in high impedance state when it is not driven. In practice, some designs pull up (or down) a floating bus with a weak resistor to avoid being a potentially problematic dangling net. The strength of the pull-up or pull-down resistor should be chosen to be just right to not have the bus stuck at one logic level. Given this complication, warnings should be issued in this situation.

Furthermore, drivers to a bus should be checked for mutual exclusion of their enable signals (in other words, at most, one driver of the bus can be active) at any time. A conclusive check is computationally expensive and belongs to the domain of formal verification. A compromise is to check mutual exclusion during simulation runs. Then, of course, the result is data dependent and indicates only partial validation. It is also desirable to ensure that when a bus receiver is enabled, a driver is already driving, or when a driver is active, at least one receiver is listening. Again, these checks should be done with formal verification tools or during simulation runs.

**FF and latch configuration.** Sequential elements are latches and FFs. How they are connected requires particular attention. Active high latches should not drive active high latches, and positive-edge trigger FFs should not drive active high latches, because the driven latches simply pass whatever the driving latches/FFs store, behaving just like buffers. This kind of configuration should invoke warnings, but not errors, because the driven latches could probably be used for other purposes than storage, such as for scanning. Similarly, an active low latch should not be driven by another active low latch or negative-edge trigger FF. Therefore, phases of sequential elements should alternate along any path.

Checking for sequential element configuration becomes complicated when there are multiple clock domains—namely, parts of the design are driven by different clocks. Then the simple phase-alternating rule would generalize to check whether a driven sequential element retains its storage functionality. Again, it is not possible to be conclusive about whether a violation is necessarily an error; hence, a warning should be issued.

**Sequential element reset.** All sequential elements in control logic should be able to be reset or be able to be driven to a known state. The reason is that when the design is first powered on, the control logic should be in a known state. However, such a requirement is not necessary for the sequential elements in data path logic, which, if designed properly, is supposedly drivable to a known state.

---

### Rules to Check

1. All operands of an operator have equal length.
2. Conditional statements have complete and nonoverlapping cases.
3. If full_case or parallel_case directives are found, warn about the possible mismatching simulation and synthesis models.
4. Warn if an `always` block has an incomplete sensitivity list.
5. The explicit connection rule is obeyed and the use of hierarchical paths is forbidden in a design.
6. Certain constructs, although legal, will be warned. A list of such constructs consists of transistors, gate primitives, strength, delays, events, time, UDPs, and test bench constructs (for example, `force/release`, `loops`, `deassign`, `fork/join`).
7. Memory arrays should be derived from a macro library.
8. An error is issued when a combinational or a simultaneously opened latch loop is detected.
9. Issue a warning when a bus with a pull-up or pull-down resistor is detected.
10. Use runtime checks to discover mutual exclusion of bus drivers, and coordination of bus drivers and receivers.
11. Only alternating types of FF and latches should be cascaded.
12. All control logic can be reset.

## 2.2  Timing Correctness

Contrary to common belief, timing correctness, to certain degree, can be verified statically in the RTL without physical design parameters such as gate and interconnect delays. In this section, we discuss some timing problems that can be discovered at RTL.

### 2.2.1  Race Problem

The foremost common timing problem is race. A race problem can be defined as the process that results in nondeterministic outcome when several operations operate on the same entity (such as variables or nets) at the same time. Therefore, the resulting outcome is at the mercy of the simulator or the physical devices. The operations can be simultaneous multiple writes or simultaneous read and write. Race problems manifest as intermittent errors and are extremely hard to trace from the design's input/output (I/O) behavior.

Race problems have many causes. The following race problems are common in practice and can be easily detected from the HDL description. The first type is assignment to the same variable from several blocks triggered by the same clock edge or event. The following circuit shows a Verilog example of a race problem caused by the write operations (in other words, assignment) from two `always` blocks triggering on the same clock edge, for which the value of x is simulator dependent.

```
always @(posedge clock)
    x = 1'b1;

always @(posedge clock)
    x = 1'b0;
```

A less obvious multiple write is the so-called *event-counting construct* shown next. Each transition of x or y causes a write to variable `event_number`. The problem arises when x and y have transitions at the same time. In this case, whether `event_number` is incremented by two or one is simulator dependent and thus nondeterministic:

```
always @(x or y)
event_number = event_number + 1;
```

An example of a simultaneous read and write is shown next, where a change in the value of y (caused by y = y + 1) triggers the continuous assignment (`assign x = y + 2`), which in turn writes a new value to x. At the same time, x is assigned z. The problem lies in what value z would get—the x value before the continuous assignment update or after? The resulting z is nondeterministic:

```
assign x = y + 2;
always (posedge clock) begin
    y = y +1;
    z = x;
end
```

In summary, when a variable is multiply written or written and read at the same time, race problems exist.

HDL beginners often commit a common read and write race mistake in using blocking assigns in FFs, as shown in the following example:

```
module FF (D,clock,Q);
    input D;
    input clock;
    output Q;
    always @(posedge clock)
        Q = D;
endmodule

module two_FFs;
    FF m1(.D(Q2),.Q(Q1),.clock(clock));
    FF m2(.D(Q1),.Q(Q2),.clock(clock));
endmodule
```

When two such FFs are connected in series, as in module two_FFs, net Q1 is read by the second FF and is written by the first FF at the same time. What the second FF sees is uncertain and varies from simulator to simulator. The remedy is to change the blocking assigns to nonblocking assigns. A nonblocking assign reads the value and holds off writing until the end of the current simulation time, at which time all reads are already finished.

In general, to eliminate read/write race problems, nonblocking assigns should be used in place of blocking assigns, because nonblocking assigns read the values when the statement is reached, and write at the end of the current simulation time, at which time all reads have been finished. As a result, no read and write can happen at the same time. However, using nonblocking assigns does not get rid of write/write race problems: All writes still occur simultaneously at the end of the current simulation time.

### 2.2.2  Clock Gating

Glitches are unintended pulses resulting from design artifacts and have widths that vary according to physical parameters (such as parasitic capacitance and trace length) and operating conditions (such as temperature and the chip's state). As these conditions change, glitches vary their widths, and consequently may get filtered out or may cause errors at random. Hence, glitches are a major source of intermittent error. Let's study several common causes of glitches.

Clock gating is a common design practice and, if not used properly, can produce glitches on the clock tree to trigger latches and FFs falsely. Let us start with the example in Figure 2.7, in which the gating signal is the output of the FF clocked by the same clock. Assume the gating signal changes from high to low at the rising edge of the clock. Because of the delay from `clock` to `Q` in the gating FF, the gating signal arrives slightly later than the clock rising transition and thus causes a narrow glitch.

One may propose to add a delay to the clock line to ensure clock rising transitions arrive late, to get rid of the glitch. There are two problems with this solution. First, factors affecting the relative delays are abundant and hard to control. For example, layout lengths and the actual gate delays depend on layout tools and fabrication processes. Second, zero-delay simulation (all gates are treated with zero delay), often employed for the RTL, still produces the glitch, because a simulator always delays the output of an FF by an infinitesimal amount relative to the clock, because of a nonblocking assign for the output of the FF, regardless of the actual delays.

A clean solution is to use an OR gate as the gating device. In this setup, the clock transits to high first to stabilize the output of the OR gate, hence preventing glitches. Similarly, if the gating FF is negative-edge triggered, the gating device should be an AND gate. In summary, if the gating signal changes on the clock's rising edge, an OR gate should be the gating

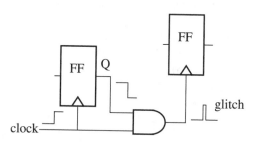

**Figure 2.7**  A gated clock produces glitches on a clock tree.

device. Conversely, if the gating signal changes on the falling edge, an AND gate should be the gating device.

Simple gates such as AND and OR should be used as gating devices because complex ones often produce glitches for certain combinations of input transitions. In addition, gating devices should be instantiated from library cells, instead of created at will as part of the RTL design, so that these timing-critical gates can be under strict control throughout the entire project.

Taking this one step further, there is a restriction on the type of latches that can be clocked by a gated clock. Based on the previous clock gating rule, a rising transition signal should be gated by an OR, then the latch receiving the gated clock should be an active low latch because the latch will correctly become opaque when the gating signal turns high to disable the clock. Similarly, for a falling gating signal, i.e., AND gating, the receiving latch should be an active high latch.

### 2.2.3   Time Zero and Zero Time Glitches

Time zero glitches refer to the transitions at the start of the simulation—at time zero. When a clock is assigned to 1 at time zero, should the transition from an unknown value, x, to 1 be counted as a transition or not? The answer depends on the simulator. A solution is to delay assignments to the clock to avoid transitions at time zero.

Closely related to the time zero glitch in name, but not in nature, is the zero time glitch. This type of glitch has zero width and is often an RTL coding artifact for which two different values are assigned to the same variable in no time. This often happens in zero-delay simulation. The following code is indicative of a finite-state machine in which, by initializing next_state to RESET, the designer wanted to make sure that next_state does not end up in an unknown state if current_state is neither ACK nor IDEL. However, if current_state is ACK (IDEL), variable next_state changes from RESET to IDLE (REQ) in no time, creating a zero time glitch in next_state and in glitch_line:

```
always @(posedge clock) begin
   next_state = RESET; // initialize next_state to RESET
   case (current_state)
      ACK: next_state = IDLE;
      IDEL: next_state = REQ;
end

assign glitch_line = next_state; // executed every time
next_state changes
```

To eliminate zero time glitches, avoid multiple writes to the same variable in one clock cycle.

### 2.2.4   Domain-Crossing Glitches

Glitches can easily form at interfaces of different clock domains. A clock domain is the group of logic that is clocked by the same clock. Interface logic of two clock domains is the logic with inputs that come from two clock domains—for example an AND gate with one input from one clock domain and another from another clock domain. If the two clock domains are not synchronized, the intervals between the transitions in the two domains can be arbitrarily small as the transitions beat against each other and hence create glitches in the interface logic. Therefore, when clock domains merge, proper synchronization should be in place. A circuit synchronizing two clock domains is shown in Figure 2.8.

---

**Rules to Check**

1. Detect race problems, such as multiple write/write and read/write race, and event-counting race.
2. The clock gate is of the right type. The rising transition should use the OR gate and the falling transition should use the AND gate.
3. For OR gating, the receiving latch is active low. For AND gating, the receiving latch is active low.
4. Clock logic should consist of only simple gates, such as AND, OR, or INVERTOR.
5. No transition should be allowed at time zero.
6. Avoid multiple writes to the same variable in one clock period to eliminate zero time glitches.
7. Signals from different clock domains should be synchronized.

---

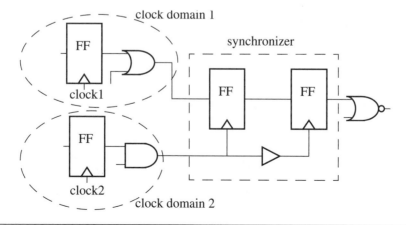

**Figure 2.8**   Synchronize two clock domains

## 2.3  Simulation Performance

### 2.3.1  Higher Level of Abstraction

It's almost universally true that the higher level of abstraction of the system, the faster it will simulate. The creation of a system goes through multiple modeling levels, from specification all the way down to the transistor level. A model of a higher abstraction level emphasizes the system's operating principles rather than implementation details, thus offering more freedom for implementation variants. As the level gets lower, operating concepts gradually transform to concrete implementation details. The highest abstraction level is specification, where only the I/O behavior is specified and there is no information about the internal system. The next level of abstraction is the algorithmic or behavioral level, where the system's functions to meet the specifications are described. This description gives an architectural view of the system as opposed to an implementational view. In Verilog, a behavioral model usually contains constructs such as if–then–else, case, loops, functions, and tasks. Further down is the RTL, in which the states of the system have been identified—the registers are defined and the functionality affecting state transitions remains at the behavioral level. The next level, gate netlist, is a complete implementation of the system specification. A design process that proceeds from a high abstraction level to lower levels is called a *top–down process*; the other way around is called a *bottom-up process*.

Here we'll look at various levels of abstraction through a design example of a simplified traffic controller in which cars cannot make turns but can only go straight across the intersection. A segment of a behavioral model is as follows, where EW and NS denote the east–west and north–south-bound traffic lights respectively:

```
initial begin // initial state of the system
   EW = GREEN;
   NS = RED;
end

always @(posedge clock)
   if ((EW == GREEN) && (NS == RED)) EW <= YELLOW;
   else if ((EW == YELLOW) && (NS == RED)) EW <= RED;
   else if ((EW == RED) && (NS == YELLOW)) EW <= GREEN;

always @(posedge clock)
   if ((NS == GREEN) && (EW == RED)) NS <= YELLOW;
   else if ((NS == YELLOW) && (EW == RED)) NS <= RED;
   else if ((NS == RED) && (EW == YELLOW)) NS <= GREEN;
```

In this model, the order of light change is described using behavioral constructs. There are no details about how such changes are to be effected, which are to be further refined in the next level (RTL) shown in the following code. An RTL model identifies the states of the system and assigned registers or FFs to represent them. In the following code segment, FF array F1, of 2 bits, represents the state of the east–west light; F2, also of 2 bits, represents north–south. In Verilog, array instantiation instantiates an array of FFs instead of having the designer do one FF at a time. In this example, each array instantiation creates two FFs, as indicated by the range `[1:0]`:

```
initial begin // initial state of the system
   EW = GREEN;
   NS = RED;
end

FlipFlop F1[1:0] (.Q(EW),.D(next_EW),.clock(clock));
FlipFlop F2[1:0] (.Q(NS),.D(next_NW),.clock(clock));

always @(EW or NS) // next_state_function, combinational
begin
   if ((EW == GREEN) && (NS == RED)) next_EW <= YELLOW;
   else if ((EW == YELLOW) && (NS == RED)) next_EW <= RED;
   else if ((EW == RED) && (NW == YELLOW)) next_EW <= GREEN;
   else next_EW = EW;

   if ((NS == GREEN) && (EW == RED)) next_NS <= YELLOW;
   else if ((NS == YELLOW) && (EW == RED)) next_NS <= RED;
   else if ((NS == RED) && (EW == YELLOW)) next_NS <= GREEN;
   else next_NS = NS;
end
```

The `always` block is very much the same as that in the behavioral model, except that it is now completely combinational because the if–then–else statements have full cases and all variables read [EW and NS], appear on the sensitivity list. The next level is gate netlist, which implements the combinational `always` block with gates. The following shows the vast complexity of the gate-level model and provides the intuition behind using models of higher levels of abstraction. The combinational `always` block translates into the following AND-OR expressions:

```
next_NS[0]=EW[0]NS[1]+NS[1]NS[0]+NS[0]EW[0]+NS[1]EW[1]EW[0];
next_NS[1]=NS[1]NS[0]+NS[1]EW[1]+NS[1]EW[0]+NS[1]EW[1]EW[0];
next_EW[1]=EW[1]EW[0]+NS[1]EW[1]+NS[0]EW[1]+NS[1]NS[0]EW[0];
next_EW[0]=EW[1]EW[0]+NS[0]EW[0]+NS[1]NS[0]EW[0];
```

If only inverters—2-input AND and OR gates—are used, the gate-level model consists of 34 lines of gate instantiations for the previous four lines of code, plus a couple more for FFs and the reset circuit. Hopefully, with this demonstration, you understand why gate-level models should be avoided for simulation performance.

## 2.3.2  Simulator Recognizable Components

Many simulators attempt to recognize some standard components, such as FFs, latches, and memory arrays, to make use of internal optimization for performance. The recognizable styles depend on the simulators; therefore, consult the simulator manual and make an effort to conform to the coding styles. In the event that you must deviate from the recommended styles, or no such recommendation is provided, you should code in a style as simple and as close to the "common" style as possible, which generally means avoiding using asynchronous signals, delays, and complex branching constructs. For FFs and latches, here are some sample "common" styles:

```
// positive-edge-triggered D-flip-flop
always @(posedge clock)
   q <= d;

// positive-edge-triggered DFF with synchronous active high
reset
always @(posedge clock)
   if(reset)
      q <= d;

// DFF with asynchronous active high reset
always @(posedge clock or posedge reset)
   if(reset)
      q <= 1'b0;
   else
      q <=d;

// active high latch
always @(clock or data)
   if (clock)
      q <= data;

// active high latch with asynchronous reset
always @(clock or data or reset)
   if (reset)
      q<=1'b0;
   else if (clock)
      q <= data;
```

Memory arrays are another tricky component: They don't have general recognizable forms. Furthermore, there are synchronous and asynchronous memory arrays—the former being faster simulationwise, and hence are preferred. Synchronous memory needs to be strobed by a clock to give output after data, address, and read/write enable are present, whereas asynchronous memory produces the output immediately once data, address, and read/write enable are present. Figure 2.9 presents a block diagram for the two types of memory.

To aid a simulator, a user directive such as a stylized comment such as

```
//my_simulator memory: clock is clk, data is in1, address is
addr1
```

can be attached to instruct the simulator to determine the clock, data, and address. The directive is usually simulator specific.

Coding finite-state machines needs special attention, and which particular style is preferred is dictated by the simulator. As a rule of thumb, separate as much combinational logic from the states as possible. For example, never encompass the next-state transition function and the sequential elements in one large sequential `always` block. Code the

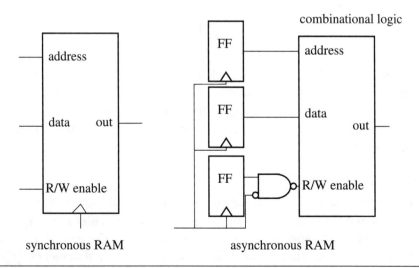

**Figure 2.9**   Synchronous and asynchronous memory

next-state transition function as a combinational `always` block using a case statement
and mnemonics for the states. An example of a next-state function is

```
//combinational next-state function. Recommended
always @(presentState) begin
    case (presentState):
        IDLE: nextState = WAIT;
        REQUEST: nextState = GRANT;
        ...
        default: $display("error");
end
```

Avoid using Verilog `@` inside an `always` block, as in the following example. When `@` is encountered, the simulation is blocked until the event of the `@` occurs; in other words, variable values
are held while waiting for the event. Therefore, an `@` construct has embedded a state. The style
using `@` mimics traversing a state diagram in which a transition is made when a clock transition occurs. Most simulators do not recognize this style as a finite-state machine:

```
//bad coding style
always @(clock) begin
    nextState = ...;
    @(clock) // on a clock edge, go to another state
        if(presentState ==..) nextState = ...;
    @(clock) // on a clock edge, go to another state
        if(presentState ==..) nextState = ...;
end
```

This style can always be recoded into a tandem of FFs for the states and a combinational
block for the next-state function. The idea is to create a state for every `@`, so that every time
a clock clicks, the state machine transits to the state corresponding to the following `@`.
Once the states are mapped with `@`s, the next-state function is written to produce the
required outputs and next state.

## 2.3.3  Vector versus Scalar

Whenever possible, use the entire bus as an operand instead of just the bits, because the
simulator takes more internal operations to access bits or ranges of bits than the entire
bus. Moreover, when a statement involves bits of the same bus, the simulator may need to
visit each bit during the execution of the statement, whereas for a statement on the whole
bus, the simulator just needs to access the bus once, even if there are multiple operations
on the whole bus. Finally, converting bit operations into bus operations often simplifies
the code. To convert bit operations to bus operations, concatenation, reduction, and mask
operators play an important role.

When bits are assigned to different values, the operation can be made into a vectored operation by concatenating the scalar operations:

```
scalar operation:
   assign bus[15:0] = a & b;
   assign bus[31:16] = x | y;

vectored operation:
   assign bus = {x | y, a & b}; // {} is the concatenation
   operator

scalar operation:
   assign output = (bus[1] & bus[2]) | (bus[3] ^ bus[0]);

vectored operation:
   assign output = (&(4'b1001 | bus)) | (^(4'b1001 & bus));
```

The first example simply concatenates the two scalar operations and assigns it to the bus. The second example uses masks to select the bits and then applies reduction operators on the selected bits. For instance, masking operation 4'b1001 | bus does bitwise OR and gives a vector (1, bus[2], bus[1], 1). Then, the reduction & operator ANDs the vector bit by bit to give bus[2]bus[1]. Similarly, 4'b1001 & bus produces vector bus[3], 0, 0, bus[0]. Then the ^ reduction operator XORs the vector bit by bit to give bus[3]^ubs[0]. Finally, the intermediate results are ORed together.

The previous conversion technique applies to any expression. A general algorithm follows.

---

**Algorithm for Converting Scalar Operations to Vectored Operations**

*Input:* a set of assignments to bits or ranges of buses

*Output:* a set of assignments with whole-bus operands

1. Repeat steps 2 and 3 for each of the assignments.
2. Express the right-hand side as a sum of products.
3. For each product term, if there are complemented variables, use a mask operation with bitwise XOR to invert the bits/ranges. Bit 1 corresponds to an inverted bit. Then select the bits/ranges of the bus in the term by bitwise ORing with a mask. Bit 0 corresponds to a selected bit. Next, reduction AND the result to produce a partial term. Finally, AND the result with the remaining literals in the product term.
4. Use the concatenation operator to make the set of assignments into a bus assignment.

To illustrate the algorithm, consider the following set of assignments, assume bus A has 2 bits and bus B has 6 bits:

```
assign A[0] = B̄[3] & B[5] & x + B[3]& y;
assign A[1] = B[4] ȳ + B[0];
```

The right-hand sides are already in sum-of-products form. Step 2 is skipped (it is always possible to transform an expression to a sum of products). In step 3, for the first product term, the partial term made of only bus bits/ranges is B̄[3]B[5]. Use mask (6'b001000 ^ B) to invert B[3], giving (B[5],B[4],B̄[3],B[2],B[1],B[0]), followed by masking with bitwise OR to select B[3] and B[5] (in other words, 6'010111 | (6'b001000 ^ B), giving (B[5],1,B[3],1,1,1)). Lastly, reduction AND to produce the partial term. Putting it all together, we have & (6'b010111 | (6'b001000 ^ B)). Finally, AND it with x. Similarly transform the remaining terms. After step 3, we have

```
assign A[0] = (& (6'b010111 | (6'b001000 ^ B))) x + (&
(6'b110111 | B)) y;
assign A[1] = (&(6'b101111 | B) )ȳ +(&(6'b111110 | B);
```

In step 4, concatenate the two assigns to make a bus assignment:

```
assign A = {(& (6'b010111 | (6'b001000 ^ B))) x +
(& (6'b110111 | B)) y,(&(6'b101111 | B) )ȳ +(&(6'b111110 | B)};
```

It's not necessary first to make all terms sum-of-products form, which is usually messy, if applications of mask, reduction, and bitwise operation can produce the required expressions.

This transformation is especially useful for error correction code (ECC) such as cyclic redundant code (CRC), for which operations on individual bits can be neatly transformed into bus operation. For instance, ECC code

```
C = A[0]^A[1]^A[2]^A[9]^A[10]^A[15]
```

can be recast into bus form:

```
C = ^(A & 16'b1000011000000111).
```

A variation of the previous conversion can prove to be useful when applied to operations on bits inside a loop, as illustrated in the following example,

```
FOR (i=0; i<=31; i = i+1)
    assign A[i] = B[i] ^ C[i];
```

can be recoded as

```
assign A = B ^ C;
```

Related to this vectorization is instantiation of an array of gates, which often occurs in memory design. Instead of instantiating one gate at a time, use array instantiation so that a simulator recognizes the inherent bus structure. An example of array instantiation is

```
FlipFlop FFs [31:0] (.Q(output),.D(input),.clock(clock));
```

where the range `[31:0]` generates 32 FFs with inputs that are connected to bus D, outputs, bus Q, and clocks `clock`.

### 2.3.4  Minimization of the Interface to Other Simulation Systems

The other systems can be C code simulations or I/Os to the host. In Verilog, it is a common practice to cosimulate a Verilog model with another C model through a programming language interface (PLI). PLIs are communication channels for values to be passed from the Verilog side to the C side and vice versa. PLIs are a major bottleneck in performance. A strategy to reduce PLI impact is to communicate a minimum amount of information and accumulate the information before it is absolutely necessary to call the PLI.

Another common cause of slow performance is displaying or dumping too much data on the host during runtime, which can easily slow down a simulation by a factor of ten or more. Thus, all display and dump code should be configured to be able to turn on and off, and should be turned on only during debug mode.

### 2.3.5  Low-Level/Component-Level Optimization

In Verilog there are data types that most simulators do not accelerate and hence should be avoided. These are time, real, named event, trireg, integer array, hierarchical path reference, UDP, strength, transistor constructs (for example, `nmos`), `force/release`, `assign/deassign`, `fork/join`, `wait`, `disable`, and `repeat`. In general, anything related to time, low-level modeling constructs, and test bench constructs should be flagged.

Remove redundant code (for example, modules/functions/tasks) that is not instantiated/ called, dangling nets, and unreferenced variables. Not all simulators remove redundant code; and for the ones that do, a longer compile time results.

**Rules to Check**

1. Encourage the use of a high-level abstraction model. Issue warnings when transistor-level and strength constructs are detected. If a design has a simulation model, ensure that the model is invoked.
2. Issue warnings if FFs or latches are not instantiated from the library.
3. Sequential components in libraries are recognized as such by the target simulators.
4. Synchronous RAM is preferred over asynchronous RAM.
5. Operations on bus bits are warned if the number of bit operands exceeds a limit.
6. Issue warnings if the number of system task or user task calls exceeds a limit.
7. Low-level constructs and non-synthesizable constructs are discouraged, such as time, real, named event, trireg, UPD, strength, transistor, `force/release`, `assign/deassign`, `fork/join`, `wait`, `disable`, and `repeat`.

### 2.3.6   Code Profiling

A code profiler is a program that is attached to a simulation and is run to collect data about the distribution of the simulation time in the circuit. From the report of a profiler, the user can determine the bottlenecks in a simulation. The profiler calculates the accumulative time spent on module definitions and instances (meaning, the total time a particular module or a particular instance of the module is simulated). It also computes the time spent on blocks (such as an `always` block), functions/tasks, timing checks, and others.

## 2.4   Portability and Maintainability

In a design team, it is essential to have a uniform style guideline so that code, which may range from tens of thousands to a couple million lines, is easy to maintain and reuse. In this section, let's take on a top-down approach in discussing portability and maintainability: first, projectwide file structure, then common code resources, and finally individual file format.

### 2.4.1   Project Code Layout

A large project often involves tens of thousands of RTL files. An obvious question is how to organize them for easy maintenance and reuse, and which files should contain what. The following is a list of guidelines found in practice:

- The design RTL file structure should reflect the structure of the top-level functional blocks. For example, if the top-level blocks in the design are M1, M2, M3, and M4, then there should be subdirectories under the main RTL source directory named M1, M2, M3, and M4 that contain RTL files for the respective blocks.

- Except for the cell library, each file should contain only one module.
- The top-level module should consist only of module instantiations and interconnects; no logic should be present. The rationale is that the top-level module represents a partition of the design's functional blocks, and thus all low-level logic should belong to one of the functional blocks.
- The RTL files may contain several models, each of which can be activated by using Verilog's `ifdef` keyword. The main reason for having more than one model embedded in the code is to enhance simulation performance. Namely, a behavioral model coexists with a structural model—the former for fast simulation and the latter for implementation and synthesis. If more than one model coexists in the code, equivalence among the models needs to be ensured, thus creating maintenance overhead. It cannot be emphasized enough that a project team must strive to minimize the number of embedded models because maintaining multiple-model equivalence has a high cost later in the project. In addition, the other models should exist only in the cell library or macro library, not in the RTL files.
- The design hierarchy represents a partition of the system's functionality and should correspond to the physical layout hierarchy. This rule makes equivalence checking between design and layout simpler. An example of a design hierarchy and its corresponding physical layout is shown in Figure 2.10.

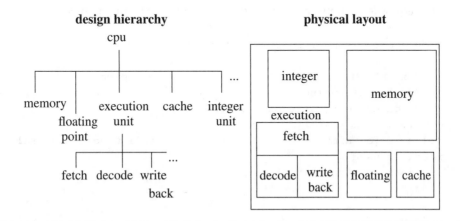

**Figure 2.10** Correspondence between design hierarchy and physical layout

• Hierarchical path access enables reading or writing to a signal directly over a design hierarchy without going through ports. Hierarchical path access should be permitted only in test benches. Therefore, all access inside the design must be done through ports. The reason for this is that hierarchical access sometimes is a necessity for a test bench because the signal monitored is only for testing purposes and may not be conveniently accessible through ports; whereas, within the design, all accesses are part of the design and thus it is reasonable to access only through the ports.

### 2.4.2   Centralized Resource

It is imperative to have centralized control over common resources. A project should have a minimum set of centralized common sources that include a cell library and a set of macro definitions as an `include` file. The macro files contain project constant definitions, memory array macro definitions, and others. All designers must instantiate gates from the project library, instead of creating their own. Similarly, memory arrays should be derived from the macro library, which expands to generate the desired structural model. No hard coding is allowed: All constants should be coded as macros. No global variables are allowed.

A cell library contains all gates and memory arrays, which may also embed other models (for example, for simulation performance). Embedding of other models is done via `ifdef`, where by defining IMPLEMENTATION_MODEL, the implementation version of the design is activated. If IMPLEMENTATION_MODEL is not defined, the faster behavioral model is selected. The equivalence between these two models needs to be maintained:

```
module adder(sum, in1, in2, carry_out);

`ifdef IMPLEMENTATION_MODEL // this is the implementation
    XOR gate1(.out(sum),.a(t1)...);
    OR gate2(.out(carry_out),.a(t2),...);
    ...
`else // for simulation performance, the following behavioral
model is used
    {carry_out,sum} = in1 + in2 ;
`endif

endmodule
```

No initialization is allowed inside a cell. Embedding initialization inside a cell adds an artifact to the true physical property of the cell, because a real physical cell does not initialize by itself. Therefore, a cell's initialization should be done by explicit reset or external loading (for example, a PLI).

### 2.4.3   RTL Design File Format

Each RTL design file should contain only one module, and the filename should match that of the module. The beginning is a section about the designer and the design (for example, name, date of creation, a description of the module, and revision history). Next is header file inclusion. Header files should contain only macro definitions, not RTL code.

In the declaration of module ports, brief descriptions about the ports are recommended. Large blocks and complex operations should have comments about their functionality and why they are coded as such. Remember that comments are not simply English translations of Verilog code: Comments should explain the intention and operation of the code. Each begin-and-end pair should have markers to identify the pair. For example,

```
begin // start of search;
...
    begin // found it
    ...
    end // end of found it
...
end // end of search.
```

Indent to show code structure. Each project should have a naming convention (for example, uppercase letters for constants/macros), a unit name should precede a module name, and markers for buses, wires, and reg (for example, first letter capitalized for bus names). An example file is presented in the following code:

```
/*****************************************
Designer: Joe Smith
Date: July 5 2003
Functionality: This module extracts the header from input stream
...
..
Revision history:
    June 11 2003: added a new counter.
    May 4 2003: fixed overflow problem.

*****************************************/
`include "project_header.h" // include macro definitions
```

```
    module InputDataEncoder (DataIn, valid, clock, ...);
    input [31:0] DataIn; // this port receives the data to be
    processed
    ...

    always @(posedge clock) begin // this block calculates ECC
       checksum = ....
          if(checksum == 1'b1) begin // find out the cause of the
          error
          ...
          end // end of cause of error
       ...
    end // end of always block
    ...
    end module
```

---

**Rules to Check**

---

1. There is a one-to-one correspondence between the logical hierarchy and the physical hierarchy.
2. Except for libraries, a file should contain only one module.
3. The name of the file should match the name of the module.
4. The top-level module should consist only of instantiations and connections.
5. The number of variant models embedded should be minimized.
6. No hierarchical paths are allowed inside the design.
7. Use library cells as much as possible, instead of writing code with the same functionality.
8. Check for project-specific file format (for example, header, line length, name convention, and indentation).

---

## 2.5 "Synthesizability," "Debugability," and General Tool Compatibility

RTL code needs to be run on tools and thus must be tool compatible. For instance, some part of the design code is synthesized, and a synthesis tool has its own subset of acceptable coding styles, called a *synthesizable subset*. Thus, that portion of code must conform to the synthesizable style. Another example includes debugging tools, which may or may not display memory content. The user needs to have some facility in place to overcome these problems.

### 2.5.1  Synthesizability

A synthesizable subset is the set of RTL constructs that is acceptable by a synthesis tool. As the tool evolves, the subset also evolves. Hence, there are no hard rules about what is synthesizable and what is not. Let's look at several common unsynthesizable constructs:

1. Asynchronous and event-based logic. Although there are special tools that can automatically generate circuits from an asynchronous RTL description, it is a safe practice to warn users when asynchronous logic is encountered, especially in a synchronous context.
2. Four-state constructs. Four-state logic includes 1, 0, x, and z, where x represents an unknown state and z represents a high-impedance state. In a real circuit, state x does not exist and a digital circuit cannot detect a z state. Thus, an RTL description that is of a four-state nature cannot be faithfully synthesized. For example, `if(out === 4'bxxzz)` cannot be implemented.
3. Constructs with delays and time. Delay is difficult to synthesize because it is a function of physical parameters and layout variations. Hence, it is not synthesizable. An example is `assign #3 a = b + c`.
4. Test bench constructs. Test bench constructs, such as `deassign`, `force/release`, `fork/joint`, `wait`, `repeat`, and `loops`, do not have direct corresponding hardware components and should be avoided during design.
5. Transistor-level models and Verilog primitive gates. RTL code is meant for gate-level synthesis and should not contain transistor-level models. Furthermore, a true transistor model should contain physical parameters such as gate width/length. Hence, a Verilog transistor model is therefore incomplete for transistor-level synthesis. Some examples are `pmos` and `nmos`. On the other hand, Verilog primitive gates, such as `nand`, should be replaced by instantiated library cells, because library cells are more project specific and are centrally controlled for projectwide consistency.

Users should consult with the synthesis tool requirements to determine what is acceptable and what is not.

### 2.5.2  Debugability

A coding style can be affected by a debugging strategy—for instance, how to dump out signal traces and how to make some nodes visible for the debugging tool. To find the root cause of a bug, some nodes values must be recorded or dumped out to trace the problem. Because the set of nodes for a bug cannot be determined a priori, the user needs to estimate the node set initially. If the bug leads to nodes that are not contained in the initial node set, the node set needs to be enlarged and the simulation needs to be rerun to dump out the values. This process repeats itself until the root cause of the bug is found. The more nodes a simulation run dumps out, the slower the simulation speed. Therefore, only very rarely will the user dump out all nodes in the circuit on the first try. Consequently, signal tracing should be divided into several levels: The higher the level, the more nodes that are traced. Usually, the tracing level is based on the design hierarchy, with the top level dumping out all nodes.

This hierarchical signal tracing strategy needs to be coded in the design, and it is implemented usually through ifdef constructs. For example, ifdef TOP_LEVEL_TRACING guards a piece of code that dumps out all nodes. If TOP_LEVEL_TRACING is defined, that piece of code is run to dump out all nodes. Similarly, ifdef MMU_TRACING dumps out only the nodes inside the MMU block.

Select dumping can be done at runtime and is achieved using plusarg. The feature plusarg allows the user to input an argument at the command line using + (hence the name plusarg). For example, we can run a simulation with variable DUMP_LEVEL set to 2 with a syntax like

```
runsim +define+DUMP_LEVEL=2 ...
```

Inside the code, the code segment triggered by DUMP_LEVEL=2 is guarded by an if statement, such as

```
if($test$plusargs(DUMP_LEVEL) == 2)
begin //$test$plusargs returns value of DUMP_LEVEL
   $display ("node values at hierarchy 2",...)
   ...
end
```

The difference between using plusarg and ifdef is that the former compiles all code for dumping and decides at runtime whether to activate the code, whereas ifdef compiles only the code that is defined and thus no selections can be made at runtime. The plusarg method uses more memory, compiles longer, and may run slower, but users do not have to recompile if they need to change node selections for dumping.

Some simulators do not allow tracing of certain data structures (for example, memory arrays and variables inside a loop). The reason for memory arrays is because of the large size of memory arrays that dumping all memory locations at each time step would cause the simulation to slow to a crawl and take up a great amount of disk space. Therefore, if users want to see the content of a memory location, they can assign the memory location to a register and dump out the register's value instead. Another way to access memory is to deduce the memory content at a given time based on prior writes to that address, which is covered in more detail in Chapter 6.

All iterations of a loop are executed in zero simulation time. Thus, the value of a variable inside a loop, when dumped out, is usually the value at the loop's exit, instead of all the values during the loop's execution. For example, if variable x increments from 1 to 9 in an execution of a loop, the value of x will be 9 when it is dumped. To access the variable's values for the entire loop simulation, users need to modify the code so that the variable's value is dumped at each loop iteration. This is easily done by using $display for the variable inside the loop. Make sure $display can be turned on or off. This method is very messy and can be costly in hardware-assisted simulation. It is a major reason why loops should be avoided.

---

**Rules to Check**

1. Issue errors for unsynthesizable constructs.
2. If loop variables and memory arrays are needed for debugging, make certain that they are observable.
3. Implement a layered dumping strategy.

---

## 2.6 Cycle-Based Simulation

Cycle-based simulation simulates the circuit only at cycle boundaries so that all transitions in a cycle but the last one are ignored. In other words, cycle-based simulators compute the steady-state behavior. The following illustrates the difference between an event-driven simulator and a cycle-based simulator. Figure 2.11 shows the transitions in a clock cycle. For instance, node c has six transitions and node d has nine transitions. On the other hand, in an event-driven simulator, a gate needs to be simulated if one of its inputs changes value. (A smart event-driven simulator may simulate only if the input transition affects the gate's output.) Therefore, gate A simulates five times; gate B, nine times; and gate C, six times, for a total of 20 times.

Because only the steady-state value is seen by the receiving FF, all the intermediate transitions can be ignored. Thus, in cycle-based simulation, the entire circuit is simulated just once, at the end of the current cycle. In this example, only three gates are simulated. In cycle-based simulation, all gates are simulated, regardless whether their inputs have

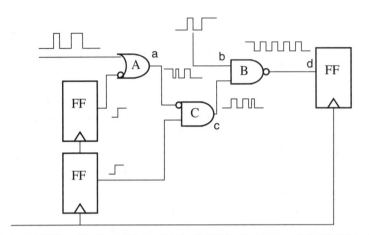

**Figure 2.11** Event-driven simulation versus cycle-based simulation

changed. Therefore, an event-driven simulator is faster if there are few transitions, typically less than 5% active nodes. In practice, experiments confirm that a cycle-based simulator is almost always five to ten times faster.

For a circuit to be simulated by a cycle-based simulator, the simulator must be able to
determine cycle boundaries (the beginning or the end of a clock cycle). It is for this reason
that an asynchronous circuit cannot be simulated by a cycle-based simulator. The following is a list of properties that a cycle-based simulated circuit should possess:

1. No event or delay constructs can be simulated using a cycle-based simulator,
   because all events except clock transitions are ignored and delays are eliminated.
2. The clock tree must be made of only simple gates, such as AND, OR, and
   INVERTOR, and the logic should be simple so that the simulator can determine
   the cycle boundary.
3. False loops should be eliminated. A false loop is a loop that consists of combinational gates, but signals cannot propagate around the loop in any one cycle. An
   example of a false loop is shown in Figure 2.12. The loop is false because one of
   the multiplexors will select input 1, breaking up the loop. A false loop presents no
   problem to an event-driven simulator, because signal propagation will eventually
   die before traversing the loop, and hence the number of gates simulated is limited. But for a cycle-based simulator, it is not as straightforward when it comes to
   determining whether it's a real combinational loop or a false loop, because
   absence of events is not used for terminating an evaluation. Furthermore, loops
   also really mess up static timing analysis.

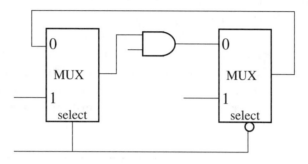

**Figure 2.12** A false loop

4. Clock gating needs special attention in cycle-based simulation, because the gating signal is not, in some simulator architectures, a part of the clock tree, but it can trigger a necessary evaluation. Let us consider the circuit segment in Figure 2.13, where the output of FF1 gates the clock of FF2. Some simulators will only consider the AND gate as a part of clock tree for FF2, and will treat FF1 and the buffer as a part of the data. Therefore, FF2 is only evaluated on transitions of clk2. However, if the output of FF1 toggles at different times than that of clk2, FF2 will not be evaluated at the transitions of FF1, but these transitions do trigger FF2 in a real circuit, causing simulation errors.

5. Every sequential component must have a port connected to the system's clock tree. This port is the clock to the sequential component and it triggers the evaluation of the component. A case in point is a frequency divider in which an output of an FF clocks another FF. If this condition is missing, incorrect simulation will result. Consider the sequential block always @(x) out = x + y, where x is not

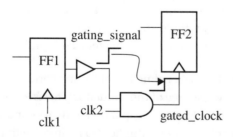

**Figure 2.13** Logic evaluation triggered by a gating signal

connected to a clock tree. In event-driven simulation, output `out` gets updated only when `x` changes. Suppose `y` changes value but `x` does not. Because `x` is not connected to a clock tree, the block will not be evaluated, and hence output `out` still has the old value of `y`. However, in cycle-based simulation, the block is evaluated every cycle, and output `out` is updated with the latest `x` and `y` values regardless whether `x` changes. Consequently, an erroneous simulation occurs. Connecting `x` to a clock tree remedies the problem, because now an event simulator will evaluate the block whenever the clock, and thus `x`, changes.

---

**Rules to Check**

1. No event or delay constructs are allowed.
2. Only gates of predefined types are allowed in a clock tree. Usually they are simple gates such as AND, OR, and INVERTOR.
3. False loops must be eliminated.
4. Clock enable signals may initiate logic evaluation.
5. All sequential components must be clocked by a signal from a clock tree.

---

## 2.7   Hardware Simulation/Emulation

Hardware simulators/emulators are special-purpose computers for accelerating circuit simulation/emulation. An emulator is like a simulator except that it is usually embedded in a system in the place of a subsystem; thus, response time is critical. Although the following discussion targets simulators, it is equally applicable to emulators. There are two categories of hardware simulators. The first kind uses FPGA chips to mimic the design. In the first kind of machine, a design is synthesized into a gate netlist, and then the FPGAs are mapped to the netlist and are programmed to the gate functionality. The second kind is based on special processors that directly simulate the circuit. Such a processor's native language includes AND, OR, multiplexing, bit operations, and other Verilog operators so that a circuit is compiled to the processor's machine code and is then simulated.

For hardware simulators there is an almost one-to-one correspondence between components in the design and the resource of the simulator. This is especially true for FPGA-type accelerators. Consequentially, loops are expanded and tasks and functions are inlined. In loop expansion, the body of the loop is duplicated $N$ times, with respective values substituted for the loop indices, $N$ being the loop count. Task/function inlining means that at every place where a task or function is called, the entire body of the task or function is inserted at the location. Therefore, a code segment of ten lines with three calls to a function of five lines expands to 7 + (3 * 5) = 22 lines.

It's imperative to make an effort to minimize the body of loop statements, tasks, and functions. The idea of loop content minimization is demonstrated in the following example, which occurs often in practice, that searches for some patterns such as memory content and operates on the found items. It is very tempting to put everything under the search loop, as seen here:

```
found = 1'b0; // not found
for(i=1;i<=100;i=i+1) begin
    if(memory[i] == 8'b10101010 && !found) begin // search
        found = 1'b1;
        read = memory[i];
        do_lots_operations(read);// big task has 200 lines
    end // end of found
end // end of loop
```

When this loop expands, the 200 lines in task `do_lots_operations` are duplicated 100 times; the resulting code has around 20,000 lines! To minimize expansion size, search and operation steps should be made separate. A disjoint search and operation optimization produces the following code:

```
found = 1'b0; // not found
for(i=1;i<=100;i=i+1) begin
    if(memory[i] == 8'b10101010 && !found) begin // search
        found = 1'b1;
        read = memory[i];
    end // end of found
end // end of loop

if(found) do_lots_operations(read); // not duplicated
```

In this code, only the search portion is duplicated 100 times; the operations are not duplicated at all. The resulting code has about 500 lines, compared with 20,000 lines in the unoptimized case. If there is more than one item for which to search, the items can be packed into an array inside the search loop and later passed to the operation. This technique finds much application, especially in test benches.

Almost all hardware simulators delegate computing of C/C++ language routines, such as user tasks written with PLIs, to the host. When this happens, overhead time in passing the

C/C++ routines to the host and then receiving the results back slow down the overall simulation. This is a major bottleneck in large system cosimulation. Therefore, PLI user tasks should be avoided at all costs. One strategy to mitigate performance degradation of PLI user tasks is to group them together and communicate with the host all at once (for example, at the end of a clock cycle). This strategy works well if the execution of all PLI user tasks can be delayed to the end of the cycle. One type of PLI user task, called *midcycle PLI*, cannot be delayed to the end of the cycle. These PLI user tasks return values that are used by other parts in the design in the *same* cycle. This drastically slows down simulation. Therefore, midcycle PLI user tasks should be treated, for practical purposes, as errors for hardware simulators.

---

**Rules to Check**

1. Issue errors/warnings for unsynthesizable constructs.
2. Minimize body statement count in loops, tasks, and functions (for example, by using search and operate disjoint optimization).
3. Restrict the use of system tasks and user tasks.
4. Prohibit midcycle PLIs.

---

## 2.8   Two-State and Four-State Simulation

Four-state (0, 1, x, and z) simulation has been in common use to detect design problems such as uninitialized states and bus contention in which the states and the bus are assigned x value. Value x denotes an unknown value and value z denotes high impedance. A simulator is a four-state simulator if every node can have a four-state value. A two-state simulator uses only 0 and 1 values. A two-state simulator is faster than a four-state simulator. If a node is uninitialized or a bus is in high impedance or contention, a two-state simulator maps x and z to either 0 or 1. It is common practice to simulate a design using a four-state simulator during the power-up stage until most nodes are initialized, then the design is switched to a two-state simulator for faster speed.

Because a real circuit does not compare or operate with x or z values, the design should never have x or z values (except for the case when x, z, and ? are used for synthesis purpose, which should be used with caution because this will give rise to discrepancies between a simulation model and a synthesis model); however, a test bench can have these values. A designer should take into consideration that the design along with the test bench may undergo two-state simulation as well as four-state simulation, and hence make an effort to minimize simulation result in discrepancies between the two. Some differences are inevitable, because of the inherent pessimism of x. Consider the two-to-one multiplexor with inputs that are both 1, but select x, as shown in Figure 2.14. The algebra of unknown x is that the complement of x is also x. Therefore, the inputs to the OR gate are both x, instead of x and $\overline{x}$, producing an unknown x for the multiplexor output. This is the result of a four-state simulation. However, in reality, because both inputs of the multiplexor are 1, the output should be 1 regardless of the select value.

If this multiplexor is simulated using a two-state simulator, the output will be 1, because the select value is mapped to 0.

Although inherent differences between four-state and two-state simulation exist, the designer should not allow these differences to propagate to create functional errors. Let us consider the following example:

```
if( FlipFlopOutput == 1'b0 ) begin
    count = count + 1;
end
```

Variable `FlipFlopOutput`, if uninitialized at power-up, takes on value x and thus makes the comparison false. In Verilog, a conditional expression is, by default, x if any one argument is x. Therefore, variable count does not increment. On the other hand, a two-state simulator represents an uninitialized variable with value 0, and hence satisfies the condition and increments variable count by one. Consequently, the values of count differ by at least one in two-state simulation and four-state simulation, and this difference can further penetrate into other parts of the design. This type of functional difference can be hard to debug and can be quite misleading.

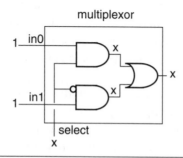

**Figure 2.14** Pessimism of four-state simulation

The proper thing to do for this type of situation is to determine whether the inputs to the block (for example, `FlipFlopOutput`) should be initialized. If yes, make sure they are initialized; otherwise, closely examine your block to see how it should behave in reality, when the inputs can be either `0` or `1`. Does it produce different results and propagate them to create different system states? If so, the designer needs to improve the design to be better immune to uninitialized states. Similar caution should be exercised about using strength.

---

**Rules to Check**

1. No `x` or `z` constants are allowed inside the design.
2. Check for comparison expressions that involve uninitialized variables.
3. All states in the design are initialized or the ensuing circuit handles unknown values properly.

---

## 2.9   Design and Use of a Linter

As we discussed earlier, for style guidelines there is no clear division between error and warning. An error for a false loop in a cycle-based environment may be just a warning in an event-driven environment. Furthermore, there are various degrees of severity for errors and warnings. As a project progresses, a warning can be promoted to an error, and vice versa. Therefore, when implementing a linter, one should set the tool to have a user programmable error/warning system. For instance, it could issue seven levels of error and seven levels of warning, each corresponding to the categories in this chapter. It could have a command-line option to demote a class of error or to promote a class of warning.

Giving out warnings and errors is only the first step. The second step is to assist engineers to locate the code causing the warnings and errors. For some warnings and errors (such as combinational loops), the problematic code can be interspersed in several files. A graphical interface is helpful to show the connections among the code fragments. Keep in mind that a text-based interface is still required for the tool to be integrated into a verification flow.

In addition, a single root cause can generate multiple messages. For example, a false loop involving a bus can generate multiple loop messages, with each loop involving a particular bit. Conversely, an error can mask other errors; so, removing an error can unmask others and generate more. Both situations can be discouraging and misleading; therefore, a tool needs to limit the effects of these two extremes.

Finally, the program should postprocess warning and error log files, generate statistics, classify warning and error according to severity, and submit reports to responsible parties.

In summary, a good linting package has three components: a checker/linter that detects violations, a locator that assists users to locate the code causing the warnings and errors, and a report generator that creates reports and statistics.

## 2.10   Summary

Designing with verification in mind prevents errors from happening before the design goes into verification. One way to achieve this is to have a set of coding guidelines and a tool to enforce it. There are five general and three ad hoc categories—namely, functional, timing, performance, maintainability, and tool compatibility; plus cycle-based, hardware, and two-state simulation. A robust enforcement package should include a checker to detect violations, a locator to identify the problematic code, and a report generator to gather statistics. Errors and warnings are issued with predefined levels of severity and can be promoted or demoted as the project progresses.

## 2.11   Problems

**1.** Find as many bad coding styles as possible in the following RTL code:

```
reg [7:0] N, out;
reg [31:0] P;
reg [63:0] G;
reg [1:0] Y;
reg clock;
wire clk, enable;

assign clk = enable & clock;

always @(posedge clk) begin
   Y = G[31:9] ^ N;
   case (Y)
      2'b00: Out = N;
      2'b01: Out = ~N;
   endcase
end
```

**2.** What would a circuit be like if it were synthesized from the following code?

```
always @(a or b) begin
   if(b == 1'b1 && a==1'b0) c = 1'b1;
   else if(b==1'b1 && a==1'b1) c = 1'b0;
end
```

**3.** Determine whether the following RTL descriptions are sequential or combinational.

a.
```
always @(y) begin
   s = y;
   z = s;
end
```

b.
```
always @(y) begin
   s < = y;
   z = s;
end
```

**4.** Rewrite the following code as much as possible to conform to the preferred styles discussed in the chapter.

```
and (out1, in1, in2);
or (out2, out1, in2);
2bitadder m1(carry_out,carry_in,in1,in2,out);

always @(posedge clock) begin
if(carry_out)
flag = 1'b1; // overflow
else if (~carry_out && carry_in) begin
check = 1; // a 1-bit flag
if(~out1) $display("do something");
end
else $call_PLI(carry_in, carry_out);
end
```

**5.** As discussed in the chapter, bus contention should be checked at runtime with assertion routines.

  a. Write a Verilog assertion routine detecting bus contention (in other words, more than one driver is enabled at the same time).

  b. What property does bus A have if `toh_eno` is printed in the following code? (Hint: Try A = 0...01000 and A = 010...01000.)

```
reg [31:0] A, B, R;

always @(posedge clock) begin
B = A - 32'b1;
R = A & B;
if CR == 32'b0)$display("toh.eno");
else $display("unknown");
```

This code can be succinctly put in one line:

```
assign R = A & (A - 32'b1);
```

**6.** Identify all true and false loops in the Verilog code. You should treat an `always` block and continuous assigns as single components:

```
always @(c or b or d)
   case (d)
        1'b0: a = d;
        default: e = b;
   endcase
assign b = a;

always @(e or c)
        d = e + c;
```

**7.** Can the latch be replaced with a buffer in each of the cases in Figure 2.15, assuming that all latches and FFs are driven by the same clock? Explain.

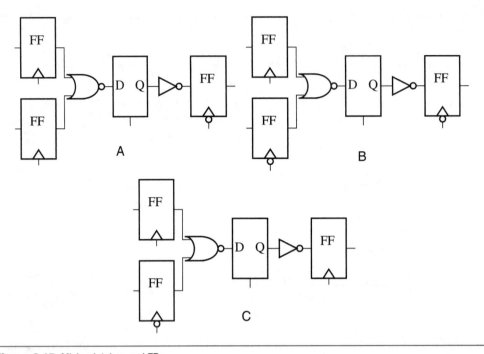

**Figure 2.15** Mixing latches and FFs

**8.** Convert the following bit operations into bus operations without part selects:

```
reg [31:0] A,B,C;
B[15:0]  = A[15:0] & C̄[15:0];
B[22:16] = Ā[22:16]^C̄[22:16] + C[22:16];
B[31:23] = Ā[31:23]^C[31:23];
```

**9.** Minimize the following loop's content. Suppose all loops must be unrolled and the routine must be expanded. What are the numbers of lines before and after you apply the optimization?

```
done = 1'b0;
For(i=0;i<100;i=i+1) begin
   if(!done && ((M[i] == 4'b1010) || (M[i] == 4'b1111))) begin
      Do_massive_computation(M[i]); //1000 lines of code here
      done = 1'b1;
   end //end of if
end //end of for
```

**10.** The following code has the so-called event-counting race. The intent is to count the total number of transitions on Cx and Cy. Redesign the code to eliminate the race condition while preserving the functionality.

```
always @(Cx or Cy)
    C = C + 1;
```

**11.** Consider the circuit in Figure 2.16. The clock tree has a multiplexor that switches clock frequencies.

    a. If $f_1 = k*f_2$, where $k$ is a positive integer, is there a $k$ value such that the output of the multiplexor is free of zero time glitches? Assume that FF1 is clocked by $f_2$.

    b. If FF1 is clocked by a waveform of frequency $f_3$, what conditions must be met to prevent zero time glitches at the output of the multiplexor?

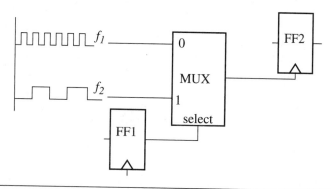

**Figure 2.16** Multiplexing clocks

# Simulator Architectures and Operations

## Chapter Highlights:

- The compilers

- The simulators

- Simulator taxonomy and comparison

- Simulator operations and applications

- Incremental compilation

- Simulator console

O nce a design is free of statically detectable errors, it is ready to be verified by simulation. To use a simulator effectively, it is imperative first to understand its architecture. Following this philosophy, this chapter is organized in two parts: The first part is devoted to architectures of simulators, and the second discusses simulator operations. In studying simulator architecture, we will examine in depth two typical simulators on the extremes of the simulator spectrum. We then discuss how simulator types in the middle portion of the spectrum can be constructed, followed by a comparative study of the various types. In the second part of the chapter, common and advanced features of simulators are presented along with their applications.

Generally, a simulator consists of three major components: a front end, a back end, and a simulation engine/control, as shown in Figure 3.1. The front end is very much standard for most simulators and is a function only of the input language. The back end performs analysis, optimization, and generation of code to simulate the input circuit, and is the main contributor to a simulator's speed. The front end and the back end form the compiler portion of a simulation system. The simulation engine takes in the generated code and computes the behavior accordingly. In this stage, the generated code has no direct knowledge of the circuit and can be in any language. If the generated code is C/C++, it first needs to be compiled using a C/C++ compiler before it is run. If the generated code is the native code of the target machine, it can be run immediately. Simulation control allows the user to interact with the operation of the simulator. An example of user control is to run a simulation in interactive mode, as in debugging a design, in which the user can set break points to pause a simulation after a number of time steps, examine variable values, and continue simulation.

## 3.1   The Compilers

The front-end portion of a compiler, consisting of a parser and an elaborator, processes the input circuit and builds an internal representation of the circuit. Specifically, a parser interprets the input according to the language's grammar and creates corresponding internal components to be used by the elaborator. For example, a module is parsed into an internal programming object that has fields for the module name, port name, port type, and a link list of all entities inside the module. The elaborator constructs a representation

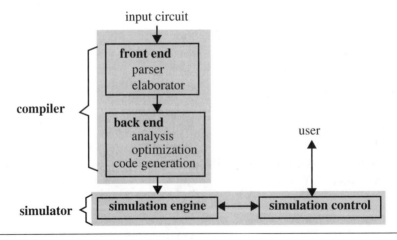

**Figure 3.1**   Major components of a simulator

of the input circuit by connecting the internal components, substituting parameters, inlining or setting up parameter passing for functions and tasks, and others. For instance, the elaborator substitutes a module instantiation with the module's definition and connects the internal objects consistent with the circuit connectivity. Sometimes the elaborator applies optimization to the internal representation. The end result from an elaborator is a complete description of the input circuit sufficient to sustain all later processing and operations.

The back end, the soul of a simulator, determines the type of the simulator. The actual operations of the analysis stage vary from one type of simulator to another. For a cycle-based simulator, it performs clock domain analysis and levelization, whereas for an FPGA-based hardware simulator, in addition to the previous analysis it also partitions, places, and routes the circuit into FPGA chips. For this reason, an in-depth discussion of analytical stage is relegated to the sections on specific simulators.

The type of simulator also dictates the construction of code generation. There are four classes of generated code: interpreted code, high-level code, native code, and emulation code. The last three are sometimes referred to as compiled code.

In an interpreted simulator, the input circuit is compiled into an intermediate language for the interpreted simulator. The interpreted simulator can be regarded as a virtual machine that reads in instructions in the intermediate language, one instruction at a time. The effect of executing the interpreted object code creates the behavior of the circuit. The diagram in Figure 3.2 depicts this interpreted simulation process. The interpreted simulator is a virtual

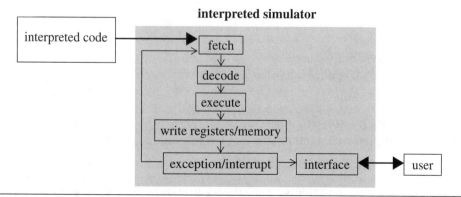

**Figure 3.2** Interpreted simulation structure and process

machine with the interpreted code as its instructions. The instructions are fetched, one by one, then are decoded and executed. The interpreted simulator has a user interface to allow data examination and execution control. An example of interpreted code and the circuit it simulates is as follows:

```
// circuit being simulated
initial
begin
clk = 1'b0;
#1 clk = ~clk;
#1 clk = ~clk;
#1 finish;
end

always @(clk)
begin
   a = b & c;
   if (a == 1'b0)
       p = q << 3;
end

// generated interpreted code
assign(clk,0);
invert(clk);
evaluate(b1);
invert(clk);
evaluate(b1);
exit();

b1: //definition of routine b1
{
    and(a,b,c);
    if(a,0) left_shift(p,q,3);
}
```

The functions in the interpreted code—assign(), invert(), evaluate()—are instructions for the interpreted simulator. Note that the stimulus or test bench is compiled with the circuit.

Interpreted code is very portable. Compiled once, the code can run on any machine that has an interpreted simulator. However, its simulation is the slowest compared with the other three kinds, because of the extra layer of execution on top of the native machine.

A compiler can also compile a circuit into a high-level language such as C/C++. To simulate, the generated C/C++ description of the circuit is compiled with a C/C++ compiler and is run just like any other C/C++ program. A sample generated C program simulating the previous circuit is shown here. By analyzing the Verilog code, it is determined that the clock toggling statement can be combined with the `always` block. In general, such a simplification may not exist. In that case, the C code of the `always` block will have to be run on a separate thread that constantly monitors changes in the clock variable. A change on `clk` will trigger an evaluation of the C code representing the `always` block:

```
main()
{
    int clk;
    int i;
    int a, b, c, p, q;
    clk = 0;
    for (i=0; i<2; i++) {
        clk = (clk == 0) ? 1 : 0 ; // clk = ~clk;
        a = b & c; // always block
        if (a==0)
            p = q << 3;
}
```

High-level code is not as portable as interpreted code because high-level code needs to be recompiled to the native language of the platform (for example, SUN workstations) every time it is simulated. This compile time can be long because the high-level code is usually very large. High-level code is portable to a certain degree, because the generated high-level code compiled from the circuit can be reused.

Native code compilation, skipping the intermediate language generation step (for example, C/C++ or interpreted code), directly produces the machine executable code for the platform. At the expense of portability, native code runs slightly faster than high-level code because of the more direct machine code optimizations. Both native code and high-level code are typically about 5 to 20 times faster than interpreted code. The major shortcoming for native code compilation is portability.

Finally, in hardware simulators/emulators/accelerators, the compiler generates the machine code for the hardware simulators/emulators. During simulation, a hardware simulator sometimes requires interaction with the host machine. An example of such an interaction is running C code in the host simultaneously with the circuit simulation in the hardware simulator (for example, PLI C code running on the host in lock step to compare result with that from the hardware simulator at the end of each cycle). (This is discussed further, later in the chapter.) This type of interaction is a major bottleneck in simulation performance. If the host–hardware interaction can be minimized, simulations on hardware are orders of magnitude faster than those in software simulators, typically in the range from 100 to 10,000 times. The disadvantages are long compilation time and capacity limitation (the maximum size of the circuit that can fit into the simulator). Figure 3.3 summarizes the four types of simulators and their simulation processes.

An interpreted simulation system has a clear separation between code of the compiled circuit and code of the simulator, with the circuit code feeding the simulator, as indicated in Figure 3.2. However, a compiled simulation system is composed of a single piece of compiled code that combines the circuit, the simulator, and a user interface program. The structure and execution flow of a compiled simulator is shown in Figure 3.4. The compiled code, the output of the compiler, has instruction memory that represents the circuit connectivity and its components' functionality, data memory that stores simulation values of nodes and variables in the circuit, and a simulation engine, sometimes called a *simulation kernel*, that manages scheduling, controls component evaluation, stores and retrieves simulation values, advances time, handles exceptions and interrupts, distributes tasks to processors in the case of a simulation using multiple processors, and other duties. To illustrate the simulation structure, consider a transition occurring at the primary input of a gate-level circuit. The simulation engine determines the fanin gates affected by the transition by following the connections of the primary input in the instruction memory, and places fanout gates in an evaluation queue. For each component in the queue, the evaluation starts by looking up its functionality in instruction memory and ends by storing the

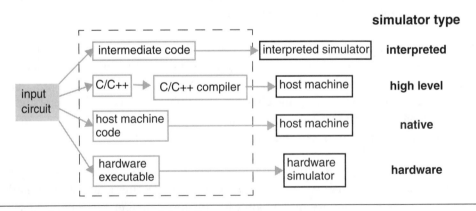

**Figure 3.3**  Summary of four types of simulation processes

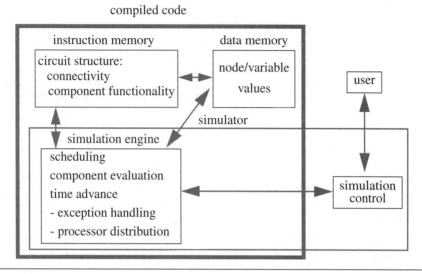

**Figure 3.4**   Compiled simulation system structure

output values in data memory. When all events have been evaluated, the simulation engine advances the time.

The simulation engine and the simulation control unit form the simulator, but the engine resides inside the compiled code, and the simulation control unit lies in a separate program. If high-level code is generated, the object code of the simulation engine is linked with the object code of the compiled circuit to produce a combined object code. If native code is generated, the final result is the executable. In running a simulation, the control unit is invoked first, then the compiled code is loaded. Through the control unit, the user directs the simulation—for example, specify the time steps a simulation is to be run, setting break points, inspecting node values, dumping out signal traces, and check-point. An example of a compiled simulator is presented in "Cycle-Based Simulators" on page 88.

## 3.2   The Simulators

Many architectures of simulators exist. In this section we first study two contrasting architectures: event driven and cycle based. Through this discussion, we introduce various concepts, terminologies, and techniques, and their benefits and shortcomings. Then we move on to an investigation of hybrid and hardware simulators.

## 3.2.1   Event-Driven Simulators

An event-driven simulator, as the name implies, evaluates a component, whether it is a gate or a block of code, only when there is an event at an input or sensitivity list of the component. An event is a change of value in a variable or a signal. If an event at a gate input causes one of its outputs to change, all the fanouts of the gate will have to be evaluated. This event ripples throughout the circuit until it causes no more events, at which time evaluation stops.

**Timing wheel/event manager.** When multiple events occur simultaneously, each of which causes further events, the simulator, being able to evaluate only one at a time, must schedule an evaluation order of the events. Events are stored in an event manager, which sorts them according to event occurrence time. Events occurring at the same time are assumed to have an arbitrary order of occurrence. Evaluations are then executed on the stored events starting from the earliest time. When the simulator is at time $T$, all events that occurred before time $T$ must have been evaluated.

---

### Example 3.1

To illustrate the interaction of event scheduling and evaluation, consider the example in Figure 3.5. Assuming each gate has a delay of one unit, the timing diagram shows five transitions or events, labeled e1, e2, e3, e4, and e5. Let us apply the event-driven idea to the circuit and derive the timing diagram. Our event manager's data structure uses a two-dimensional queue, called a *linked list*. One dimension is a queue of time slots, with each entry pointing to another queue that holds all the events occurring at that time. The first event is the falling transition at input *in* occurring at time 0, and the affected gate is A. So event e1 is stored in an event queue at time slot 0, and it is the first event to be evaluated. An evaluation on gate A produces a falling transition on its output, e2, at time 1, because of the delay of the gate. Event e2 is placed into the event queue at time slot 1. When gate A is done evaluating, e1 is deleted from the event queue. The next event is taken off the queue, which in this case is e2. The affected gates are B and C. The evaluation of gates B and C resulting from e2 create output transitions e3 and e4, both at time 2, which are then placed at time slot 2. Event e2 is deleted. Event e4 is taken from the queue, and its evaluation generates no further events because there is no fanout. Thus, e4 is deleted. Next, e3 is evaluated and is found to produce e5 at time 3, which is queued at time slot 3. Then e3 is done and deleted. Finally, e5 affects no gates, so the chain of evaluation stops. This series of evaluations gives the transitions shown in Figure 3.5.

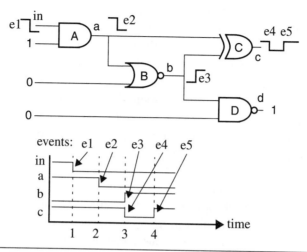

**Figure 3.5** Event-driven simulation example

In practice, the event queue is often implemented as a circular queue or a timing wheel, where the time queue wraps around itself, as shown in Figure 3.6. A timing wheel has time slots, and each time slot points to a queue that stores all events occurring at that time. Simulation progresses along the time slots. At each time slot, the events of the queue are evaluated one at a time until it is empty. During an event evaluation, if events are generated, they are inserted into the queue in the time slots at which they will occur. When all the events in the queue are examined, simulation advances to the next time slot.

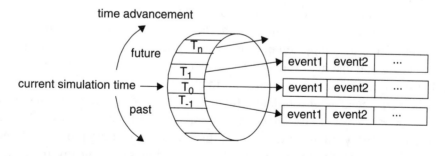

**Figure 3.6** A timing wheel: a two-dimensional queue made of a circular time queue and a linear event queue

The number of time slots in a timing wheel reflects an estimate of the maximum number of occurrences of future events, which is equal to the number of distinct delays from a primary input to a node. In practice, the size of the timing wheel is not equal to the maximum theoretical bound, but is equal to an empirical average. Therefore, it is possible that the timing wheel experiences overflow. When this happens, an overflow two-dimensional queue is created to hold the new entries. As simulation time advances, time slots on the timing wheel are freed so that new events can be placed there instead of in the overflow queue. An overflow queue is less efficient to manipulate than a timing wheel. Thus, choosing the right-size timing wheel has a palpable impact on simulation performance.

**Scheduling semantics.** So far, we have assumed that all events in the same time slot are processed in an arbitrary order. However, events at the same time have to be prioritized according to IEEE Verilog standards. In Verilog, events at a simulation time are stratified into five layers of events in the following order of processing:

    1. Active
    2. Inactive
    3. Nonblocking assign update
    4. Monitor
    5. Future events.

Active events at the same simulation time are processed in an arbitrary order. The processing of all the active events is called a *simulation cycle*. Inactive events are processed only after all active events have been processed. An example of an inactive event is an explicit zero-delay assignment (#0 x = y), which occurs at the current simulation time but is processed after all active events at the current simulation time have been processed. A nonblocking assignment executes in two steps. First it samples the values of the right-side variables. Then it updates the values to the left-side variables. A nonblocking assign update event is the updating step of a nonblocking assignment and it is executed only after both active and inactive events at the current simulation time have been processed. Note that the sampling step is an active event and thus is executed at the moment the nonblocking statement is encountered. Monitor events are generated by system tasks $monitor and $strobe, which are executed as the last events at the current simulation time to capture steady values of variables at the current simulation time. Finally, events that are to occur in the future are future events. Each time slot, in reality, points to four subqueues corresponding to the four groups of events. Within each subqueue, the order of events is arbitrary.

---

### Example 3.2

In the following example, the assignments occur at the same simulation time. The non-blocking assignment y <= x executes in two steps. First it samples the value of x, and then it updates variable y. Because the two `always` blocks occur at the same simulation time, the two blocking assignments x = a and x = b are active events, and the order in which they are executed is arbitrary. Therefore, the sampled x value of the nonblocking assignment can be either a or b. But the nonblocking assignment is *updated* after all blocking assignments are processed, including blocking assignment y = c, as stipulated by the IEEE standard. Therefore, the value of y is either a or b, but never c:

```
always @(posedge clock)
begin
    x = a;
end

always @(posedge clock)
begin
    x = b;
    y <= x;
    y = c;
  end
```

It should be noted that the execution order of blocking assignment x = b and y = c is as written, because they are procedural statements in the same `always` block.

---

**Update and evaluation events.** When an event is placed in a queue, it only means that the event may happen. Whether it actually *will* happen has to be evaluated. In the context of simulator design, events are further conceptually categorized into update events and evaluation events. An update event occurs when a variable or a node changes its value. When an update event has occurred, all processes sensitive to the variable or node are triggered and must be evaluated. This evaluation process is called an *evaluation event*. If an evaluation event of a process changes the values of some variables, then update events are generated for the affected variables. Therefore, an update event causes evaluations of the processes sensitive to it, which in turn may produce update events for their output variables.

The update event simply replaces the existing value of a variable or node with the new value. The evaluation event essentially simulates the gates or blocks sensitive to the update event. If the affected gates or blocks have no delays, the variables and nodes in the gates or blocks are computed at the current simulation time. If there are changes, update events are scheduled. If a gate or block has delays, then the result of the evaluation will be known only at a future time. When this happens, the evaluation schedules future events. These future events need to be validated at future times. It is possible that some of these scheduled future events may be canceled, as the following example demonstrates.

---

**Example 3.3**

Consider the AND gate in Figure 3.7, with two inputs that have delays as shown. At time 1, input a has an update event, a rising transition. This event triggers an evaluation event for the AND gate. Because the delay from input a to output c is 3, a future event, say E1, for output c at time 4 is scheduled. It appears that at time 4, a rising transition may occur at output c. Next, at time 2, a falling transition at input b occurs, causing an evaluation event for the AND gate. Because the delay from input b to output c is only 1, a future event, say E2, for output c is scheduled at time 3. When time advances to 3, we evaluate event E2 and conclude that output c is 0. At time 4, evaluating event E1 shows that output c remains at 0, because a falling transition at input b at time 2 travels along the faster path to the output and stabilizes the output to 0 before the rising transition can take effect. Therefore, event E1 is suppressed or canceled.

---

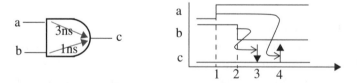

**Figure 3.7** A predicated event is canceled because of a later event in a gate with asymmetric delays

---

Suppose that a future event is scheduled at time $T$. We now examine how the event is validated. At time $T$, the event is taken off the timing wheel. From the event's content, we get the gate that was predicted to produce this event. Because all events up to the current time at all the inputs of the gate have already happened, all input waveforms are known up to the current simulation time. So we can compute the output transition at $T$, using the gate's logic functionality and internal delay information. If a gate has internal delays, then only the portions of the input waveforms that are still "in flight" inside the gate need to be used. For instance, to compute the transition at the output of the AND gate in Figure 3.7, we only need to know the value at input a at time $T - 3$ns, and the value at input b at time $T - 1$ns. Input values at these times have just propagated to the output. In other words, only the portions of input waveforms determined by the gate delays, not the entire history, are required for validation. The computed output transition occurs if it is validated.

Let us illustrate this using the example in Figure 3.7. At time 1, the rising transition at input a schedules an event at time 4; and at time 2, the falling transition at input b schedules another event at time 3. Let us validate these two events. When the simulation advances to time 3, we first determine the values at the output that came from inputs a and b respectively. The value from input a is the value at input a at time $3 - 3 = 0$. In other words, $T$ minus the delay from input a to output c. The value at input a at time −1 is 0. The value propagating from input b is the value at input b at time $3 - 1 = 2$, which is 0. (We compute a transition at the moment just after time $T$.) Therefore, the output is 0 at time 3. When the simulation advances to time 4, we validate the event predicted by the rising transition at input a back in time 1. The output value propagating from input a is the value at input a at time $4 - 3 = 1$, which is 1. On the other hand, the output value from input b is the value at input b at time $4 - 1 = 3$, which is 0. Therefore, the resulting output value at time 4 is 0. That is, the scheduled event at time 4 is canceled and the output remains 0.

This event validation algorithm is summarized in the following list. If the gate has multiple outputs, the outputs are validated one at a time:

1. Let the present time be $T$, g have n inputs, and the functionality of the gate g be f().
2. For each input $x_i$ of g, let the value $x_i$ at time $T - d_i$ be $y_i$, where $d_i$ is the delay from $x_i$ to the output of g.
3. The output value of g is f($y_1, \ldots, y_n$).

**Event propagation.** When an event has been confirmed to happen, all fanout gates or blocks sensitive to the event must be examined for event propagation. At this stage, all fanouts should be assumed to propagate this event to their outputs and should be confirmed later at the event validation stage. It is dangerous to delete a future event at the current simulation time when it appears not to happen in the future. For example, an input of a 2-input AND gate goes high while the other one stays at 0 does not necessarily mean that the rising transition will not cause a rising transition at the output. A case in point is the AND gate in Figure 3.7 with input a rising, followed by b rising 1ns later. The b transition, although it arrives later, allows the transition from input a to propagate to the output.

A complicated component, with multiple paths from an input to an output, may produce more than one event at the output when it is triggered by a single input update event. In this case, all potential events must be scheduled. The time of occurrence of these potential events is then computed, and the events are inserted into the respective slots in the timing wheel.

Event propagation requires a knowledge of fanouts. It is interesting to note that fanouts can change during simulation. It is true that fanouts of a node never change during the simulation of a gate-level circuit. However, this is not true in an RTL simulation, and such code does occur in test benches. An example is the following block with two event operators @, assuming x is an output of gate A and y is an output of gate B:

```
gate A(.out(x),...);
gate B(.out(y),...);
always
begin
    @x
    a = b;
    @y
    b = c;
end
```

The block waits at @x for any event on x. So gate A has this block as a fanout. After an event from x has occurred, the block proceeds until it hits @y, where it waits for an event on y. Now gate A no longer has the block as a fanout. Instead, gate B has it as a fanout (the fanouts of gates A and B change during simulation). Therefore, fanout lists must be updated as a simulation progresses.

**Time advancement and oscillation detection.** Events are deleted from the event queues after they are evaluated. When no event for the current simulation time remains, the simulation time is advanced. Time advancement can also be controlled by users through simulation control, during which users can pause the simulation at a specified time or when a condition is met.

Event-driven simulators accept circuits with combinational loops. If a combinational loop is stable, meaning after a number of iterations it settles down to a steady-state value, eventually all events for the current time will be exhausted. Then, time advances. If an oscillation occurs, event queues will never be empty. To detect oscillation, the simulator keeps track of the number of iterations processed before time is advanced. If a maximum number of iterations is exceeded without time advancement, the simulator declares that an oscillation has occurred.

FFs and latches are treated as black-box components, as opposed to logic with combinational loops. In so doing, no oscillation detection mechanism is necessary for these components. This idea can be extended to larger logic blocks with combinational loops: These loops are "black boxed" as individual components, with evaluation and oscillation detection a part of the components' function. In this way, the centralized event scheduler is relieved of its burden, and simulation performance benefits.

**Event-driven scheduling algorithm.** Let's summarize the previous discussion on simulation processes based on update and evaluation events as follows:

---

### A simulation cycle for event-driven simulator

```
while (there are events) {
  if (no events for current time) advance simulation time.
  Foreach (event at the current time) {
    remove the event and process as follows:
    if (event is update)
      update the variables or nodes.
      schedule evaluation events for the affected processes.
    else // event is evaluation
      evaluate the processes
      schedule update events for outputs that change
  } // end of Foreach
} // end of while
```

---

This event-driven algorithm follows events in the timing wheel to evaluate gates. For zero-delay simulation, performance can be enhanced if the evaluation follows the gates instead of the events. During zero-delay simulation, all gates have zero delays. If only steady-state values matter, the simulation can be executed based on the gates instead of the events. To illustrate this, consider the circuit with zero delays in Figure 3.8. If the evaluation follows the events, then the evaluating gate A with event e1 produces event e2. Event e2 in turn causes gate B to be evaluated to give event e3. Event e2 also causes gate C to be evaluated

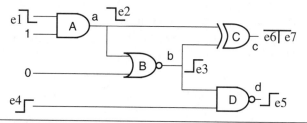

**Figure 3.8** Excess evaluation of events in a zero-delay model

to produce event e6. Event e3 causes gate C to give event e7. Event e3 causes gate D to be evaluated; so does e4. Both events produce event e5. There is a total of six evaluations. However, if the evaluation follows the gates, in the order A, B, C, and D, then each gate is evaluated only once, because all inputs to a gate are ready by the time it is being evaluated. There is a total of four evaluations—a savings of 33%. Therefore, for zero-delay simulation, gate-oriented evaluation (if following the correct order) enhances performance over event-oriented evaluation. An algorithm to prune unnecessary evaluations is relegated to a later session on leveled event processing. See "Leveled event processing for zero delay simulation" on page 101.

## 3.2.2   Cycle-Based Simulators

The other extreme in the simulator spectrum is cycle-based simulators. To motivate the need for cycle-based simulators, consider the combinational logic that computes the next-state function for a finite-state machine. Every time the FFs change, many events are generated in the combinational logic, but only the steady state is latched at the next clock edge; evaluations of all intermediate events are wasted. To avoid evaluating transient events, cycle-based simulators take in the steady-state values of the FFs at the current cycle, regardless of whether a state bit has changed, and apply the next-state function combinational logic to compute the inputs to the FFs for the next cycle. In other words, the combinational logic is evaluated at each clock boundary and a gate is evaluated once, regardless of whether its inputs see events. Therefore, for a circuit to be simulated by a cycle-based simulator, the circuit must have clearly defined clocks and their associated boundaries. Consequently, asynchronous circuits and circuits with combinational loops cannot be simulated by cycle-based simulators. Furthermore, because only steady states are computed, all delays in the circuit are ignored in cycle-based simulation. All components are assumed to have zero delays.

**Leveling.**   To compute steady-state values, gates must be evaluated in a proper order. In the circuit in Figure 3.9, the FFs and the inputs change at a clock transition, and we want to simulate the steady state after the clock transition. If we evaluate gates in the order A, D, B, and C, as an event-driven simulator would schedule it, gate D would use the old value at output B because B has not yet been evaluated, which produces an incorrect value. A correct evaluation order for cycle-based simulation must guarantee that a gate is evaluated only after all its inputs have already been evaluated. One way to visualize the ordering is to arrange components into levels. The first level consists of components with inputs that are directly connected to primary inputs or FF outputs. The nth level contains components connected directly to primary inputs, FF outputs, or outputs of components from level N − 1 and lower. Such order can be obtained using the so-called *topological sort*.

A circuit can be modeled as a directed graph, where the vertices are circuit components and the directed edges are connections. A connection from an output of gate A to an input

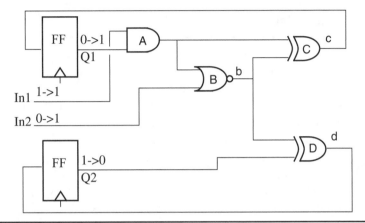

**Figure 3.9**  Proper levelization for gate evaluation

of gate B is represented as an arrowed edge from A to B. FFs are conceptually removed, leaving their outputs as primary inputs.

Topological sort, based on depth-first search (DFS), starts on primary inputs and FF outputs and returns an ordered list of nodes. DFS arriving at a node first visits all nodes that can be reached from an outgoing edge of the node before visiting other nodes that can be reached from another outgoing edge of the node. When representing a graph as $G(V, E)$, $V$ is the set of vertices and $E$ is the set of edges. A topological sort algorithm is shown on page 91. The resulting list is the order for cycle-based evaluation. $N$ records when a node is visited. When a node is visited, $N$ is stored in the node's entry time attribute. When a node is done visiting, $N$ is stored in the node's exit time attribute. $N$ is included for the sole purpose of showing how nodes are visited, and hence is not necessarily a part of the algorithm. DFS is the algorithm minus the list insertion step (last statement).

## Example 3.4

Let us topologically sort the circuit in Figure 3.9, which has the graph representation shown in Figure 3.10, where the FFs are removed and their outputs are labeled Q1 and Q2 respectively, and the vertices represent gates. We start the sort from input In2, which we mark visited and tag with $N = 1$. In the figure, the first number is the entry time and the second number is the exit time. Following the arrow, we arrive at B, which has two fanouts. We mark B visited and tag B with $N = 2$ and follow a fanout to D, which is then marked visited and tagged with $N = 3$. Because D has no fanout, D is done visiting; we return from D. On exiting from D, we tag D with exit time 4 and insert it in List. Following the other fanout of B, we arrive at C, which is marked and tagged with $N = 5$. Because C has no fanout, we return

*continues*

---

**Example 3.4    (Continued)**

from C, tag it with exit time $N = 6$, and insert it in the front of List. On returning to B, we find that all its fanouts have been visited, so we exit from B, tag it with exit time 7, and insert it in the front of List. Finally, we return to node In2, tag it with exit time 8, and insert it in List. This completes the first traversal. However, there are still unvisited nodes, such as A, so we start from another primary output, say, In1. We tag In1 with entry time 9 and follow its sole fanout to A. On arriving at A, we tag it with entry time 10. Because both fanouts of A have already been marked visited, we return from A, tag its exit time with $N = 11$, and insert it in front of List. On returning to In1, we tag it with exit time 12 and insert it in List. The remaining unvisited nodes, In2 and Q2, having no unvisited fanouts, are marked, tagged, and inserted in List. The final List is shown in Figure 3.10. The edges that were traversed in DFS are in bold and form a forest—a collection of trees. These trees are called *DFS trees.*

---

We need to understand why the algorithm guarantees that a gate be placed after all its fanin gates. When a gate's fanin is reached, either the gate has been visited or not. If the gate has already been visited, the gate must have been already inserted in List. Because insertion to List is always in the front of List, the fanin will be inserted before the gate when the fanin is inserted. If the gate has not yet been visited, the gate will be finished visiting first before the fanin can be inserted into List. This is because all fanouts of the fanin must return from VISIT, as indicated in the for_each loop of the VISIT part of the algorithm. Hence, the fanin is again before the gate in List. This argument applies to any fanin. Therefore, the order in List guarantees that a gate be after all its fanins.

So far, we have discussed levelization in the context of a gate model. The algorithm applies to RTL models as well. First, the RTL model must be leveled by constructing a graph representing the model. The key in constructing such a graph lies in identifying the fanins and fanouts of the RTL constructs. The RTL constructs include gates, blocking and non-blocking assignments, system and user tasks, and monitors. The order produced by the

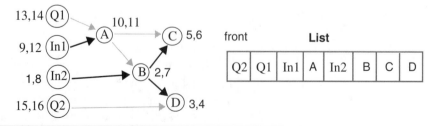

**Figure 3.10** Topological sort to get a proper evaluation order

---

**Topological Sort Algorithm and Depth First Search**

---

Input: G(V, E)

Output: a queue of ordered nodes, List

Initialization: N=1 // used to record entry and exit time of a node

TopologicalSort (G) {
  while (node v in V is not marked visited) VISIT(v);
}

VISIT(v) {
  mark v visited;
  v.entry = N; N = N + 1; // record node entry time
  for_each (u = fanout of v)
    if (u is not marked visited) VISIT(u);
  v.exit = N; N = N + 1;
  insert u in *front* of List; // this line is only for topological sort
}

---

topological sort only mandates that fanins be evaluated before the gate itself. The stratified event queue in the IEEE 1364 standard further imposes an ordering constraint on RTL constructs, such as nonblocking assignments, system and user tasks, and monitors. Let us call this group of constructs *end-time group*. This constraint requires that the end-time group of constructs be evaluated last and, within the group, nonblocking assignments are executed first, followed by system and user tasks, and finally monitors. Therefore, gates, continuous assigns, and blocking statements are scheduled according to the order in which the gates are leveled. After they have been evaluated, the end-time group is evaluated. User/system tasks with outputs that are used by other parts of the circuit before the end of the cycle must be executed at the time they are called, instead of moved to the end of the cycle.

---

**Example 3.5**

---

Schedule the following RTL code for cycle-based simulation:

```
always @(posedge clk)
begin
    a = b;
    c <= a;
    $myPLI(a,b,d); // d is an output
    $strobe("a=%d,b=%d,c=%d",a,b,c);
    e = d;
end
```

*continues*

**Example 3.5    (Continued)**

```
        assign x = a << 2;
        assign y = c;


        gate gate1(.in1(x), .in2(y),...);
```

A graph showing the connection relationship among the constructs is presented in Figure 3.11. The `always` block is to be evaluated first, followed by `assign x = a <<2`, `assign y = c`, and ended with the gate evaluation. The end-time group consists of statements `c<=a;` and `$strobe();`. The user-defined system task `$myPLI()` has an output `d` that is used by another statement, `e=d`. Thus, `$myPLI` has to be scheduled as it is. Within the end-time group, `c<=a` is executed before `$strobe()`. Therefore, the order of execution is as follows:

1. `a=b;`
2. `$myPLI(a,b,d);`
3. `e=d;`
4. `assign x = a <<2; assign y = c;`
5. `gate gate1(.in1(x),.in2(y),...);`
6. `c<=a;`
7. `$strobe("a=%d,b=%d,c=%d",a,b,c);`

**Combinational loop detection.** As a side product, topological sort detects loops. (More precisely, DFS detects loops.) In the case of circuit simulation, a combinational loop exists if the topological sort detects a loop in the circuit graph with FFs and latches removed. Before discussing loop detection, let us first define some terms. Vertex A is called a descendant of vertex B if A can be reached, following directed edges, from B in a DFS tree. Vertex B is called an ancestor of A. A back edge is an edge that goes from a descendant to an ancestor. A loop exists if and only if there is a back edge.

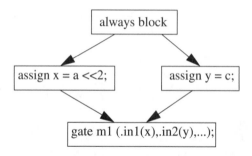

**Figure 3.11** Scheduling RTL code for cycle-based simulation

---

### Example 3.6

Use DFS to find loops in the graph of Figure 3.12. Apply a topological sort (DFS) to the graph in Figure 3.12. The bold edges form the DFS tree. Node a is a descendant of nodes e, c, f, and b. Edges (a,b) and (a,c) are the only back edges. Therefore, there are loops, and they are (a,b,f,c,e,a) and (a,c,e,a). In this example, only entry times are recorded as shown.

---

Although applying a topological sort to a circuit with combinational loops will produce a list of ordered nodes, the order can no longer guarantee that all fanins of any gate be placed before the gate itself. This is because any two gates in a loop have each other as a transitive fanin. So no matter how they position in the list, one gate will be evaluated before the other, which, as a transitive fanin, must have been evaluated first. An impossible situation results. Therefore, cycle-based simulators cannot simulate correctly circuits with combinational loops. (By grouping loops into strongly connected components (SCCs), a circuit with combinational loops can be transformed into one without loops and hence can be addressed by a cycle-based simulator if the SCCs are simulated separately. A later section on levelized compiled simulators discusses this in more detail.)

**Clock domain analysis.** So far we have assumed that the entire circuit has only one clock and that all logic between FFs and latches is evaluated once every cycle. When a circuit has multiple clocks, not all logic has to be evaluated at every clock transition—only the part that is triggered by the clock transition. That is, we have to determine the part of the circuit that requires evaluation at each clock's transition. This task is called *clock domain analysis*, and the set of components that are evaluated at a clock's transition is the domain of the clock.

To find a clock's domain, we first identify all FFs and latches triggered by it. For these FFs and latches to store the correct values at the triggering transition of the clock, all logic converging at their inputs must be evaluated just before the clock transition. A clock triggering

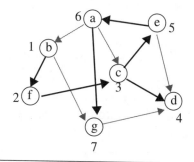

**Figure 3.12** Detect loops using DFS

transition is the one that locks a value into an FF or a latch (for example, a rising transition for a positive-trigger FF and a falling transition for a high-sensitive latch). Therefore, a clock can potentially have two clock domains, one for a rising transition and another for a falling transition. We see later that these two domains are not necessarily identical. To find the logic cone for a given FF or latch, we back trace from its input until we arrive at either a primary input, an FF, or an opaque latch. The reason why we stop at an opaque latch but not a transparent latch is that only an opaque latch's output value is a steady-state value of the latch. If back tracing eventually stops, which is guaranteed if no combinational or latch loops were encountered, then all the logic traversed forms a clock domain. This procedure is then applied to both transitions of every clock. After all the clock domains are determined, they are leveled and evaluated at their associated clock transitions. The following summarizes this domain partitioning algorithm:

---

**Clock Domain Partitioning Algorithm**

1. Identify all FFs and latches triggered by C.
2. For an FF or latch, back trace from its input until it arrives at a primary input, an FF, or an opaque latch. The traversed logic is a part of a clock domain.
3. Repeat step 2 for each of the FFs and latches.
4. The union of all traversed logic of positive-triggered FFs and low transparent latches is the rising transition clock domain, and is similar for the falling transition clock domain.

---

**Example 3.7**

Determine clock domains for the circuit in Figure 3.13. Let us apply this algorithm to the circuit in Figure 3.13, which has two clocks, C1 and C2. C2 is 90 degrees out of phase with respect to C1. Because back tracing continues on transparent latches but stops on opaque latches, we need to know the relative phases of the clocks to determine whether a latch is opaque or transparent. We will determine the rising transition domain of clock C1 for gate G1 and the falling transition domain of C1 for G6. Just before a rising transition on C1, C2 is low; so, G3 is transparent and thus will not stop the rising transition back tracing. Back tracing from the D input of G1, we stop at G5 and G8 because these two gates are FFs and thus have steady-state values. The traversed logic consists of G2, G4, and G3, with functionality equivalent to that of a buffer. Therefore, the domain of rising C1 for G1 is G2, G4, and G3.

Because G6's triggering transition is falling, its input needs to be evaluated just before a falling transition. Just before a falling transition of C1, C2 is high, making G3 opaque. Therefore, this back tracing stops at G3 and G8. The traversed logic is just G7. Therefore, the domain of falling C1 for G6 is G7.

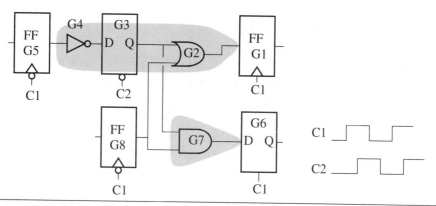

**Figure 3.13** Clock domain partitioning of a circuit with multiple clocks

**Clock tree processing.** After all clock domains have been identified and leveled, evaluation of the clock domains is triggered by clock transitions. When a clock transition occurs, its corresponding clock domain is executed according to its levelized order. When it finishes, it returns and waits on the next clock transition or a user interrupt. Before calculating clock transitions, the clock logic or clock tree that supplies the clock waveforms has to be identified first. A clock tree is the part of the circuit that is bounded by clock pins and primary inputs, and is usually made up of only simple combinational gates. A sample clock tree is shown in Figure 3.14.

**Execution order.** The functional procedures discussed earlier are executed in the following sequence. After the input circuit is parsed and elaborated, its clock tree is traced out and the clock waveforms are computed. Then, for each clock transition polarity (falling and rising), its domain is carved out and levelized. At the various stages, optimization may be

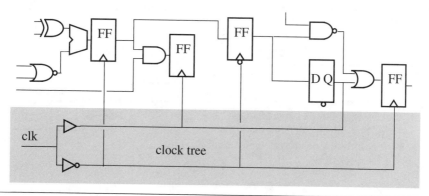

**Figure 3.14** A sample clock tree

applied to increase code performance. An example is to prune logic that is not driving any output, and to collapse buffers. At this point, the result is ready for code generation.

**Code generation and simulation control.** Code generation produces code that executes to simulate the circuit with cycle-to-cycle accuracy. As mentioned earlier in the chapter, there are four types of generated code: interpreted, high level, native, and emulator. Here we look at code generation by creating high-level code. Generated code performs all the operations of a simulator and thus has all the components described in Figure 3.4, such as clock waveforms, circuit connectivity, component functionality, storage space for circuit node values, a time advancement mechanism, and a facility to read and write node values.

For the user interface, embedded in generated code are communication channels to simulation control, through which the user can specify the length of the circuit simulation, pause the simulation when a condition is met, and inspect node values. The communication channels can be memory locations of variables that simulation control can deposit values (for example, number of cycles to simulate), or wait statements where the simulation waits on commands from the user.

Furthermore, system/user tasks can be implemented as either part of the compiled code or as a dynamically loadable library. With the former, the generated code contains the object code of the tasks. A call to a task is a jump to the task's entry point. With the latter, the generated code simply has in the place of a system/user task, a call to simulation control with the name of the task. The simulation control, when called, dynamically loads the code segment of the task from the library and executes the task. In this implementation, system/user tasks are compiled independently of the circuit. Figure 3.15 shows the flow of generated cycle-based code.

To start the simulation, the user invokes the simulation control to load the generated code, specifies a number of conditions and parameters (for example, the number of cycles to

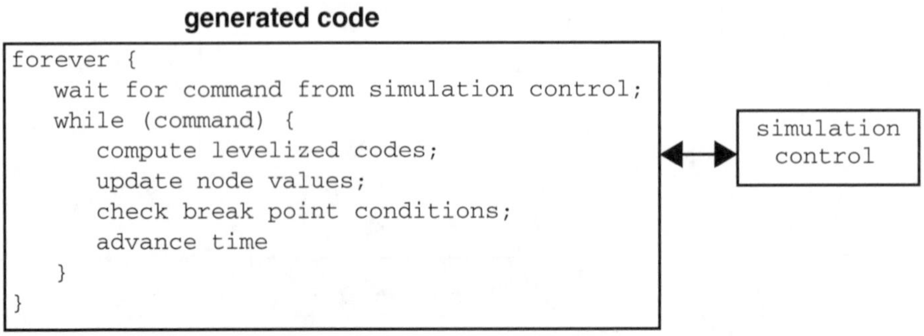

**generated code**

```
forever {
    wait for command from simulation control;
    while (command) {
        compute levelized codes;
        update node values;
        check break point conditions;
        advance time
    }
}
```

```
simulation
control
```

**Figure 3.15** Flow of generated cycle-based code

simulate, or a break point condition such as `stop if Node1.value == 1'b1`, and starts the execution. The generated code runs until one of the specified conditions or parameters fires, or the end of the simulation is reached. During the simulation, communication between the simulation job and the simulation control may take place—for instance, displaying error messages from the simulation to the console or handling system/user task calls using a dynamically loaded library. When the simulation is paused, the user may inquire about node values, may step cycle by cycle, or may save the simulation state to be restored in a later session.

---

### Example 3.8

Generate cycle-based pseudo C code from Verilog. Let us generate high-level code for the following simple circuit. The first line generates a clock waveform of period 2 of 50% duty cycle. The following `always` block is executed on every rising transition of the clock:

```
always clock <= #1 ~clock; // generate a clock of period 2

always @(posedge clock)
begin //block executes on rising of clock
   a = b + c;
   $MyTask(a,b); //call user task MyTask.
end
```

Sample generated pseudo C code is as follows. The clock waveform is not explicitly shown in the generated code, but its effect is incorporated through the use of time *T*. In this example, we assume that the simulation control can do only two things: load the generated code and allow the user to set the number of cycles to be simulated, which is stored in variable `CycleLimit`. To simulate, the user first loads the compiled generated code through the simulation control, sets the number of cycles to be simulated, and starts the simulation. The user task `MyTask` is implemented in a dynamically loadable library and thus is computed by calling the simulation control. The simulation exits when the number of cycles exceeds `CycleLimit`:

```
T = 0;
while ( T<= CycleLimit ) {
   a = b + c;
   call_simulation_control("MyTask(a,b)");
   T = T + 1; // advance time
}
```

It should be stressed that not only cycle-based simulation, but event-driven simulation, can have compiled high-level simulation. Consider the following event-driven code. Assume the processing function has a zero delay:

*continues*

**Example 3.8    (Continued)**

```
always
begin
    #1 trigger = trigger_function(...);
end

always @(posedge trigger)
begin
    // processing function
    ...
end
```

Pseudo C code for this event-driven model is as follows. The code has two threads, each simulating an `always` block. The interthread communication is implemented using condition variables c and d. UNIX function `cond_wait(&c,&m)` waits for variable c to change and, during the waiting period, it also releases `mutex` m. UNIX function `cond_signal(&c)` notifies all threads waiting on c.

```
// thread 1: first always block
while(){
    // these codes are here to ensure thread 2 is finished
    // before another transition occurs.
    lock_mutex(&m2);
    while(!done) {
        cond_wait(&d,&m2);//wait for posedge done, release
        mutex m2
    }
    done = 0;
    unlock_mutex(&m2);

    lock_mutex(&m1);
    trigger = trigger_function() ;
    cond_signal(&c); // signal that a transition has occurred.
    unlock_mutex(&m1);
} // end of while
```

**Example 3.8    (Continued)**

```
      //thread 2: the second always block
      while () {
          lock_mutex(&m1);
          while(ClockEdge(trigger)!= RISING) { // wait for a rising edge
              lock_mutex(&m2);
              done = 1; // evaluation done for falling edge
              cond_signal(&d); // signal evaluation done
              unlock_mutex(&m2);

              cond_wait(&c, &m1); // wait for a rising edge, release
              mutex m1
          } // end of ClockEdge check

          // a rising clock transition has arrived; evaluate the
          always block.
          // processing function here...

          lock_mutex(&m2);
          done = 1; // evaluation for rising edge done
          cond_signal(&d); // signal evaluation done
          unlock_mutex(&m2);

          unlock_mutex(&m1);
      } // end of while
```

The first thread waits for thread 2 to finish before sending out a triggering signal. It waits on done to become 1. When it becomes 1, thread 1 computes `trigger_function()` and signals an update of the value on `trigger` by calling `cond_signal(&c)`. Then, thread 1 goes back to `cond_wait(&d, &m2)`, waiting for done acknowledgement from thread 2. In the beginning of the `while` loop, thread 2 waits for `trigger` for a rising transition. When a transition comes, thread 2 determines whether it is rising or falling. If it is falling, it sets done to 1 and continues to wait. If it is rising, it processes the function and alerts thread 1 at the end by calling `cond_signal(&d)`. UNIX functions `lock_mutex(&m)` and `unlock_mutex(&m)` are used to guarantee that the variables sandwiched between the pair can be changed by, at most, one thread at any time.

**Example 3.9**

Compare the performance of event-driven and cycle-based simulators. In this example, we will put together all the steps in event-driven simulation and cycle-based simulation by estimating the relative performance of the two types of simulators. For simplicity, our computation model assumes that the following operations are equal in complexity: insertion to a queue, deletion from a queue, search of a fanin or a fanout, comparison of two node values, retrieval of a gate's delay, and computation of logic function with two inputs. Furthermore, we take the average gate input, fanin, and fanout count to be 4.

For event-driven simulation, the following operations are performed:

1. At time $T$, take an event from the timing wheel. The operation cost is 1.
2. To validate the event, retrieve the gate's fanins. Assume an average of four fanins, the cost is 4.
3. Once the fanin values are known, compute the gate output. Because the computational cost is 1 for two inputs, it takes three operations to compute a gate with four fanins, because it takes two operations to compute the two intermediate results from four fanins, and one operation to get the output from the intermediate results.
4. Retrieve the output value from the previous evaluation and compare it with the current value to determine whether there is a transition. The operation cost is 1.
5. If there is an output transition, determine the fanout gates. The operation cost is 4.
6. For each fanout, insert an event in the timing wheel at the time equal to the delay of the fanout gate. The operation cost is 1. For four fanouts, the cost is 4.

The total number of operations for an event is the sum of the costs in the previous steps, which is 17. For a circuit with 15% of its nodes toggled, with each node having 2.5 transitions per clock cycle, the number of operations that an event-driven simulator needs to perform for a clock cycle is $0.15 * 2.5 * 17 = 6.375$ per circuit node. In a cycle-based simulator, each gate is computed once, and leveling eliminates the search for fanins. Each gate has an average of four inputs and thus costs three operations to compute its output. Therefore, our crude estimate predicts that an event-driven simulator is about 2.12 times slower than a cycle-based simulator based on the assumed transition statistics.

## 3.2.3  Hybrid Simulators

So far, we have discussed two simulators at the extremes: event-driven and cycle-based simulators. The former is versatile but slow; the latter is fast but restrictive. It is possible to construct a simulator tailored for specific requirements that lays in the middle of the simulator spectrum. Simulator characteristics (such as event driven, levelized, compiled, interpreted, and centralized or distributed event schedule) are independent of each other and hence can be chosen individually to concoct a hybrid simulator. A simulator can be designed to be levelized, event driven in nature, but with some components compiled, and can have a centralized event scheduler. On the other hand, a simulator can be constructed so that each of its clock domains is levelized and it runs like a cycle-based simulator, but the interaction among the domains is event driven. In the following sections we will look at several typical hybrid simulator designs.

**Compiled event-driven simulator.** With this combination, components of the circuit are compiled code. However, triggering the evaluation of a component and thus execution of its compiled code is dictated by the events among the components. The granularity of components in this case is not necessarily a single gate or RTL construct. It can be a group of gates or RTL constructs. The finer the granularity of the component, the closer the simulator is to being a true event-driven simulator.

**Leveled event processing for zero-delay simulation.** Levelization is not a technique used solely for cycle-based simulation. It can also be applied to event-driven simulation. In the previous discussion of event-driven simulation, we assumed an arbitrary order of execution of events occurring at the same time, except for nonblocking assignments and monitors, which are executed at the end of the current simulation time, as specified in the IEEE Verilog standard. Now we will see how event prioritization can improve event-driven simulator performance.

Consider simulating the zero-delay circuit and input events in Figure 3.16. If we follow the usual event-driven scheduling algorithm, the event evaluation order would be e1, e2, e3, and e4, followed by the output events on gates A, B, D, and C. The widths of the pulses are exaggerated to illustrate the number of evaluations induced by them. They are zero in actual simulation. A gate is evaluated once for every input event; therefore, there is a total of nine evaluations. Because all glitches have zero widths, oftentimes the user is only

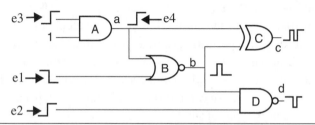

**Figure 3.16** Leveling events improves performance

interested in the steady-state values. Therefore, only steady-state values are of concern, and levelization can prune away transient evaluations. For example, gate D can compute its steady-state output by delaying its evaluation until gate B is done evaluating. In general, a gate should be evaluated only after all its input gates have finished, which is exactly what a levelized order guarantees. If a leveled order is imposed on event scheduling, gate A is evaluated first, followed by B, then D, and finally C, giving a total of only four evaluations.

To incorporate levelization into event scheduling in a zero-delay simulation, the circuit is first leveled and then the event queues are sorted by level. At simulation time $T$, events from the level 1 queue are evaluated first. If events are generated at a component's output, the events are inserted in the queues corresponding to the level of the fanout gates. Events are evaluated one level at a time. When queues at all levels are finished, time advances. In a timing wheel, each time slot is modified to hold queues, one for each level. A levelized event driven for a zero-delay simulating scheduling algorithm is as follows:

---

### Zero Delay Levelized Event Driven Scheduling Algorithm

1. The current simulation time is $T$.
2. for_each level in the levelized circuit
3. while (event queue at level L and time $T$ is not empty)
   fetch an event
   validate the event and update simulation states
   predict events for fanout gates
   get fanout gates' levels
   insert the predicted events into queues of the levels.
4. Advance the simulation time and go to step 2.
5. Exit when there are no more events or the simulation is terminated.

---

This levelized event-scheduling technique resembles cycle-based simulation, but the main difference is that cycle-based simulation evaluates all circuit components whereas this levelized event-driven algorithm evaluates only the ones with input events. For instance, if events e1 and e3 in the previous example are absent, a levelized event-driven simulator will only evaluate gate D, whereas a cycle-based simulator will evaluate all four gates. The similarity between these two types of simulator is that, if a gate is evaluated, it is evaluated just once in both types of simulators.

**Compiling combinational loops for cycle-based simulation.** In a previous discussion, we noted that a cycle-based simulator cannot accept a circuit with combinational loops. This restriction can be relaxed if a hybrid method is used. If we think a little deeper about cycle-based simulation, we would notice that FFs and latches are essentially combinational loops, but they are simulated in cycle-based simulators. Thus, we can conclude that combinational loops can be simulated by a cycle-based simulator if they can be encapsulated into macro models. To extend the scope of cycle-based simulation, all combinational loops should be encapsulated in macro models, and the macro models should be simulated using an event-driven simulator, so that the circuit with the macro models can be simulated with a cycle-based simulator. The next question is how to find and isolate all combinational loops. The following algorithm isolates all loops and guarantees that the resulting circuit with the loops encapsulated is loop free.

An SCC of a directed graph is a maximal subgraph such that every node can be reached from any other node. The SCC is an expanded notion of a loop. A directed acyclic graph (DAC) is a tree or forest, and hence is free of SCCs. It is known that any directed graph can be decomposed into a DAC and SCCs. In other words, this decomposition breaks any directed graph into looping components and straight components. Once a graph is decomposed in such a way and the SCCs are encapsulated, the resulting graph is a DAC, and is loop free. SCCs can be determined by applying DFS, as follows:

---

**Algorithm for Finding SCCs**

Input: Graph G

Output: A collection of SCCs in G

1. Execute DFS on G and record the exit times for the nodes.
2. Reverse the edges of G and apply DFS to this graph, selecting nodes in the order of decreasing exit number during the `while` loop step.
3. The vertices of a DFS tree from step 2 form an SCC.

---

In step 2, "reversing an edge" means making the head of the edge the tail, and vice versa. When applying DFS to this graph, in the `while` loop of the DFS algorithm (see page 91), select unvisited nodes in the order of decreasing exit numbers, which were derived in step 1.

**Example 3.10**

Identify SCCs. Apply the SCC-finding algorithm to the graph in Figure 3.17. In A, the numbers underlined are the exit times of the nodes from the DFS in step 1. B is obtained from A with edges reversed. To apply DFS to this derived graph, the first unvisited node is node d, because it has the largest exit number. This search ends after it has visited node a, node b, and node h. The next unvisited node is node f, because it has the largest exit number in the remaining unvisited nodes. This search concludes the DFS after it has visited nodes g and e. The bold edges in B are the DFS trees from this DFS. The nodes of a DFS tree form an SCC. Therefore, {a,b,d,h} is an SCC, as is {e,f,g}. Making an SCC into a composite node, the resulting graph is a DAC, shown in C.

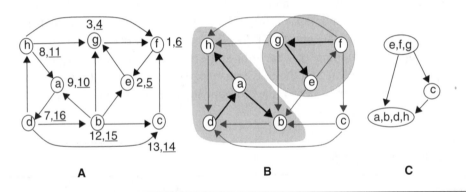

**Figure 3.17** Finding SCCs. (A) Result of a DFS. The underlined numbers are exit times. (B) DFS on the graph derived by reversing the edges. (C) Resulting DAC by treating SCCs as composite nodes.

When the SCCs are found, they are compiled individually and are treated as macro models. The resulting circuit is free of combinational loops and can be simulated with a cycle-based simulator. During simulation, if an SCC macro model is encountered, its compiled code is called to perform its own evaluation, very much like FFs and latches are simulated.

**Example 3.11**

Transform a circuit with combinational loops into a circuit that can be cycle-based simulated. The circuit in Figure 3.18 has a combinational loop and therefore cannot be simulated in a cycle-based simulator. Here we transform it to be cycle-based simulated by first
identifying all SCCs and then "black boxing" them. The SCC is identified as the part of the
circuit in the shaded box in B. By making the shaded box a component, the circuit can be
cycle-based simulated. The functionality of the black box is compiled and the pseudocode
is shown in C, where variable N counts the number of iterations before output q stabilizes
(in other words, q = old_q). If a preset limit is exceeded, an oscillation error is reported.

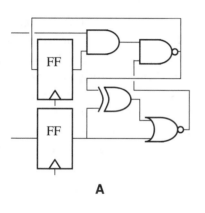

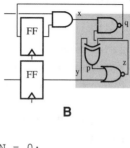

B

```
N = 0;
do {
    N = N + 1;
    old_q = q;
    p = y XOR old_q;
```

C

A

**Figure 3.18** Macro modeling a combinational loop for cycle-based simulation. (A) A circuit with a
combinational loop. (B) After the loop is isolated and compiled, the resulting circuit can be cycle-based
simulated. (C) High-level compiled code for the encapsulated loop.

**Distributed event management.** A major cause of slow simulation performance in event-
driven simulators is the centralized event manager: the timing wheel and its maintenance.
To mitigate this problem, the centralized event manager is replaced by multiple distributed local event managers embedded in partitioned domains, each of which is in charge of
its own domain. Cross-domain events are handled by a global event manager. An event

domain can be a clock domain in circuits with multiple clocks, or an instance at the unit granularity level. A key indicator for an effective event domain partition is minimal cross-domain events. Such a simulator gains performance when run on a multiple-processor machine.

### 3.2.4   Hardware Simulators and Emulators

Hardware simulators and emulators are computers that are specially designed for running simulations. An emulator is just a simulator with an interface connected to a hardware system as a substitute for the circuit being simulated. When the system operates in real time, the emulator takes in input signals, computes the outputs, and responds with the results, all in real time. Hence, emulators have more stringent response time requirements than simulators. Besides, there are no major differences between the two. For this reason, we only discuss hardware simulators in this section.

There are two types of hardware simulators, classified by their underlying hardware computing components: one is FPGA based and the other is processor array based. The FPGAs or processors can be connected in any network configuration, but the common ones are two- or three-dimensional mesh or torus, where processors are placed on grids and connected along the grid lines. A torus is a mesh with the ends wrapped around (see Figure 3.19). Another common configuration is through a central switch, such as a crossbar or butterfly switch, or a simple bus. The user interface is often through a host machine that in turn is connected to the simulator. A block diagram of a generic hardware simulator using a central switch is shown in Figure 3.20.

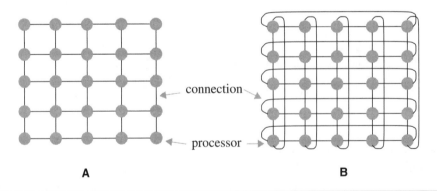

**Figure 3.19** A two-dimensional mesh (A) and torus (B)

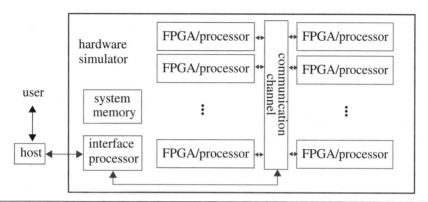

**Figure 3.20** Architecture of a general hardware simulator

In an FPGA-based architecture, each FPGA chip has a network of prewired blocks of look-up tables and a bank of FFs. A look-up table can be programmed to be a Boolean function, and the look-up tables can be programmed to connect or bypass the FFs. If connected, the FPGA chip operates as a finite-state machine. If bypassed, the FPGA chip operates as a combinational circuit. The look-up tables can be programmed to mimic any combinational logic of a predetermined number of inputs and outputs. To run a circuit on an FPGA-based simulator, the circuit must first be compiled. A compiler partitions the circuit into pieces, each fitting into an FPGA chip. The partitioned subcircuits are then synthesized into the look-up tables (that is, generating the contents in the look-up tables such that the look-up tables together produce the function of the partitioned subcircuits). Then the partitioned netlist is placed and routed on the FPGA chips. In other words, we assign the subcircuits to the chips and connect the chips in a way that preserves the connectivity in the original circuit.

Similarly for a processor array-based architecture, the input circuit is partitioned into subcircuits so that each piece fits the instruction memory of a processor. Besides instruction memory, each processor also has data memory. The code running on a processor can be either event-driven or cycle-based code. After the code is loaded into the processors, the processors simulate their respective portions of the circuit. At predefined times, the processors propagate and synchronize their results, in effect simulating the input and output flow of signals among the partitioned subcircuits. In cycle-based simulation, synchronization occurs at the end of the cycle, whereas in event-driven simulation, it occurs when cross-domain events happen.

In summary, a hardware simulator compiler follows very much the same flow as a software simulator compiler. The major difference is that it has a partitioner to break down a large circuit and a scheduler to set up communication among the computing resources. Hardware compilation can generate event-driven code or compiled code. In the event-driven case, event management can be centralized to a processor or it can be distributed among processors. Furthermore, hardware simulators can only verify logical functionality but not timing properties, because delays from a network of FPGAs or processors do not correlate with those in the design.

After compilation is done, the compiled image is downloaded via the interface processor to the hardware. Besides downloading compiled images, the interface processor also uploads simulation results, controls the simulator in interactive mode, calls the host to execute system functions or tasks, and passes the results back to the simulator. Each processor has its own instruction and data memory. In addition, there may be system memory that can be used to model the memory arrays in the circuit.

## 3.3 Simulator Taxonomy and Comparison

Although a new type of simulator results with each combination of the simulator architectural features, there are only a few in use. Let's study their relative strengths and limitations, and survey their application arenas.

### 3.3.1 Two-State and Four-State Simulators

In a two-state simulator, a node can have the value 0 or 1, whereas in a four-state simulator it can have 0, 1, $x$, and $z$, where $x$ denotes an unknown value and $z$ denotes high impedance. An $x$ value results when a node is uninitialized or two sources are driving the node to opposite values at the same time. A $z$ value results when all drivers on a bus are brought to high inpedance. If an $x$ or $z$ value is encountered in a two-state simulation, it is mapped to either 1 or 0. For the following discussion, let's assume that the $x$ and $z$ values are mapped to 0.

The algebra of $x$ and $z$ can be summarized as follows: If an input is either $x$ or $z$ and other inputs do not have a controlling value, the output is $x$. If an input has a controlling value, the output is then determined by the controlling value. A controlling input value determines the output of the gate independent of other input values. For example, 1 is the controlling value of the OR gate, and 0 is that of the AND gate. The complement of $x$ is $x$. The complement of $z$ is $x$.

The algebra of $x$ and $z$ can produce pessimism in simulation results. A well-known example is that of a multiplexor. The following RTL code is a multiplexor. When s = 0, output y = j. When s = 1, y = i.

```
assign n = i & s;
assign m = j & (~s);
assign y = n | m;
```

If s takes on an unknown value $x$, and inputs i and j are both 1, then one might reason that, because output y is either i or j, and i = j = 1, y should be 1 regardless of the value of s. However, if output y is computed according to the algebra of $x$ and $z$, a different value results. Because both s and ~s are of value $x$—the complement of $x$ is $x$—and both i and j are 1, n and m have $x$ value, giving an $x$ value to y. The cause of this pessimistic result is the result of the rule that the complement of $x$ is $x$. Making the complement of $x$ to be $\bar{x}$ only complicates matters.

In two-state simulation, high impedance, bus contention, and a zero value on a bus are all mapped to 0. Sometimes, the simulator needs to distinguish these situations, such as when detecting errors on a bus and determining whether a true zero or high impedance is read from a bus. Knowing the value of all bus drivers' enable pin and the input values distinguishes these situations. Bus contention occurs if more than one driver is enabled, and inputs to the drivers are opposite. The bus is in high impedance if no bus driver is being enabled and there is neither a pull-up nor a pull-down. The bus has a true zero value if only one driver is enabled.

A two-state simulator is faster, because evaluations are shorter and storage is smaller with two values. A four-state simulator is normally used to simulate the power-up period, when many states are uninitialized. After a while, a well-designed system will be free of unknown states. Then, a two-state simulator can switch to replace the four-state simulator and can continue the simulation at a faster speed.

## 3.3.2 Zero- versus Unit-Delay Simulators

A zero-delay simulator ignores all delays in the circuit and is used mainly for functional verification. A unit-delay simulator assumes that all gates have a delay of one. Unit-delay simulation generates orders of magnitude more events than zero-delay simulation, because all glitches that are collapsed into a single transition in a zero-delay model may now occur at different times (see "Leveled event processing for zero-delay simulation" on page 101). Therefore, zero-delay simulators run much faster. Unit-delay simulation aims at hazard and race detection. The unit delay is introduced to "spread" out transitions so that glitches and race problems are revealed. A design with realistic delays, back annotated from layout information, provides more accurate timing information but runs even slower because more events may surface as the delays spread out the glitches further. Note the separations between transitions arising from three delay models—zero, unit, and real delay—as shown in Figure 3.21. Unit-delay simulation is very useful for detecting reset problems and logic where the RTL and gates do not match, because it is much faster to simulate than a full-delay model.

### 3.3.3 Event-Driven versus Cycle-Based Simulators

The main cause of slow performance in event-driven simulators is centralized event management, whereas the potential performance drawback in cycle-based simulators is the indiscriminate simulation of all components, regardless of input excitation. Empirical data have shown that unless switching activity inside a circuit is less than 1% (the percentage of nodes switching), cycle-based simulators are faster. In practice, the average switching activity is around 10 to 20%, which translates to 5 to 20 times acceleration. However,

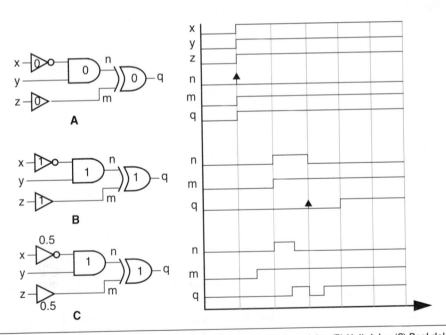

**Figure 3.21**  Transition spreading as a result of a delay model. (A) Zero-delay (B) Unit delay (C) Real delay

cycle-based simulators have a more stringent coding style. They cannot simulate asynchronous circuits, timing behavior of circuits with delays, and some test bench constructs, just to name a few. Moreover, it takes longer to compile a circuit for cycle-based simulation. Therefore, cycle-based simulators are ideal for functional verification, whereas event-driven simulators are more suitable for timing verification and prototyping.

### 3.3.4 Interpreted versus Compiled Simulators

Compiled simulators have better simulation performance but take longer to compile, are less interactive, and are less portable. Interpreted simulators find usage in prototyping, where speedy compilation and accessibility to internal nodes are a premium. Compiled simulation optimizes much for performance and destroys more user-entered structure. Interpreted simulation also finds its strength in intellectual property (IP) procurement. An IP design can have various levels of implementation—from an RTL netlist to a simulatable binary model to a layout of the design. If an IP only offers simulation functionality but not its implementation details (for example, a functional model of a hardware core), the seller can ship the IP in a precompiled format compatible with a standard interface (such as SWIFT), so that the content is protected yet the user can execute the precompiled code using any interpreted simulator accepting the standard interface.

### 3.3.5 Hardware Simulators

Hardware simulators are often a few hundred to a few thousand times faster than software simulators. However, improper design of a test bench (for example, extensive interaction between the host and the simulator) or improper use (for example, heavy dumping of signals) can drastically decrease the performance to that of software simulation. Therefore, a major bottleneck for which to watch when using a hardware simulator is the interaction between the hardware simulator and the host. To minimize interaction, data exchange is often cached and flushed all at once, when it is full. Aside from performance, hardware simulators fall short of their software counterparts in the scope of design constructs that can be simulated, inability to verify timing properties, and ease of debugging.

The critical characteristics of a hardware simulator are capacity, speed, compilation time, and debugability. With today's technology, processor-based simulators have a higher capacity than FPGA-based simulators, but they are more complex to design. Performance-wise, they are comparable. For hardware simulators, compilation time is a major factor for application consideration, because it can be 5 to 50 times longer than software compilation. The bottleneck comes from partitioning the design, placing components, and routing signals among the components. Oftentimes, large designs require intervention from the user to complete compilation. Consequently, hardware compilers should be able to compile incrementally. Between FPGA and processor array accelerators, FPGA accelerators have much slower compile time, and processor array accelerators have compile times approaching to those of software simulators. Designs run on a hardware simulator should be optimized to have a minimum amount of output data. Finally, because hardware simulators are a rare resource shared by many, in addition to their limited circuit node visibility, debugging directly on a hardware simulator is counterproductive. A solution to this is to save the state of the simulation, when errors are detected, and load the saved image to a software simulator, where debugging can conveniently proceed.

Table 3.1 summarizes the relative effects of several simulator architectures on performance, capacity, compile time, debugability, and portability. If a feature has no directly significant impact, it is labeled as NDI. Four grades are used: best, better, NDI, and worse.

## 3.4   Simulator Operations and Applications

This section studies typical features in a simulator and where these features are applicable. This section is not meant to be a substitute for simulator manuals, but rather it serves to introduce the concepts and commands that are available in a simulator. It is not feasible to cover all simulator commands, because of their enormous number and variation over simulators and time. However, simulator commands are just an embodiment of more fundamental concepts. It is these concepts that are precisely the focus of our study in this

**Table 3.1**    Relative Effects of Simulator Architectural Features on Five Qualities of Simulator

| Feature | Event driven | Interpreted | High level | Native | Levelized | Hardware |
|---|---|---|---|---|---|---|
| Performance | Worse | Worse | Better | Better | Better | Best |
| Capacity | Better | Better | Best | Better | NDI | Worse |
| Compile time | Best | Best | Better | Better | Better | Worse |
| Debugability | Best | Best | Worse | Better | NDI | Worse |
| Portability | Better | Best | Better | Worse | NDI | Worse |

section. Therefore, the commands in this section are pseudocommands. Besides explaining functions in command categories, we will also discuss how commands are applied.

### 3.4.1 The Basic Simulation File Structure

Every simulator has a directory structure for input, output, and command files. The input directory, which usually has subdirectories, holds HDL design files, include files, library files, Makefiles, compiled images, and sometimes C/C++ files for PLIs. The HDL design directory often has subdirectories corresponding to the functional units of the design. Within a functional unit subdirectory are design files that contain RTL code, along with macros, parameters, and constant definitions, which reside in include files. The library file contains cell definitions, such as FFs and encoders. A cell is defined to be the module lowest in the design hierarchy and there is no module instantiation inside it. Makefiles perform various tasks such as compiling C/C++ code and linking object code to produce executables, expanding macros to generate HDL files, and compiling the design for simulation. Compiled images are files produced by the simulator's compiler and are input to the simulator. The output directory, possibly having a couple layers of subdirectories, contains log files generated from simulation runs, signal tracing files, error logs, and others. The command directory has script files for simulation compilation, simulation run, debugging options, and others. An example simulation directory organization is shown in Figure 3.22.

To guide a simulator or compiler to search for a file or a directory, the information is passed through runtime options. For example, to specify an input file or to designate an output file, a full path to the file is specified on the command line, or the directory holding the files is passed as option arguments and the simulator or compiler searches the files. If

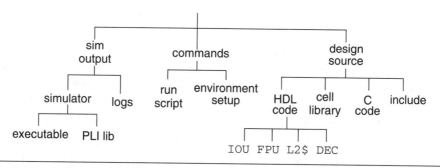

**Figure 3.22** Simulation directory structure

directories are passed, the compiler searches files in the current working directory and then the specified directories. A typical command line for compilation may look like

```
compile -f filelist -y srcDirectory +option1 +define+MyVar=1
-output=logFile -o sim
```

where `filelist` contains paths to the HDL design files and include directories. The following is an example:

```
+incdir+/home/design/include/
+incdir+/home/design/macro/
/home/design/adder.v
/home/design/multiplier.v
...
```

The first two lines specify the include directories to be `/home/design/include` and `/home/design/macro/` so that when an include file is encountered during compilation, these two directories will be searched. The remaining files are HDL design files. The argument after `-y` is the directory for library cells. The next item, `+option1`, can be any option known to the compiler. The next item, `+define+MyVar=1`, sets the value of the compile time variable `MyVar` to `1`, so that whenever `MyVar` is encountered in the source files during compilation, it is replaced by `1`. The next item designates `logFile` to be the output file. Finally, the last item specifies the name of the compiled image to be `sim`. After compilation, the simulator can be invoked using a command such as

```
simulate -image sim +option2
```

where the simulator loads the compiled image file `sim` and takes in runtime option `option2`.

### 3.4.2 Performance and Debugging

In this section we discuss simulator options for enhancing performance and debugability of the circuit. Options for performance and for debugability have opposite effects on simulation: High performance means less debugability and vice versa. This is because to increase simulation speed, the circuit representation often needs to be restructured. For instance, buffers and inverters are combined with other gates and are eliminated, bus bits are aggregated, redundant logic is pruned, and blocks with the same sensitivity are merged. Consequently, the eliminated nodes are not observable, and the resulting structure is not easily recognizable to the user, making the circuit more difficult to debug.

Most simulators have several levels of performance optimization. We assume the highest level means the highest performance and hence the lowest degree of debugability. Debugability usually refers to how the user may inquire or manipulate circuit nodes or variables during simulation runtime or through user PLI tasks. Different modules in a design can be tailored to have different levels of optimization so that the well-tested modules can be optimized to the greatest extent. At different levels of optimization, the corresponding debugging restrictions imposed at each level vary. An example guideline follows. At the highest level, nodes or variables can only be read. At the next level, values of nodes and variables can be modified, and delays of gates can be altered. Changing a node value can be done, for example, by using Verilog's `force` construct, or PLI's `tf_put`, or by assigning to a new value during an interactive simulation session. At the lowest level, all performance optimizations are disabled, and everything is readable, writable, and traceable. Traceable means that the circuit structure can be traversed through PLI routines (for example, inquiring about fanouts or fanins of a node through PLI's `acc_next_driver` or VPI's `vpi_iterate`). Obviously, to enable traceability, the simulator must maintain a mechanism to support the PLI or the VPI routines, which slows down performance. If a node or a variable is accessed at a level not permissible by the optimization option (for example, if it is assigned a new value while the highest performance option is specified), an error will result. An example compile command with tailored optimization options is as follows:

```
compile -f filelist +optimize+level2+file=ALU.v +optimize+level1
```

where the first optimization option specifies that file `ALU.v` be optimized at level 2 and the rest optimized at level 1.

To debug a circuit, viewing signal waveforms is a necessity. A common practice is to dump out signal traces during a simulation run and view them later with a waveform viewer. Using this method, the user can debug off-line and free up the simulator for others. Unless all resources have been exhausted, it is inefficient to dump out all signal traces in the design, especially when the design is large. Instead, only a portion of the design is selected for dumping, and this selection can be made by the user during compilation or simulation. To implement selective dumping at compilation, Verilog's `ifdef` guards a dumping code segment that can be activated to dump signals in a functional unit. When the variable of `ifdef` is defined, the dumping code is activated. If the code is not activated, signals from the unit are not dumped. For example, to create selective dumping for functional unit ALU, the following code is used:

```
`ifdef DEBUG_ALU
    $dumpvar(0, alu);
`endif
```

System task `$dumpvar` dumps out all node values inside module `alu` in value change dump (VCD) format. The first argument, `0`, means that all levels of hierarchy inside `alu` are dumped. To activate this task at compile time, the following command is used, which defines variable `DEBUG_ALU`:

```
compile -f filelist +define+DEBUG_ALU ...
```

Because variable `DEBUG_ALU` is defined, the code `$dumpvar(alu)` is compiled with the rest of the circuit, and dumping is activated. Dumping can also be activated during simulation runtime and it is done via `plusarg` (short for +argument). Change the previous `ifdef` to the following `if` statement:

```
if($test$plusargs(debug_alu == 1))
    $dumpvar(0, alu);
```

where task `$test$plusargs` checks the value of argument `debug_alu`. If it is equal to 1, the following line will be executed. To invoke this dumping at runtime, the simulator is invoked with the plus argument `+debug_alu+1`:

```
simulate -image sim +debug_alu+1
```

`plusarg +debug_alu+1` defines the value of the argument to be 1, and hence turns on dumping of `alu`.

The differences between compilation time and simulation time selection are the size of the compiled image and the ability to select dumping based on actual simulation results. If a dumping code is implemented as a compilation time option, the decision to dump (or not) must be made at compilation time. Once compiled, it cannot be changed without recompilation. The advantage is that, if selected not to dump, the resulting compiled image is smaller. On the hand, if it is implemented as a simulation time option, what to dump can be decided when a bug shows up, without recompiling. The disadvantage is that the code has already been compiled, even though it is selected not to dump.

Table 3.2 summarizes the effects of simulator options on compilation and simulation speed, as well as debugging capability.

### 3.4.3 Timing Verification

To verify timing properties, a delay model for the circuit must first be chosen. One delay model is a zero-delay model, in which all gates and blocks, specified explicitly or not, are assumed to have zero delays. This delay model does not reveal much timing property

**Table 3.2**    Effects of Simulator Options on Compilation and Simulation Speed and Debugability

| Option Type | Effects |
| --- | --- |
| Enable read, write, and connectivity trace | Slow down compilation and simulation but increase debugging capability |
| Enable two-state simulation | Speed up both compilation and simulation but decrease debugability |
| Disable timing checks | Speed up simulation but decrease debugability |
| Use a zero-delay or a unit-delay model | Speed up both compilation and simulation but decrease debugability |
| Perform structural optimization (combine bits, eliminate buffers) | Slow down compilation, speed up simulation, and decrease debugability |
| Enable interactive simulation | Slow down compilation and simulation but increase debugability |

about the circuit and thus is used mainly for functional verification. A zero-delay model produces the fastest simulation speed compared with other delay models. Another model is a unit-delay model for which all gates and blocks have a delay of one, and all specified delays are converted to unit delays. This delay model is not realistic, but it is a reasonable compromise between a realistic but slow delay model and the zero-delay model. Its main application is in detecting hazards. Finally, a full-delay model allows each gate, block, and interconnect to have its own delay. The delay values are usually extracted from the layout of the design and are back annotated to the design for timing simulation. This full model has the most accurate timing information, but it runs the slowest. It is used for timing closure verification after functional verification is completed.

To build a full-delay model, delay information on gates and interconnects is computed based on the cell library and RC extractions from the design's layout. The delay numbers used in timing simulation are interconnect delays and gate propagation delays.

Interconnect delays are calculated from the interconnect's physical dimension and the resistive and capacitive parameters of the IC fabrication process. Gate delay is determined by three variables: input transition speed, delay equation of the gate, and output capacitive load. A steeper input transition produces a smaller gate delay. A larger capacitive load causes a larger gate delay. A delay equation of a gate takes in an input transition speed and an output load, and produces the gate's delay and the output transition speed. A gate's delay equation is obtained by characterizing the gate using a transistor-level simulator, such as SPICE. The characterization process simulates and measures the gate's propagation delays and output transition speeds for a range of input transition speeds and output capacitive loads. The measures are then fit into a set of equations.

To calculate a gate's delay in a layout, the gate's input and output capacitance are first extracted from the layout. Next, the input transition speed is calculated by computing the output transition speed of the driver on the gate's input capacitance using the driver's delay equation. With this input speed and the gate's output capacitance, the gate's propagation delay is calculated using the gate's delay equation. This iterative process is captured in Figure 3.23.

The calculated delays numbers, gate and interconnect, are then stored in standard delay file (SDF) format. The exact format can be found in the *OVL Standard Delay File (SDF) Format Manual*. These delays are then written and annotated to the gate or block models using Verilog's # `delay` construct or `specify/endspecify` construct.

A delay model can be selected as an option in the command line or as a compiler directive. When both are present, the former takes precedence over the latter. The exact syntax for

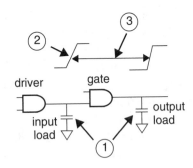

(1) capacitance extracted from layout

(2) input speed calculation

(3) propagation delay based on input speed and output load

**Figure 3.23** Calculating gate propagation delay from a delay equation

delay model selection is not an IEEE standard, and thus is simulator specific. An example of delay model selection is as follows:

```
compile -f filelist -unit_delay_model // command-line option
selecting unit delay model
```

or as compiler directive `**`use_unit_delay_model**` inside the HLD code. Both keywords `unit_delay_model` and `use_unit_delay_model` are understood by the compiler to choose the unit-delay model.

So far, we have assumed a single number of gate delays. As a part of the IEEE standard, a delay can have three possible values: minimum, typical, and maximum. A gate with a delay triplet is declared as follows, in which its minimum is `1`; typical, `1.2`; and maximum, `1.5`:

```
buffer #(1:1.2:1.5) gate1(...);
```

Which delay value is to be used in simulation can be selected by passing a simulator-specific option to the compiler, such as

```
compile -f filelist -typical_delay
// command option selecting typical delay among maximum,
typical, and minimum delays
```

Once a delay model is selected, a simulator can be configured to verify various timing properties. Some common properties are timing checks, race check, and narrow pulse check. There are eight IEEE standard built-in timing checks in Verilog: `$setup`, `$hold`, `$setuphold`, `$width`, `$period`, `$skew`, `$nochange`, and `$recovery` (based on IEEE 1364–1995). These timing checks perform three tasks: (1) determine the elapsed time between two events, (2) compare the elapsed time with a specified limit, and (3) report a timing error if the limit is violated. For example, `$setup(data_in, psedge clock, 1)` compares the time elapsed between a transition of signal `data_in` and a rising edge of `clock`. If the elapsed time is less than one unit of time, a violation will be reported. The same applies to `$hold` and `$setuphold`. `$width` checks for pulses with a width narrower than a specified limit—glitch detection. `$period` flags an error if the period of the signal is smaller than a specified limit. `$skew` issues an error if the time interval between two transitions (the skew) is greater than a specified limit. Finally, `$recover` checks for the recovery time of a changed signal, whereas `$nochange` checks for a steady-state value of a signal within a time interval. For instance, `$nochange (posedge clock, data, 0, 0)` issues an error if `data` changes while `clock` is rising. For a more detailed description of

these checks, please refer to IEEE 1364-1995 or later version. A simulator can be configured to perform timing checks on selected modules. For example, the following command
passes in a file, `timing_file`, that specifies which modules should be skipped for timing
checks or which block delays should be replaced with zero delays:

```
compile -f filelist -timing timing_file
```

A typical format for `timing_file` is

```
<module path> <timing specification>
```

an example of which is

```
top_module.* no_timing_checks,
```

meaning all submodules under `top_module` should be skipped for timing checks.

In a real circuit, every transition has either a nonzero rise time or a nonzero fall time, and
consequently it is possible that the finite rise and fall times shape a narrow pulse so that it
does not have enough energy to propagate through a gate, as seen in Figure 3.24. This phenomenon is called *narrow pulse filtering*.

RTL simulators combine rise and fall times with gate propagation delay, and use the combined delay as the overall gate delay. Effectively, all transitions have zero rise and fall times.
Most simulators have a mechanism to detect narrow pulses. First, let us define some

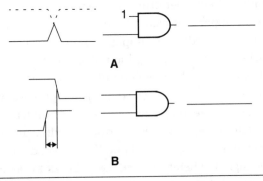

**Figure 3.24** Effect of nonzero rise and fall times on narrow pulses. (A) A narrow pulse is filtered out. (B) Two
closely spaced transitions fail to propagate the glitch.

terms. The gate delay measured from an input transition to an output transition is called the *transport delay*. The minimum width a pulse must have to propagate to an output is called the *inertial delay*. The narrow pulse filtering effect is modeled by *inertial delay*. A common practice is to filter out automatically pulses of a width less than or equal to the delay of the gate (transport delay = inertial delay). To override this, the user can pass, in compile time, options specifying a limit on the minimum pulse width, in terms of a percentage of gate delay. Furthermore, this option can be applied to selected modules or paths. An example command follows:

```
compile -f filelist -pulse_width_limit=50 -pathpulse ...
```

where `-pulse_width_limit=50` sets the minimum pulse width to be 50% of the gate delay, and `-pathpulse` enables module path-specific pulse control rules. Module path-specific pulse control rules specify pulse widths for paths inside a module. The rules are embedded in RTL code within `specparam` with a keyword such as `PATHPULSE$ = 5`, meaning the minimum pulse width for the module is five units of time.

When a pulse violates a pulse control restriction (for example, a pulse width is narrower than the inertial delay) the output of the gate becomes unknown. When this situation occurs, the time that the output becomes unknown can be determined using two methods. The first method, called *alert on output transition*, sets the time of unknown output to be the time when the first edge of the input pulse appears at the output. The rationale is that this is the time the output recognizes the input pulse. The second method, called *alert on detection*, sets the time of unknown output to be at the moment the input pulse is determined to be in violation. The rationale here is that this is the time the violation occurs. Most simulators allow the user to choose either method of reporting. Figure 3.25 illustrates the two reporting methods. The pulse at input `in` of the invertor has a width of 2, whereas the gate has an inertial delay of 3. Therefore, this pulse will flag an error. The transport delay of the gate is 3. View A in Figure 3.25 illustrates the rule of method 1. It produces an unknown output (shaded region) when the first transition of the input pulse has propagated to the output, which happens at time 5. The unknown value lasts until the second transition gets to the output, which occurs at time 7. View B illustrates the rule of method 2. It sends the output to unknown once the pulse is detected to be narrow. The detection time is when the second edge of the pulse arrives at the input, which is at time 3. This unknown value persists until the second edge of the pulse has reached the output at time 7.

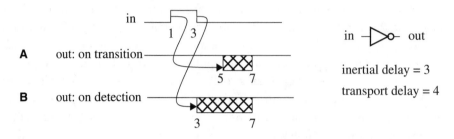

**Figure 3.25** Two different alert systems: (A) on first output transition and (B) on pulse violation detection

### 3.4.4   Design Profiling

Design profiling gathers information on how simulation time is distributed throughout the design and underlying supporting operating system (OS). The main purpose of using design profiling is to find simulation bottlenecks and optimize them for performance improvement. Activating profiling slows down simulation speed.

Profiling results can be collected at various levels of the design hierarchy. A profiling result in a design scope is sometimes called a *view*. One view is the overall summary of computing times spent on the design, the OS kernel, PLI calls, and signal dumping. An example of an overall view is shown in Table 3.3, where the design took 313.2 seconds, about 36% of the total simulation time. OS kernel time is the time spent on calling OS system tasks such as those for file I/O. PLI task time is that used by PLI tasks. Signal trace dumping is a major consumption of simulation time (for example, dumping VCD files).

**Table 3.3**   Example Simulation Profiling Summary

| Scope | Total Time, sec | Time, % |
|---|---|---|
| Design | 313.2 | 36 |
| OS kernel | 72.4 | 8 |
| PLI tasks | 155.7 | 17 |
| Signal trace dumping | 322.9 | 39 |

Inside a design view, there can be more scope. Examples include the module view, where runtime distribution statistics on modules are collected, and the construct view, where statistics on `always` blocks, continuous assignments, functions/tasks, timing checks, UDPs, and other constructs are gathered. In the construct view, each construct is identified by filename and line number. An example of a construct view is shown in Table 3.4. For example, 2.9% of time is spent on an `always` block in file `chip.v` on lines 122 to 244.

To activate design profiling, an argument is passed to the compiler so that the mechanism to collect the statistics can be constructed and compiled during compilation, such as

```
compile -f filelist +profiling ...
```

### 3.4.5  Two-State and Four-State

Two-state simulation is faster but four-state simulation detects unknown and high-impedance states better. Some simulators allow users to specify at compilation time with an option such as `+two_state` whether two-state or four-state simulation is to be executed. When simulating in two state, some simulators convert the entire design to a two-state version by replacing `x` and `z` with `0`, and ignoring the unconvertible constructs. Therefore, for these simulators, the result from the unconvertible constructs may be

**Table 3.4**   Profiling Statistics of Constructs of a Design

| Construct | Instance | Time, % |
|-----------|----------|---------|
| `always` block | chip.v: 122-244 | 2.9 |
|  | chip.v: 332-456 | 3.3 |
| `initial` block | reset.v: 100-144 | 2.5 |
| Function | ecc.v: 320-544 | 2.0 |
|  | m ask.v: 124-235 | 1.3 |
| Task | cache.v: 212-326 | 1.9 |
| Timing check | pipeline.v: 32 | 0.4 |

wrong. Some simulators, on the other hand, preserve certain constructs that are inherently four state. For these simulators, the acceleration is less. Therefore, when coding for simulation performance, it is important to know what constructs are inherently four state. The following is a list of four-state constructs.

1. Strength data types. Verilog data types `tri1` and `tri0` model nets with implicit resistive pull-up and pull-down. If no driver is driving `tri1`, the value of `tri1` is 1 with strength of pull. Similarly, `tri0` is 0 if it is not driven. Data type `trireg` models a net with charge storage with three storage strengths. These three data types should be preserved in two-state simulation; otherwise, the wrong result will occur. This is because in two-state simulation, there is no concept of strength. All strengths are the same. Therefore, when converting to two state, the implicit pull-up in `tri1` is mapped to 1 and hence causes bus contention when `tri1` is driven to 0. Consequently, all strength-related constructs should be preserved. Some such constructs are pull-ups, pull-downs, and primitives such as `tran`, `rtran`, and their relatives, which propagate signals with strength. Also parameters with Xs and Zs should be preserved.

2. Four-state expressions. Verilog operators such as `===`, `!===`, `casex`, and `casez` operate on four-state expressions and hence should be preserved.

3. User-defined four-state data type. Some simulators allow users to define the four-state date type. An example is shown here, where wire w is defined through the stylized comment to be a four-state wire and hence should be preserved:

```
wire /* four_state */ w;
```

4. Any constructs connected to the previous four-state constructs or variables that propagate four-state data should be considered secondary four-state constructs and hence preserved. Consider the following:

```
wire /* four_state */ a;
assign b = a;
buffer gate1(.in(b), .out(c));
```

where wire a is declared as a four-state variable using a simulator `pragma`. Wires b and c should be preserved as four state because they form a conductive path for wire a. Any four-state value coming from wire a will be propagated to wires b and c.

To preserve four-state constructs, simulators allow the user to select modules to be simulated in four state or two state, and the selections are made through a configuration file. A configuration file contains module identifications and designation of four state or two state. For example, the following line specifies that module `mod1` and `mod2` be simulated in four-state mode:

```
module {mod1, mod2} {four-state}.
```

The configuration file `4state.config` is then passed to the compiler on the command line:

```
compiler -f filelist +2_state +issue_2_state_warning
+4_state_config+4state.config
```

which invokes a two-state simulation compilation, issues warnings on constructs that may cause simulation differences arising from conversion of four-state constructs to two-state constructs (as indicated by `+issue_2_state_warning`), and simulates some modules in four-state mode as specified in configuration file `4state.config`.

### 3.4.6 Cosimulation with Encapsulated Models

Encapsulated models, arising mainly from IPs, and reused and shared libraries, are precompiled object models that offer simulation capability while concealing their functionality. An encapsulated model has an HDL wrapper that defines the model's interface and design parameters. An encapsulated model also provides view ports through which the user can access certain internal nodes for read or write (for example, loading to memory inside the model or reading some control and status registers). To use an encapsulated model, it is first linked with the simulator and is then instantiated in the design through its wrapper interface. A simulator communicates with an encapsulated model through its wrapper interface. Two standard interfaces are open model interface (OMI) and SWIFT. To use an encapsulated model, the following steps are required:

1. Install the encapsulated model.
2. Link the simulator with an interface to the encapsulated model. The interface passes and retrieves values through the model ports.
3. Modify the library path to include the installed directory.
4. Instantiate the model wrapper in the design, then compile and simulate.

Hardware emulators can also be interfaced as an encapsulated object. Instead of having the wrapper interface talking to precompiled code, the wrapper communicates with the hardware emulator itself.

Figure 3.26 shows simulation with two encapsulated models: One is precompiled object code and the other is a hardware emulator. For the hardware emulator, sometimes an

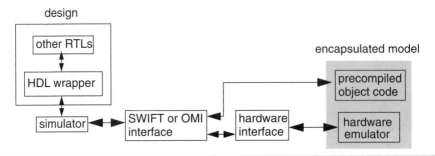

**Figure 3.26** Cosimulation with an encapsulated model

additional interface is needed between the emulator and the standard interface. Whenever the interface wrapper is encountered during simulation, the wrapper simply collects the inputs and passes them to the encapsulated model, which executes and returns the outputs to the interface wrapper, which in turn passes up to the design.

## 3.5    Incremental Compilation

Incremental compilation compiles just the portion of the circuit that has been modified, and it reuses existing compiled code for the remaining circuit. The savings in compilation time can be significant when compiling a large circuit for hardware simulation. The principle behind incremental compilation is that every file has a signature file that compares with the file to detect changes. If no changes are detected, the file is not compiled and its compiled image is linked with other images to generate the final code. Because images of individual files can be shared, it is good practice to create a central repository for images paired with their signature files. When a user compiles, the central repository is searched for signature and image files. The signature files are then compared. If some signature files match, meaning there are no changes in the design files corresponding to these signature files, the associated binary code is retrieved and reused. For the mismatched files, compilation is required. This flow of incremental compilation is shown in Figure 3.27.

Changes in file content certainly causes recompilation. However, it is possible that factors other than changes in file content can trigger recompilation. Such factors include command-line options that have changed, referenced modules that have changed, and the simulator version that has changed. In other words, if file content has not changed, but the options used to compile its image have been changed, then the archived image cannot be reused. Similarly, if the compiled image contains information related to some reference

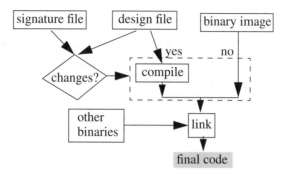

**Figure 3.27** Flow of incremental compilation

modules or simulator features, then a change in a reference module or the simulator's version outdates the archived image. An example command line for a compiler with incremental compilation is as follows:

```
compile -f filelist -shareDirectory=centralLib [other options]
```

where `centralLib` contains the archived binary images.

## 3.6 Simulator Console

Simulations run in two modes: batch mode and interactive mode. Batch mode starts the execution immediately after the simulator is invoked, whereas the interactive mode creates a session awaiting commands from the user. The software program responsible for interactive mode is called the *simulator console*.

In batch mode, runtime arguments are passed to the simulator at invocation and they generally belong to the following categories: parameter passing, library path, checking control, command file, output control, and miscellaneous simulator information. Runtime arguments were discussed in detail in the previous sections. To contrast with interactive mode, let's briefly summarize here. Parameters can be passed using `plusarg` (for example, `+myFlag+1` sets variable `myFlag` to 1) such that system task `$test$plusargs("myFlag=%d",value)` inside the code assigns 1 to variable *value*. A library can be a precompiled PLI user task library or the binary code of an encapsulated model. Its location can be specified through a command-line option. An argument of checking control turns on or off certain checks such as timing checks. A command file contains runtime commands and the commands are executed in the order they are listed in the file. A command file contains simulator control instructions that direct how many cycles to run, when to pause to print out variables, and when to "check point" or exit. An output control argument tells the simulator what and where to log (for example, log type I warnings to file *warnings*). Finally, miscellaneous simulation information can be the version of the simulator. The following is a sample runtime command:

```
simulator -command_file list_of_commands +myFlag+1 -log log_file
-PLI_lib /home/design/pli
```

where file `list_of_commands` contains runtime commands, argument `myFlag` is set to 1, log file is `log_file`, and the path to the PLI library is `/home/design/pli`.

In interactive mode, the simulator console takes over immediately after the simulator is invoked. The user can enter commands to direct the simulator. Interactive mode is mainly used for debugging. Generally, the interactive commands fall into the following types:

1. Break point. The user sets a break point, possibly with some condition. The simulator will pause when the break point is hit, at which time the user can examine variables. Related commands can delete and display break points.
2. Variable inquiry. The user reads from or writes to variables, dumps out signal traces to a file, or displays signal waveforms. In addition, connectivity of the design can be displayed and traversed. As said previously, the extent of visibility can be controlled at the expense of simulation performance.
3. Simulation drive. The user runs the simulation for a number of time steps or to the end. The user can also restart a simulation from a halted simulation.
4. Check pointing. The user saves the state of simulation and restores it later in the simulator or in a different one. This is useful when the simulation is first run on a fast simulator—say, a hardware simulator—to detect errors. Once an error is found, the state of the simulation is saved and is restored to a second simulator for debugging. The situation is often that the first simulator is in high demand and hence cannot afford to be in debug mode, and it has limited visibility to circuit nodes.
5. Simulator information. The user inquires about the current simulation time, the current clock cycle, memory usage, simulator version, and other miscellaneous information.

## 3.7 Summary

In this chapter we discussed simulation compilers and various architectures of simulators. While studying simulation compilers, we listed the typical components in the front end and the typical output formats from the back end. In particular, we illustrated interpreted code and high-level code. We also studied a compiled simulation system structure and its interaction during simulation.

We examined different types of simulator architectures with an emphasis on event-driven and cycle-based simulators. For event-driven simulators, we discussed the timing wheel, event queue, event prediction, event propagation, event validation, time advancement, and oscillation detection, and ended the discussion with an event-driven scheduling algorithm. The next topic we emphasized was cycle-based simulation. We introduced leveling and the topological sort algorithm to level. The algorithm was further applied to combinational loop detection. Next, we studied clock domain analysis, clock tree processing, execution order, code generation, and simulation control. We considered event-driven and cycle-based simulators to be the extremes of the simulator spectrum. We then discussed hybrid simulators that employ a combination of key concepts from event-driven and cycle-based simulation—namely, leveled event processing and cycle-based simulation of combinational loops. We also touched on the subject of hardware simulator construction.

We then listed common simulator terms and contrasted them. We covered two-state, four-state, zero-delay, unit-delay, event-driven, cycle-based, interpreted, compiled, FPGA-based, and process array-based hardware simulators. In the final section we studied simulation operation and application. The purpose of this section was to introduce key concepts and to show you the commands and features available in a typical simulator. We covered simulation file structure, command-line arguments, options for selective performance optimization, and signal dumping. We summarized the effects of simulator options on performance and debugability.

We then discussed what a simulator is equipped for timing verification. First we introduced three delay models and three-value delay. Then we enumerated the IEEE standard built-in timing checks and explained their function and how they can be invoked selectively. Next, in studying pulse control, we talked about transport and inertial delay, and described output behavior on pulse width violation. In the next section we illustrated how design profiling reveals simulation bottlenecks. We concluded the chapter by examining some special aspects of simulation: inherent four-state constructs, cosimulation with encapsulated models, incremental compilation, and interactive simulation commands.

## 3.8 Problems

1. During a design project, the team has access to three types of simulators—interpreted, cycle based, and hardware—but can only use one type of simulator during a development phase. Each type of simulator has its own compilation time and simulation speed, as shown in Table 3.5. The design's revision frequency and the number of diagnostic tests run in each revision are shown in Figure 3.28, in which there are four stages: prototyping, alpha, beta, and product. The relative magnitudes of the numbers reflect industrial reality. During the early stage of design, bugs abound and revisions are often, but test cases are few. During the production stage, the design is stable and lots of tests are run before it is manufactured. During a stage, the design is revised the number of times equal to the revision frequency. And for each revision, the design is compiled once and simulated on diagnostic tests with a total number of cycles that is determined by test size. Determine which simulator should be used for each development phase so that the overall verification time—compilation plus simulation—is minimal.

2. Here we compare the number of operations between an event-driven simulator and a cycle-based simulator for an XOR parity network made of 2-input XOR gates. A number of bits have even parity if the number of 1s is even; otherwise, they are odd. A three-level XOR parity network is shown in Figure 3.29. Calculate the number of events in an N-level XOR network if all inputs switch, counting zero-width glitches. If one event evaluation and one gate evaluation are considered an equal operation, what is the relative speed of an event-driven simulator versus a cycle-based simulator for this N-level XOR network for large N?

3. In this exercise, you are to generate a C program simulating the circuit shown in Figure 3.30 for the input stimuli. Print values of output x and y after each application of an input vector.

**Table 3.5**   Simulator Characteristics

| Simulator type | Compilation time, sec | Simulation speed, cycle/sec |
|---|---|---|
| Interpreted simulator | 200 | 2 |
| Cycle-based simulator | 1,500 | 20 |
| Hardware simulator | 36,000 | 10,000 |

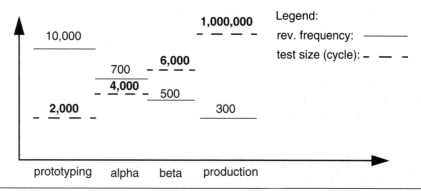

**Figure 3.28** Revision frequency and diagnostic test size in a design project

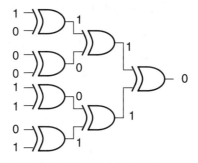

**Figure 3.29** An XOR parity circuit for comparing simulation operations

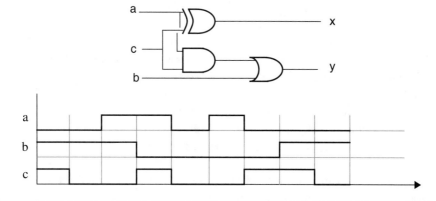

**Figure 3.30** Circuit for generating C code

**4.** If the probability is 0.5 that the value of a node is 1, what is the probability of a transition propagating from an input to the output of a 2-input AND gate? How about a 2-input OR gate? A 2-input XOR gate? N-input AND or OR gate?

**5.** In this problem, calculate event counts for the 4-bit ripple carry adder in Figure 3.31 using different delay models, when 0111 is added to 1010 with carry-in equal to 1. Assume all inputs to the adder are zero before the addition.

  a. Assume each gate has zero delay. Calculate the maximum total number of events possible in the 4-bit adder, counting zero-width glitches.
  b. Repeat step a with event levelization. What is the saving?
  c. Repeat step a with a unit-delay model.
  d. Tabulate the results from a, b, and c.

**6.** Use DFS to determine all loops in the following Verilog RTL code:

```
always @(a or b or c) begin
if(a) d = ~b;
else d = b + c;
end

always @(d or h)
e = d | h;

always @(a or e or h) begin
if(a) c = e & h;
else b = ~e | h;
end

always @(d or c or b) out = d & c ^ b;
```

  a. Convert the RTL code to a graph. Each always block is represented by a node.
  b. Use DFS to find all loops in the graph. Are there any false loops (loops that can never propagate signals round-trip in the same cycle)?

**7.** Simulate the circuit in Figure 3.32 using the event-driven scheme. Construct timing diagrams, a timing wheel, and show how events get queued and dequeued. Assume that each gate has a unit delay. Ignore zero width glitches.

**8.** Apply a topological sort to the circuit in Figure 3.32. Show an order of the gates so that the evaluation is correct for steady-state evaluation.

**9.** Find all SCCs in the circuit in Figure 3.33 and write C code to simulate the SCCs with an iteration limit equal to five.

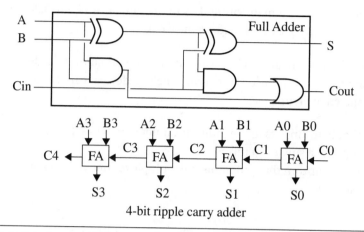

Figure 3.31 A 4-bit ripple carry adder for event evaluation

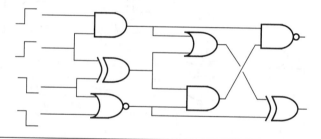

Figure 3.32 A unit-delay circuit for event-driven simulation

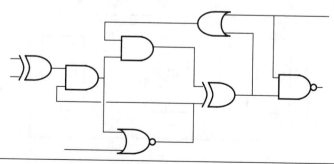

Figure 3.33 Compiling a circuit with combinational loops

**10.** Determine the clock domains of the FFs in the circuit in Figure 3.34. Assume that all FFs and latches are connected to the same clock.

**11.** The basic element in an FPGA-based hardware simulator is a device that can be programmed to any Boolean function. Such a device can be a multiplexor network, as shown in Figure 3.35.

a. What should the inputs to the multiplexor network be so that its function is the sum output of a single-bit full adder?

b. Derive an algorithm that programs the multiplexor network to any function of three variables.

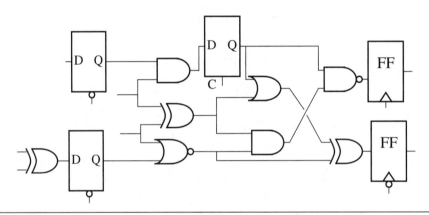

**Figure 3.34** A mixed FF and latch design for clock domain analysis

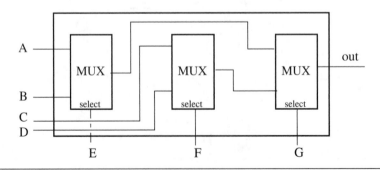

**Figure 3.35** A multiplexor network for a hardware simulator

12. Write a C/C++/Java (or your favorite language) program that evaluates the circuit in Figure 3.32. Does your program need to search for fanins or fanouts during simulation?

13. We compare an implementation difference between a compiled two-state simulator and compiled four-state simulator. The circuit is a bus with two drivers and one receiver, as shown in Figure 3.36.

   a. Write a C/C++/Java program (or your favorite language) to simulate the circuit in two states.
   b. Repeat step a for four states. Note that you need to use 2 bits to code a node value.
   c. How many signals are required to be examined to determine the value on the bus for step a and step b?

14. Dumping signal traces can drastically decrease simulation performance. One strategy is to dump only the minimum necessary signals and compute the rest off-line. This strategy is useful, especially for hardware simulators, whose purpose would be defeated if a large amount of signal traces are dumped. What is the minimum set of signals to dump?

15. Compile and simulate a design.

   a. Compile the design with and without optimization. Record the compilation times.
   b. Invoke the simulator with `plusargs` and dump signal traces.
   c. If the simulator has a profiling feature, turn it on and examine the profile.
   d. Go into interactive mode, set the break points, examine the variable values, and step until it finishes.

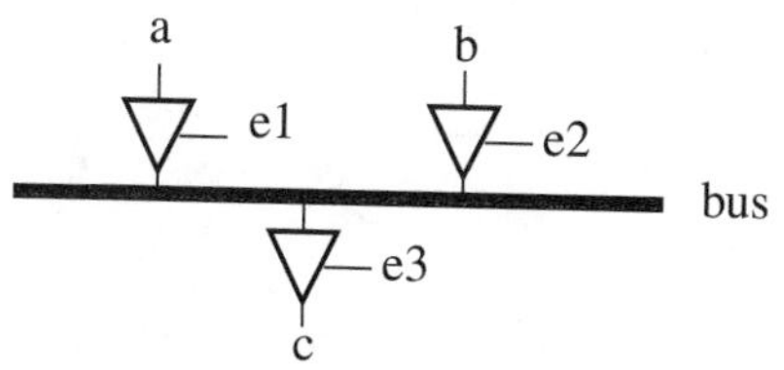

**Figure 3.36** A bus for two-state and four-state simulation

**16.** Compile a design using an incremental compilation feature and study its triggering behavior.

    a. Modify a module's functionality and invoke incremental compilation. Record the files that get recompiled.

    b. Alter a module's coding style but preserve its functionality. Does incremental compilation recompile the file? How about when you just change comments in the code?

**17.** Construct a large design, as simple as a long buffer chain and as complex as a CPU. Measure simulation performance for the following scenarios.

    a. Compile and simulate with various levels of optimization, and record and plot simulation speeds.

    b. Experiment with dumping various parts of the circuit. Plot simulation speed as a function of nodes dumped.

CHAPTER 4

# Test Bench Organization and Design

## Chapter Highlights

- Anatomy of a test bench and a test environment
- Initialization mechanism
- Clock generation and synchronization
- Stimulus generation
- Response assessment
- Verification utility
- Test bench-to-design interface
- Common practical techniques and methodologies

## 4.1 Anatomy of a Test Bench and a Test Environment

To simulate a design, an external apparatus called a *test bench* is often required to mimic the environment in which the design will reside. Among other functionality, the main purpose of a test bench, written in HDL, is to supply input waveforms to the design and to monitor its response. Because a test bench is not manufactured as the design, it has far fewer coding style restrictions. Together with the perception that test benches are discarded once the design is verified, the structures of a test bench are often at the mercy of verification engineers. Consequently, test benches frequently generate wrong stimuli, compare with wrong results, or miss corner cases, eventually diverting valuable engineering time to debugging the test benches instead of the design. Furthermore, without well-organized guidelines, test benches can be a nightmare to maintain and hence are not reusable. Therefore, to have easily maintainable and reliable test benches, it is important to understand organizations and designs of test benches.

Through the process of creating a test bench for a simple circuit, we will see the major components of a test bench as a prelude to a systematic test bench architecture. Consider verifying a circuit computing the remainder for ECC CRC-8 with generator $x^8 + x^2 + x + 1$, whose coefficients are binary. The remainder from the generator is a polynomial of degree, at most, seven, and thus can be represented by a byte, whose bits are the remainder's coefficients. For example, $x^7 + x^5 + x^2 + x + 1$ is represented by 10100111. The circuit accepts a binary stream representing an arbitrary polynomial and computes the remainder divided by the generator. An example of binary division with this generator is shown in Figure 4.1. The input polynomial is $x^{14} + x^{13} + x^{11} + x^9 + x^7 + x^5 + x^2 + 1$. The remainder is 10110100, which is polynomial $x^7 + x^5 + x^4 + x^2$. The shift and subtract (exclusive OR in Boolean domain) operations are implemented using shift registers and XOR gates. A circuit computing a remainder using this generator is shown in Figure 4.2. The input stream is fed, MSB first, to port in of the shift register, which has a zero initial state. After the input is completely shifted in, the outputs of the FFs give the remainder. As a sanity check, shift in, MSB first, and byte 100000111, which is the generator itself. After eight cycles, the MSB appears at Q7 and is XORed with the last three 1 bits to produce all zeros at the FF outputs, giving a zero remainder, as expected when the generator is divided by itself.

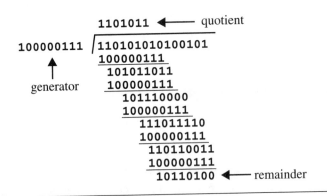

**Figure 4.1**   Compute remainder of CRC-8

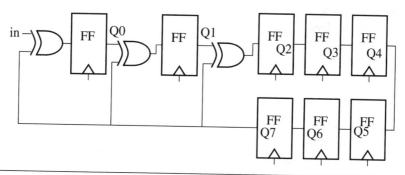

**Figure 4.2**  A circuit computing the remainder of CRC-8

Let us call the RTL module for this circuit CRC8, with input ports CLK and IN, and output ports Q0, . . . , Q7. The following line instantiates the design:

```
//design module instantiation
CRC8 DUV (.IN(in), .CLK(clk), .Q0(q0),..., .Q7(q7));
```

To set up a test bench for CRC8, we need to apply a bit stream to the circuit, which can be done by storing bits in an array and feeding the design from the array. Verilog code for this method of input application is shown here. At each rising edge of clk, the *i*th element of array input_array is fed to the input port in:

```
//apply input to CRC8 remainder circuit
initial i = size_of_input_polynomial;

always @(posedge clk) begin
   if (i! = 0) begin
    in <= input_array[i];
    //input_array is binary representation of input polynomial
    // LSB is input_array[0]
    i <= i - 1;
   end // end of if
end
```

When the array is all shifted in i.e. i == 0, the outputs of the design are compared with the expected result. The expected result, the remainder, can be computed by dividing the input array with the generator using the modulo operator %. This remainder is compared with the outputs of the design. If they are unequal, a task is called to print out the error. Sample code for checking the response is

```
// compute and compare with expected result
remainder = input_array % 8'b10000111 ;
if(remainder != {q7,q6,...,q0}) print_error();
```

Right after the circuit is powered up, the FFs should be initialized to zeros. This is done using an `initial` block. In this block, the FF outputs are initialized using assignments through hierarchical paths, such as DUV.Q0:

```
// initialize states to zero
initial begin
   DUV.Q0 = 1'b0;
   ...
   DUV.Q7 = 1'b0;
end
```

Finally, a clock waveform is generated to drive the design:

```
always clk <= #1 ~clk; // generate clock waveform
```

Putting these pieces together, with clock initialization, we have a test bench for CRC8:

```
//design module instantiation
CRC8 DUV (.IN(in), .CLK(clk), .Q0(q0),..., .Q7(q7));

// initialize states to zero and other variables
initial begin
   i = size_of_input_polynomial;
   clk = 0;

   DUV.q0 = 1'b0;
   ...
   DUV.q7 = 1'b0;
end

//apply input to CRC8 remainder circuit
always @ (posedge clk) begin
   if (i! = 0) begin
      in <= input_array[i];
      //input_array is binary representation of input polynomial
```

```
        // LSB is input_array[0]
        i <= i - 1;
    end // end of if
end

// compute and compare with expected result
remainder = input_array % 8'b10000111 ;
if(remainder != {q7,q6,...,q0}) print_error();

always clk <= #1 ~clk; // generate clock waveform
```

From this simple example, we see the six major components of a test bench, sometimes called a *test harness*:

1. Initialization
2. Input stimuli
3. Response assessment
4. Test bench-to-design interface
5. Clock generation and synchronization
6. Verification utility

The test bench-to-design interface, in the example, consists of access to the design signals through primary inputs, outputs, and hierarchical paths. Clock generation and synchronization, in the example, is just the clock generation. Verification utility is a collection of functions, tasks, and modules shared by various parts of the test bench, such as task `print_error()` in the example. The diagram in Figure 4.3 illustrates the organization and the components of a test bench.

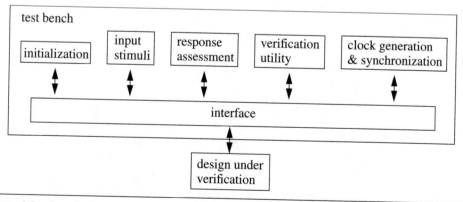

**Figure 4.3**   Organization and major components of a test bench

In the previous example we assumed that the property to verify was that the CRC8 circuit calculates the correct remainder, given the initial state is zero. Another property that we could verify is that the circuit initializes correctly. (In this example, initialization is provided by the test bench. In a real design, initialization can be a part of the design. In that case, verifying correct initialization is more meaningful.) Still another property to verify is correct remainder calculation when input polynomials are "pipelined" into the circuit. Each of these properties, or scenarios, is called a test case. To illustrate the concept of a test case further, consider verifying an ALU. One test case can be verifying integer computations, while another test case verifies Boolean operations. Each of these test cases can have its own unique initial values, input stimuli, and expected responses. For example, to test integer addition and subtraction, input vectors could be chosen to cause overflow or underflow along with other corner cases; whereas in testing Boolean operations, certain bit patterns could be selected (such as alternating ones and zeros, or even or odd parity). Because multiple test cases use the same test bench, to maximize portability of a test bench, test cases must be separated from the test bench. To modify our example to extract the test case from the test bench, the initial values and input_array can be read from an external file (for example, using readmemb or readmemh) that contains a test case. A more in-depth discussion of this topic is relegated to a later section. Figure 4.4 shows a typical verification environment.

## 4.2   Initialization Mechanism

Initialization assigns values to state elements such as FFs and memories. Although a completed design has a circuitry that initializes on power-on, initialization is often done through a test bench for the reason that the initialization circuit may not have been designed at the time, or a simulation is to run starting from a time long after power-on so that simulating through the initialization stage will consume too much time, or the simulation emulates an exception condition that a normal operation of the design will not reach from its legal initial state.

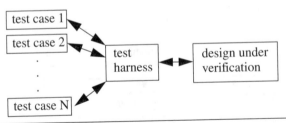

**Figure 4.4**   A typical verification environment with the separation of the test harness and the test cases

Initializing can be done using test bench RTL code or PLIs. The former is usually reserved for designs with a small number of state elements or one in which accessing of state elements is relatively simple. The latter, for large designs in which the structure of the design is traversed to locate state elements (for example, enumerate all instantiated gates and blocks) is used to determine the sequential ones and apply corresponding initial values.

## 4.2.1   RTL Initialization

Initialization code, usually enclosed in `initial` blocks, assigns values to state elements using hierarchical paths. Although initial values can, in theory, propagate through the design's primary inputs to the state elements, in practice the propagation sequences are difficult and time-consuming to generate.

One must decide whether to initialize a state element's input or output. In most cases, outputs are initialized. If, for some reason, inputs are initialized, you should make certain that the values are not overwritten by the input logic cones. Furthermore, the time when initialization takes place needs to be determined, because not all initializations should occur at the same time or at the very first clock edge. For example, one cannot initialize a latch without knowing when the latch becomes transparent, because initializing when it is transparent runs the risk of having the initial values being overwritten. An example of initialization using test bench RTL code is shown here, in which some state elements in modules `alu` and `usb` are assigned initial values, and `usb` is initialized ten units of time later than `alu`:

```
initial
begin
    top.design.alu.multipler.ff9.Q = 1'b1;
    top.design.alu.multipler.ff9.Q_bar = 1'b0;
    ...
    #10;
    top.design.usb.xmit.latch3.Q = 1'b0;
    top.design.usb.xmit.latch3.Q_b = 1'b1;
    ...
end
```

Hard coding initial values makes the code unable to be reused if different initial values are often required for other test cases or the circuit requires initialization at a different time. To make it reusable, the code should be encapsulated inside a task with an input for initial values. In this way, whenever initialization is needed, the task is called. Applying this principle to the previous example, we can split the code into two tasks, one for `alu` and the other for `usb`:

```
task initialize_alu;
input [N:0] init_value;
begin
    top.design.alu.multipler.ff9.Q = init_value[0];
    top.design.alu.multipler.ff9.Q_bar = init_value[1];
    . . .
end
endtask

task initialize_usb;
input [M:0] init_value;
begin
    top.design.usb.xmit.latch3.Q = init_value[0];
    top.design.usb.xmit.latch3.Q_b = init_value[1];
    . . .
end
endtask
```

Now, to initialize, simply call the tasks with initial values:

```
initial
begin
    v1 = ...;
    v2 = ...;
    initialize_alu(v1);
    #10;
    initialize_usb(v2);
    . . .
end
```

To see the power of reusable code, let's make use of these tasks and readily implement a random initialization routine using Verilog's system task `$random(seed)`. It is highly recommended that the seeded version be used for repeatability. Note that `$random(seed)` returns a value with the size of an integer. So, if the input vector is of a different size, the returned value is scaled by the ratio of the maximum value of the input to that of the integer—namely, $2^{(a/b)}$—where a is the number of bits of the input and b is that of the integer. If the returned value is not scaled properly (such as linearly), the resulting distribution is changed:

```
task random_init;
input [K:0] seed;
begin
   reg [31:0] v1;
   reg [31:0] v2;
   v1 = $random(seed);
   v2 = $random(v1);
   initialize_alu(v1);
   #10;
   initialize_usb(v2);
   ...
end
endtask
```

Initializing memory is usually done by loading memory content stored in a file using system tasks $readmemh or $readmemb. System task $readmemh expects values in hex whereas $readmemb uses binary. For example,

```
$readmemh("filename",top.design.data_memory,`START_ADDRESS,
`END_ADDRESS);
```

Initialization codes cannot be synthesized (at the time of this writing) and usually are not supposed to be part of a library cell's functionality. Thus, they should never be placed inside any library cells. Another reason for not putting initialization code inside a library cell is that different instantiations of the same cell may require different initial values. Besides, sometimes an instantiation without initialization is used as a means to verify initialization circuits. In any case, embedded initialization has detrimental effects and must be avoided all the time.

## 4.2.2 PLI Initialization

The aforementioned initialization method has several disadvantages. First, it produces a large amount of code in a large design, because an assignment is needed for every state element and memory. Second, explicitly specifying the state elements is not portable because the design can change, decreasing or increasing the number of state elements. To overcome these problems, a better mechanism should search for state elements and memory, and initialize them. Verilog does not offer a facility to perform this search directly, but PLI or VPI has routines to do just that, and more. Here we look at how sequential elements can be initialized using PLIs.

For simplicity, let's consider only FFs. However, the algorithm can be generalized to other types of state elements. First, all gates in the design are enumerated and each gate is checked for sequential or combinational type. If a sequential gate is found, all its outputs are then iterated, and each output is assigned an initial value, which is, for example, retrieved from a database. These procedures are facilitated by the acc routines of PLI. For a complete description of PLI and its applications, refer to *The Verilog PLI Handbook: A User's Guide and Comprehensive Reference on the Verilog Programming Language Interface* by Stuart Sutherland. C pseudocode for initializing FFs using PLI is as follows:

```
void initialize_flip_flop ( ) {
    ...
   db = fopen("database","r");
   module = acc_fetch_by_name("my_design");
   cell = NULL:
   while(cell = acc_next_cell(module, cell ){
      if(cell is sequential){
         port = acc_next_port(cell, port);
       if (port is output) {
            get_init_value(db, port, &value);
            acc_set_value(port, &value, &delay);
         }
      }
   }
}
```

The first acc routine gets the handle of the design by name, assuming the module name of the design is my_design. The second acc routine, acc_next_cell, iterates through all cells in the design (in other words, each call to this acc routine returns a cell in the design until it runs out, and then it returns NULL to exit the while loop). If the cell is sequential, acc_next_port iterates all its ports. For each port, its polarity is checked. If it is an output, the initial value for the output is retrieved from a database using function get_init_value. Function get_init_value is a user function that searches the initial value for a port in database db. Once found, the value is assigned to the port using acc routine acc_set_value.

This C code is compiled into object code and is linked to a Verilog user-defined task, say $init_FFs. The thread of execution during a simulation run is shown in Figure 4.5. The RTL code runs on a simulator and the C object code runs on the host. When $init_FFs is encountered during simulation, the simulation is paused and control is passed to the host

RTL code                          PLI C code

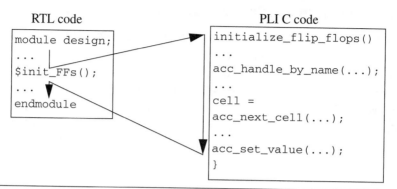

**Figure 4.5**   Flow of execution of a circuit using a PLI to initialize FFs

system to execute the compiled PLI C code, in which all cells in the design are traversed using PLI's acc routines and sequential gates are processed. Outputs of each sequential cell are iterated and assigned initial values stored in a database. When the PLI C code is finished, all FF outputs have been initialized. Then the control is passed back to the simulator, and simulation resumes.

### 4.2.3   Initialization at Time Zero

Initializing a variable at time zero requires care, because the change from an unknown value before time zero to the initialized value at time zero can constitute a transition for some simulators, whereas other simulators will not pick up this transition. Therefore, inconsistent results are produced from simulator to simulator. To avoid this, one strategy is to initialize at a positive time, and treat this time as the time of origin. Sometimes it is safer to initialize explicitly to the unknown value x at time zero, followed by an initialization to the desired values at a later time. For example,

```
task init_later;
input [N:0] value;
begin
   design.usb.xmit.Q = 1'bx; // avoid time zero event
   ...
   #1;
   design.usb.xmit.Q = value[0]; // now real initialization
   ...
end
endtask
```

## 4.3    Clock Generation and Synchronization

Clocks are the main synchronizing signals to which all other signals reference. In a majority of situations, clock waveforms are deterministic and periodic. Therefore, if you know how to write RTL code for one period, the complete waveform can be generated by embedding the one period inside a loop. To create one period of clock, there are two methods: explicit and toggle. With the explicit method, you specify rising and falling times of the clock and assign a value to it at each of the transition times. With the toggle method, you also specify rising and falling times of the clock, but invert the clock at the transition times.

### 4.3.1    Explicit and Toggle Methods

Consider generating the clock waveform in Figure 4.6. First, one period of a clock is generated. Using the explicit method, one period of clock has the RTL codes

```
#1 clock = 1'b1;
#1 clock = 1'b0;
#2 clock = 1'b1;
#2 clock = 1'b0;
```

The delays before the assignments are delay intervals between successive transitions. Putting this period inside a loop and initializing the clock produces the complete waveform, as shown here:

```
initial clock = 1'b0;
always begin
   #1 clock = 1'b1;
   #1 clock = 1'b0;
   #2 clock = 1'b1;
   #2 clock = 1'b0;
end
```

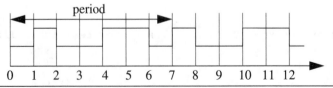

**Figure 4.6**    Generating a clock signal

The same clock waveform can also be generated using the toggle method:

```
initial clock = 1'b0;
always begin
    #1 clock = ~clock; // rising
    #1 clock = ~clock; // falling
    #2 clock = ~clock; // rising
    #2 clock = ~clock; // falling
end
```

This toggle method can be difficult to see the value of `clock` at a given time. Thus, comments indicating rising or falling transitions are recommended. Furthermore, if the clock is left uninitialized, the clock will not toggle and simply stays at the unknown value x—a potential pitfall that the explicit method avoids. On the other hand, the toggle method is easy to change the phase, or the initial value, of the clock by simply initializing the clock to a different value while keeping the toggle statements intact. Changing a clock's initial value is more complicated with the explicit method; all `assign` values have to be modified.

Note that we used blocking assignment operator =, but nonblocking operator <= could also be used in this example.

### 4.3.2  Absolute Transition Delay

In the previous example, the delays are interval delays between successive transitions, and there are situations when absolute transition times are desired. To do so, nonblocking intraassignment delay can be used. The following code representing the waveform shown in Figure 4.6 uses nonblocking intraassignment delays:

```
initial begin
    clock <= #0 1'b0;
    clock <= #1 1'b1;
    clock <= #2 1'b0;
    clock <= #4 1'b1;
    clock <= #6 1'b0;
    clock <= #7 1'b1;
    ...
end
```

Statement `clock <= #2 1'b0` assigns `1'b0` to `clock` two units of time from the current time. Because this statement does not block, the next statement is simulated immediately. The next statement, `clock <= #4 1'b1`, assigns 1 to `clock` four units of time from the current time. Therefore, all delays in the nonblocking assignments refer to the current time.

Because the delays are absolute transition times, all transitions have to be explicitly specified, as opposed to embedding one period in a loop. Hence, generating waveforms using absolute times is used only for aperiodic waveforms such as reset signals. It's important to note that if the nonblocking assignments are replaced by blocking assignments, the next statement must wait until the current statement is executed, meaning the delays now have become interval delays.

### 4.3.3   Time Zero Clock Transition

Similar to the time zero initialization problem, the very first transition of `clock` at time zero may be perceived as a transition because `clock` has an unknown value before time zero and gets assigned to a value at time zero. Whether this time zero transition is perceived is simulator dependent, and thus care must be exercised to deal with either scenario. One way to avoid this ambiguity is to initialize `clock` explicitly to unknown value x at time zero, hence eliminating the time zero transition, and start the clock at a later time.

### 4.3.4   Time Unit and Resolution

During verification, the clock period or duty cycle may change. It is beneficial to use parameters to represent the delays, instead of hard coding them. For example, to generate a clock starting with zero that has a 50% duty cycle, one may code as follows:

```
define PERIOD 4
initial clock = 1'b0;
always #('PERIOD/2) clock = ~clock;
```

Caution should be exercised when PERIOD is not evenly divided by two. If PERIOD is odd, the result is truncated. If integer division is replaced by real division, the result is rounded off according to the specified resolution.

In Verilog, the unit for delay (for example, #3) is specified using 'timescale, which is declared as

```
'timescale unit/resolution
```

where `unit` is the unit of measurement for time and `resolution` is the precision of time. For example, with 'timescale 1.0ns/100ps, #(4/3.0) clock = 1'b1 means at 1300 ps, `clock` is assigned to 1. Note that although 4/3.0 gives 1333ps, it is rounded off to 1300ps because the resolution is declared to be 100ps.

### 4.3.5 Clock Multiplier and Divider

A complex chip often uses multiple clocks generated from a phase lock loop (PLL) block. The analog behavior of PLL is not modeled in HDL, but is abstracted to generate clock waveforms using delays and assigns. When multiple clock waveforms are generated, their relationship needs to be determined (for example, whether their transitions are independent, whether some clocks are derived from others). A clock can be derived from another via a frequency divider or multiplier. If the frequency ratio is an integer, it is easy to generate a derived clock from the base clock. For frequency division, the derived clock can be generated from base clock without knowing the base clock's frequency. To divide `base_clock` N times to get `derived_clock`, trigger a transition on `derived_clock` for every N transitions of `base_clock`:

```
initial i = 0;

always @( base_clock ) begin
i = i % N;
if (i == 0) derived_clock = ~derived_clock;
i = i + 1;
end
```

Multiplying a clock frequency by N can be achieved using Verilog's `repeat` statement once the base clock's frequency is known (for example, for every positive or negative transition of the base clock, repeatedly generate 2N transitions for the derived clock). For example,

```
always @(posedge base_clock) begin
repeat (2N) clock = #(period/(2N)) ~clock;
end
```

If the period of the base clock is not known or is changing constantly, and/or the ratio is not an integer, a different technique is required. First, the base clock's period is measured on the fly and then the derived clock is generated using

```
forever clock = #(period/(2N)) ~clock;
```

A sample code to implement this general clock divider/multiplier is as follows:

```
// measure the first period of base_clock
initial begin
derived_clock = 1'b0; // assume starting 0
@(posedge base_clock) T1 = $realtime;
```

```
@(posedge base_clock) T2 = $realtime;
period = T2 - T1;
T1 = T2;
->start; // start generating derived_clock
end

// continuously measure base_clock's period
always @(start)
forever
@(posedge base_clock) begin
T2 = $realtime;
period = T2 - T1;
T1 = T2;
end

// generate derived_clock N times the frequency of base_clock
always @(start)
forever derived_clock = #(period/(2N)) ~ derived_clock;
```

Make sure that the proper time scale is used so that the division of the period by 2N has the correct precision.

## 4.3.6   Clock Independence and Jitter

If the clocks are independent, each of them should be modeled with its own `always` statement. For example, the following code generates two independent clocks:

```
initial clock1 = 1'b0;
always clock1 = #1 ~clock1;

initial clock2 = 1'b0;
always clock2 = #2 ~clock2;
```

An incorrect way to produce these two clocks is to use one clock to generate the other, as illustrated here:

```
initial clock1 = 1'b0;
always clock1 = #1 ~clock1;

initial clock2 = 1'b0;
always @(negedge clock1) clock2 = ~clock2;
```

Although any waveform viewer shows that this set of clocks has the same transitions as the previous set, they are fundamentally different. `clock2` of the second set is synchronized with `clock1`.

To emphasize more the relative independence of two clocks, jitter can be introduced to one of the clocks to simulate the nondeterministic relative phase. This can be done with a random generator:

```
initial clock1 = 1'b0;
always clock1 = #1 ~clock1;
jitter = $random(seed) % RANGE;
assign clock1_jittered = #(jitter) clock1;
```

The modulo operator `%` returns the remainder when divided by RANGE, and thus restricts the range of jitter. Clock `clock1_jittered` is a version of `clock1` with edges that are randomly shifted in the range `[0, RANGE]`. All random functions/tasks should be called with a seed so that the result can be reproduced later when errors are found. Verilog offers a variety of random distributions, such as uniform, normal, and Poisson.

### 4.3.7   Clock Synchronization and Delta Delay

Because independent waveforms are not locked in phase, they can drift relative to each other. Jittering models this phase-drifting phenomenon. When two independent waveforms arrive at the same gate, glitches can come and go depending on the input's relative phase, creating intermittent behavior. Therefore, independent waveforms should be first synchronized before propagation. A synchronizer uses a signal, the synchronizing signal, to trigger sampling of another to create a dependency between them, hence removing the uncertainty in their relative phase. The following code is a simple synchronizer. On every transition of signal `fast_clock`, signal `clock1` is sampled. The synchronizer is essentially a latch; thus, nonblocking `assign` is used to avoid a race condition:

```
always (fast_clock)
    clock_synchronized <= clock1;
```

If the synchronizing signal's transitions do not align with those of the synchronized signal, some transitions will be missed, as shown in Figure 4.7. Because transitions will be missed if the synchronizing signal has a lower frequency, the signal with the highest frequency is usually chosen as the synchronizing signal.

Because of the nonblocking `assign` in the synchronizer, the synchronized signal and the synchronizing signal are not exactly aligned but are separated by an infinitesimal

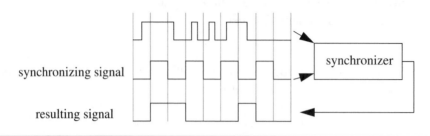

**Figure 4.7**   Intermittent glitches and synchronizing independent signals

amount. This is the result of a characteristic of nonblock assignment. At simulation time, nonblocking assignments are evaluated after all regular events scheduled for the time are evaluated. For example, in the following RTL code, the value of v is sampled when the nonblocking statement is encountered, but the actual assignment to x happens only after all blocking statements are evaluated. In this case, x gets the value of v after the blocking assignment y = x is executed. Thus, y gets the old value of x (the value of the previous clock cycle). Therefore, even though the two assignments are evaluated at the same simulation time, y always lags x by one clock cycle:

```
always @(posedge clock) begin
x <= v;
y = x;
end
```

In the previous synchronizer example, the synchronized signal, clock_synchronized, always lags synchronizing clock1 by an infinitesimal amount. This amount is from the time clock1 is sampled to the time clock_synchronized is actually assigned. Nevertheless, the two clocks' transitions always have identical simulation times. However small this infinitesimal amount is, it sometimes can cause other signals to lag by a cycle, as illustrated in the previous example.

### 4.3.8  Clock Generator Organization

A central clock module generates various clock waveforms from the same clock source. During circuit implementation, the clock source is a PLL, and various clock signals are derived from the PLL output. In RTL, clock generation should be encapsulated as a module with parameters to change clock frequencies, phases, and jitter ranges, among others. A typical block diagram for a clock generation network is shown in Figure 4.8. The primary

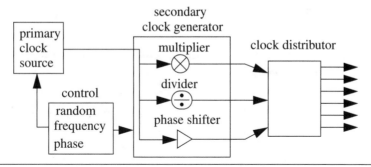

**Figure 4.8**   A typical clock generation network

clock source is a simple waveform generator, modeling a PLL. The secondary clock generation block derives clocks of various frequencies and phases with frequency multipliers, dividers, and phase shifters. Frequency, phase, and random distribution are controlled through parameters and variables. Finally, the clock distributor multiplexes the clocks to appropriate clock domains. In RTL, the clock generation network is

```
module clock_generation_network (clk1, clk2,...);
output clk1, clk2, ...;
parameter DISTRIBUTION 1;
parameter JITTER_RANGE 2;
...
primary_clock pc (.clock_src(csrc));
clock_gen cg(.in_clock(csrc),.freq1(freq1),.phase1(.ph1),...);
clock_distributor cd (.outclk1(clk1),..., .inclk1(freq1),...);

endmodule
```

## 4.4   Stimulus Generation

Many ways exist to apply input vectors to the design under verification. Let's start with the most basic method: Assign vectors to the primary inputs of the design synchronously. A block diagram illustrating this method is shown in Figure 4.9. Vectors are stored in stimuli memory. Triggered by a stimulus clock, memory is read one vector at a time and is

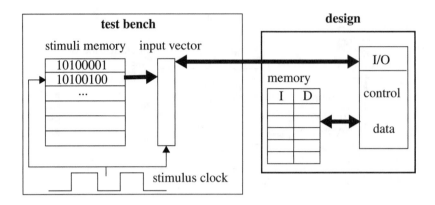

**Figure 4.9**    Apply input stimuli through primary inputs

assigned to an array. The array is then applied to the design's primary inputs. A sample RTL test bench implementing this method is as follows:

```
reg [M:0] input_vectors [N:0] // stimulus memory
reg [M:0] vector; // input vector

initial
begin
   $load_memory (input_vectors, "stimuli_file");
   i = 0;
end

always @(posedge stimulus_clock)
begin
   if(apply_input == TRUE) begin
      vector = input_vectors[i];
      design.address <= vector[31:0];
      ...
      design.data <= vector[M:M-31];

      i = i + 1;
   end
end
```

The initial block loads memory `input_vectors` from file `stimuli_file`, which holds all stimulus values, and resets counter `i`. Variable `apply_input` is controlled elsewhere in the test bench and is TRUE when the design is ready to accept stimuli.

Clock `stimulus_clock`, not part of the design, models the external clock that the design is seeing, and should be synchronized to the design's clock. Furthermore, the assignments to the design's primary inputs should be nonblocking; otherwise, race problems may result, as demonstrated in the following simple example. The design is a DFF and the stimulus clock is the same as the FF clock:

```
// design under verification, a D-flip-flop
always @(posedge clock)
   Q <= D;

// test bench that applies stimuli to design's primary inputs
always @(posedge clock)
begin
   D = vector[0]; // creates race
   i = i + 1;
end
```

On the positive transition of the clock, the following two events are scheduled: the sampling of value D to be applied to Q, and the assignment to D from `vector[0]`. The execution order of these two events is not specified in the Verilog IEEE standard and hence is at the mercy of the simulator. This order of execution affects the value D sampled. If sampling of D is executed first, Q will be assigned the value of D from the previous cycle. If assignment to D is first, Q will have D's current value; hence, a race problem has occurred. Note that the use of a nonblocking assignment in the design does not prevent this race problem. However, using a nonblocking assignment in the test bench eliminates it.

The `always` block in the previous example explicitly ties the stimulus memory to the design's primary inputs. This is not very flexible or portable, because the `always` block cannot be used to apply stimuli from other memory or to apply the same memory to another design. To remedy this, the code can be encapsulated as a task that has an input to accept a vector so that whenever a stimulus needs to be applied, the task is called with the stimulus as the input. Furthermore, one such task is associated with each design so that to stimulate the design, the design's task is invoked. In this way, stimulus application is separate from

particular memory or design, and it is self-contained, reusable, and portable. The following
example demonstrates this technique in applying a reset sequence:

```
// task applying an input to cpu's PIs
task cpu_apply_input;
input [N:0] vector;
begin
   cpu.address = vector[0];
   ...
   cpu.data = vector[j];
   ...
   cpu.interrupt = vector[k];
   ...
end
endtask

initial // use the cpu task to apply inputs
begin
   #10;
   cpu_apply_input(64'b0); // reset
   ...
   #50;
   cpu_apply_input(v1); // now start input vector sequence
   #60;
   cpu_apply_input(v2);
   ...
end
```

Besides loading stimulus vectors from a file to memory and iterating the memory to apply
input vectors, stimuli can be generated on the fly via a task, a function, a PLI user task, or a
random generator. Code using these methods is similar and can be encapsulated in a task
such as cpu_apply_vector, as in the previous example. Let's look at some sample implementations.

In the first implementation, the task accepts an index indicating the vector is to be applied
to the input and assigns the vector to it. The indexed vector is obtained through another
task generate_vector:

```
task apply_vector_task; //apply vectors from a task
input index;
output [N:0] primary_input;
reg [N:0] vector;
```

```
begin
  generate_vector(index, vector); // generate indexed vector
  primary_input <= vector; // apply vector
end
endtask


always @(start) // on start, call apply_vector_task
begin
  #10;
  apply_vector_task(i, top.cpu.inputs);
  i = i + 1; // next vector
end
```

In the second implementation, the vector-generating task is replaced by a PLI user task, which computes the indexed vector in C/C++. The user task, when encountered in RTL, calls the C/C++ routine and returns with the vector:

```
task apply_vector_PLI; //apply vectors from a PLI
input index;
output [N:0] primary_input;
reg [N:0] vector;
begin
  $gen_vector_PLI(index, vector); // generate indexed vector
  primary_input <= vector; // apply to
end
endtask
```

In the third implementation, random vectors are generated using Verilog's system task $random:

```
task apply_vector_random; //apply random vectors
input [32:0] seed;
output [N:0] primary_input;
begin
  primary_input <= $random(seed); // gen. random vector
end
endtask
```

## 4.4.1  Asynchronous Stimuli Application

All the previous examples clock input vectors into the design's primary inputs. There are situations when no clock is available and input stimuli are applied based on other events. An example is CPU data and instruction memory that sends bytes only when the CPU requests them. Another example is an asynchronous I/O specification, which often exists during the early stage of design, such as the one shown in Figure 4.10. When the sender is armed, 4 bytes of data are sent to two receivers. When the sender finishes transmitting, a done event is sent to the receivers, which in turn disable the ready signals and proceed to process the data.

Let us write a stimulus generator to model the specification in Figure 4.10. There are two components in the test bench. The first one decides whether both receivers are ready and, if so, activates an arming signal. The other component sends out blocks of data when armed and notifies the receivers when done. Lowering or raising signals ready1 and ready2 is the receivers' responsibility. When the receivers are ready to get data, they enable, lower the ready lines, and snoop on the bus for data. In addition, the receivers also monitor signal done. When signal done is activated, the receivers wrap up data retrieval and disable the ready signals by raising them:

```
always // detects whether receivers are ready
begin
   @(ready1 or ready2)
   arm = ready1 | ready2; // arm when both are ready
end

always @(negedge arm)
begin
   transmit_data();
   -> done; // notify receivers after data are transmitted
end
```

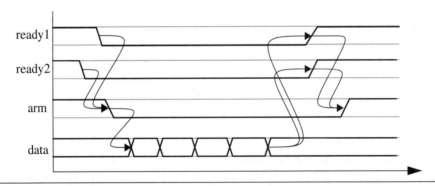

**Figure 4.10** Multiple-agent asynchronous waveform specification

### 4.4.2  Instruction Code or Programmed Stimuli

In large designs, working at the bit level is difficult to manage. Vectors at the bit level are not intuitive to the functions they represent. Therefore, functional coverage (that is, how much the design is tested) is hard to determine. In addition, generating random bit-level vectors is prone to generating illegal operations. An alternative is to elevate stimuli to the instruction code level. This method is especially attractive for designs that operate from instructions, such as processors.

Using the programmed stimuli method, the memory holding the stimulus vectors is a part of the design—instruction and data memory. A test, described by a programming language such as C/C++ or Assembly, is compiled into the design's machine code and is then loaded into instruction and data memory. After the memory is loaded, the design is started to run off the memory. This process is depicted in Figure 4.11.

A major advantage for programmed code verification is its convenience of generating large tests. These tests can be as close as possible to real applications intended to run on the design. The main drawback is that a tool chain is required. The tool chain consists of a compiler, an assembler, a loader, and possibly a disassembler and a debugger.

To determine whether a simulation ends correctly, a high-level software simulator of the design is constructed to run the same instruction code. The results from the high-level simulator are then compared with the RTL ending states. If a high-level simulator is not available or is difficult to design, checking the RTL simulation results can be difficult with this method. This strategy of running a high-level design simulator along with the design's RTL model is sometimes called *cosimulation*.

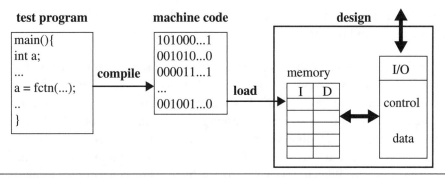

**Figure 4.11** Verification using programmed code

Simulations based on programmed code usually run for many cycles, because the programs or test cases are often large. Therefore, debugging is a challenge.

## 4.5  Response Assessment

The general principle in assessing simulation results consists of two parts. The first part monitors the design's nodes during simulation, and the second part compares the node values with expected values. A discrepancy warrants an in-depth investigation, and the cause could be a real design bug or a bug in the specification. Absence of discrepancies could have many meanings. It could mean that the erroneous nodes were not monitored, bugs were not exercised by the input stimuli, there are bugs in the expected values that mask the real problems, or the design is indeed free of bugs.

Comparing simulation results with expected values can be done on the fly or off-line. With the on-the-fly method, node values are gathered during the simulation and are compared with the expected values. With the off-line method, called *postprocessing*, node values are dumped out to a file during simulation, and the file is processed to find discrepancies after the simulation is finished. In this section, we start with postprocessing method, followed by self-checking routines, and conclude with techniques for checking temporal specifications.

### 4.5.1  Design State Dumping

State dumping prints node values to a file during simulation. With dumping on, simulation speed is drastically decreased, from 2 to 20 times. Thus, in practice, simulation performance requirements prohibit dumping out all nodes in the design. With access only to a part of design, it is necessary to use other measures to locate bugs effectively. Some such measures are to restrict dumping to certain areas where bugs are more likely to occur, turn on dumping only within a time window, dump out signals only at a specified depth from the top module, and sample signals only when they change. These features—scope, interval, depth, and sampling, should be configured in a well-designed dumping routine.

Scope refers to the range of signals to be printed out. A scope can be the Verilog scopes (such as module, task, function, and named block) or a user-defined scope, which can be a group of functionally similar modules. To say that a scope is to be "dumped out" is to print out all signal values within the scope. For example, if the scope is a module, then dumping out that scope prints out all node values inside that module. In a user-defined scope, arbitrary modules can be grouped together as a scope. In practice, these modules usually form a functional unit. The following code is a user-defined scope consisting of modules A and B. Dumping of this scope is activated by defining PRINT_IMAGE:

```
`ifdef PRINT_IMAGE
    $display("top.A.n1 = %b, ... ", top.A.n1, ...);
    ...
    $display("top.B.m1 = %b, ... ", top.B.m1, ...);
    ...
`endif
```

Explicitly indicating the nodes to be printed is a tedious task and is error prone. A better way is to write a user task, $dump_node, that takes in a path to a module and dumps out all the nodes inside the module. For example,

```
`ifdef PRINT_IMAGE
    $dump_nodes(top.A);
    $dump_nodes(top.B);
`endif
```

There are two ways that the scope of dumping can be controlled. The first method uses the ifdef directive, as shown in the previous example. In this case, the decision to dump (or not) is made before compilation time. If it is not defined, the code within the directive is excluded from compilation. Thus, if the ifdef directives are disabled, the compiled image is smaller and thus runs faster; however, the decision to disable dumping cannot be reversed later during simulation runs. The other method uses a runtime argument to turn dumping on or off through the use of plusargs. In this case, the entire design together with the dumping routines is compiled, resulting in a larger image. However, the set of signals to be dumped can be adjusted during simulation, resulting in dumping fewer irrelevant signals and better overall simulation performance. An example of using plusargs is

```
if($test$plusargs("dump_on"))
begin
    $dump_nodes(top.A);
    $dump_nodes(top.B);
end
```

To turn on dumping during simulation, the following command can be used:

```
simulate +dump_on ...
```

The plusargs +dump_on is checked by $test$plusargs("dump_on"). In this case, $test$plusargs returns true, and modules A and B are dumped out.

We have been referring to dumping out nodes inside a module. What does this really mean? If a module has instantiations of other modules, are the nodes inside these modules to be dumped out too? This brings up the question of depth. A convention is that depth is one plus the maximum number of module boundary a node has to cross downward from the dumping scope. Using $dump_nodes(top.A) as an example, the dumping scope is module A. Because all nodes inside module A can be reached without going through any module boundary, all these nodes have depth 1. If module A has a module instantiation, such as instance1, all nodes inside module mod1 are one module boundary down from the dumping scope A. Therefore, nodes inside instance1 are of depth 2:

```
module A; // dumping scope
    ...
    mod1 instance1(...); // one level down
    ...
endmodule

module mod1(...)
begin
    ...
endmodule
```

If depth N is specified in dumping, it means all signals of depth not greater than N, as opposed to only the signal of depth N. If nodes of all depth are to be dumped, the depth is defined to be 0. With the concept of depth introduced, $dump_nodes can be enhanced to include an argument for depth, and you can only dump out signals of the required depth (for example, $dump_nodes(top.A, depth)).

Oftentimes, signal traces are needed only during a certain time interval. For large designs (for example, an entire CPU), it is rare that signals are dumped out right from the start of the simulation to the end of the simulation. Tracing signals for the entire simulation run, even when it is needed only for a small time interval, slows down simulation drastically and wastes a large amount of disk space. Therefore, dumping tasks or routines should have parameters to turn on and off. For example, $dump_nodes can be modified as shown so that signals are dumped when variable dump_on is 1:

```
$dump_nodes(top.A, depth, dump_on);
```

With this on/off switch, nodes can be traced contingent on circuit state. For example, if a forbidden state is reached, dump_on can be set to 1 to start tracing problems:

```
$dump_nodes(top.A, 0, bad_state_reached);
```

When a signal is dumped, how often should it be sampled? A common practice is to sample signals with respect to a clock. For example,

```
always @(clock)
    $display("signal = %d", signal);
```

There are several problems with this sampling technique. First, if the signal's transition does not line up with the clock transitions, the signal's change will show up only at the next clock transition; that is, the dump data will not show the time the signal changes, but will show the time the change is sampled by the clock. Second, if the signal does not change often, many samples would be redundant. Third, in a design with multiple clocks, choosing a reference clock creates confusion. To solve these problems, a signal should be sampled only when it changes. Verilog provides such a facility called $monitor. $monitor takes in a list of signals as arguments and prints out the signals whenever they change. Furthermore, $monitor can be turned on with $monitoron and off with $monitoroff. Dumping changes only also increases simulation performance, because the I/O file is very slow compared with the simulation. An example of using $monitor is as follows:

```
initial monitor($time,"a=%b,b=%d,...",A,B,...);
// whenever signal A, B, ... changes, time stamp and the
// signal list will be printed
...
if(initialization) $monitoroff;
// turn off display during initialization
...
if(start_design) $monitoron;
// turn on display in running the design
...
```

$monitor is executed whenever any variable, except $time, changes. $monitor needs to be instantiated just once, and it will be in effect for the entire simulation run, as opposed to $display, which is called whenever something is to be printed. Therefore, monitor should reside in an initial block.

By default, signals are printed to the screen in the format specified (for example, in $display or $monitor). When the amount of data is large, formats more compact than text are used. These formats are usually proprietary and require a software program, called a *waveform viewer*, to be viewed. Therefore, text format is still used for small-scale signal tracing, mainly for its convenience. A common text format is VCD (or value change dump).

This format records only the times and values of signal changes. A simplified sample VCD file is

```
. . .
$var reg 8   *@   busA   $end
$var reg 8   *#   busB   $end
. . .
#100
$dumpvars
bx*$
bx*#
. . .
$end
#200
00101111*@
11110000*#
. . .
#300
10101010*@
00001010*#
. . .
```

The first two lines associate buses busA and busB with symbols *@ and *# respectively. The reason for this abbreviation is to use shorter names for the original names so that file size can be reduced. The first time stamp is at time #100. The keyword $dumpvars lists all the variables and their initial values. Note that the short names are used instead of the original names. The future time stamps start with # (for example, #200 and #300). The lines following #200 are the values for *@ and *#. For example, at time 200, busA has value 00101111 and busB has value 11110000.

IEEE Verilog standard 1364-1995 provides a suite of VCD dumping system tasks. These tasks are $dumpfile, $dumpvars, $dumpon, $dumpoff, $dumpall, $dumplimit, and $dumpflush. $dumpfile ("file") specifies the name of the dump file. $dumpon and $dumpoff turn on and off dumping. $dumpvars(depth, scope) dumps out variables in the scope at the specified depth. If it is used without arguments, all variables are dumped. $dumplimit(size) limits the size of the dump file. $dumpall creates a check point or a snapshot of all selected variables. $dumpflush empties the VCD file buffer to the dump file. Sample code using the dump system tasks is as follows:

```
initial
begin
   $dumpvars(0,top.design); // dump everything in design
   $dumplimit(100,000,000); // limit file size to 100MB
end

always @(start_dump)
begin
   $dumpon;
   repeat (1000) @(posedge clock)// dump 1000 cycles
   $dumpoff;
end
```

### 4.5.2   Golden Response

Signals are dumped out for either detecting bugs or debugging bugs. To detect bugs, the signal traces are visually inspected, viewed with a waveform viewer, or compared with a known, good output. Visual inspection is suitable only for dealing with a small number of signals. Viewing with a waveform viewer is the main route for debugging a problem when the user knows where to look for clues and causes, and thus the scope of interest is narrow enough to handle manually. In detecting bugs, especially in large designs, the entire set of dumped traces needs to be examined, and thus manual inspection is not feasible. A common method to detect bugs is to compare signal traces with a set of known, good reference traces. These reference traces are also called *the golden response*. Comparing a set of traces and the golden response can be done using the UNIX diff command if the format is text; otherwise, a waveform comparator must be used. The comparison is fully automatic and can easily handle large numbers of traces.

A golden response can be generated directly from the specification or can be generated by a different model of the design (for example, a nonsynthesizable higher level model or a C/C++ model). If there are differences between a set of traces and the golden response, the causes can be design bugs or bugs in the golden response, that is, bug in specification.

The following design demonstrates how a golden response is generated. This design has two 32-bit inputs and one 32-bit output. If we suppose this design is a reference model, then the values of its variables can be used as a golden response. If we deem it sufficient to compare only the inputs and the output, then the $fdisplay statement creates a golden response. To create a golden response, the design is simulated on a number of tests and each run prints out the I/O values, which constitute a golden response. In this example, the file named `gold_file` contains a golden response.

```
module reference_model (in1, in2, out1);
input [31:0] in1;
input [31:0] in2;
output [31:0] out1;
integer fp;
  ...
`ifdef GEN_GOLDEN_FILE
  initial fp = $fopen ("gold_file");
`endif


...
`ifdef GEN_GOLDEN_FILE
always @(posedge clock)
  $fdisplay(fp, $time, "in1 = %d, in2 = %d, out1 = %d", in1,
in2, out1);
`endif


  ...
endmodule
```

A sample output produced by simulating above is

```
...
1012 in1 = 121, in2 = 92, out1 = 213
1032 in1= 32, in2 = 98, out1 = 124
...
```

What variables should be printed in a golden file? Are I/O ports sufficient? The answer depends on the design. I/O ports are the minimum. Printing every node is overkill and often is not feasible because the reference model and the design usually do not have the same internal nodes. If state variables are well defined in the reference model, they should be included in the golden response. Besides variable selection, the time stamp should be included in most cases. Furthermore, the time window in which traces should be considered valid is another judgment the user needs to make. The wider the window, the more coverage it provides, and the more time and disk space it consumes.

Golden responses do not usually remain invariant throughout the design cycle. Specifications can change and, more often than not, the dumping formats or variables change. In either case, the golden responses need to be regenerated. Moreover, golden responses very often need to be augmented to record new stimuli and responses, and there are practically infinitely many stimulus-and-response pairs. Even if a design's responses match the golden responses, they may not meet other design requirements not dictated by golden responses (examples of which include power, area, and testability constraints). This means that the design will change and may alter its responses, even the reference model's responses. In summary, in practice, golden files need to be updated constantly as the project progresses.

An advantage of using golden files is its simplicity for comparison. There are several drawbacks. The first one is maintenance. If a bug is found in the supposedly correct design, the design is changed to meet other constraints, the specifications are changed, or the printing formats or variables are changed, all golden files must be updated. In a large design, thousands of golden files are typical. Furthermore, golden files can be very large, especially in long simulation runs with many signals dumped out; gigabytes of golden files are commonplace. Their large size costs disk space and presents problems for maintenance.

**Self-checking codes.**   Dumping out signals and comparing with golden files have costly performance hits; file I/O can easily slow down a simulation by a factor of 10x or more. Even when there are no bugs in the design, large amounts of data are still dumped out. To avoid dumping signals, checking is moved to the test bench so that signals are monitored and compared constantly against expected behavior. This technique is called self-checking. The general structure of self-checking codes consists of two components: detection and alert. The detection component compares monitored signals with expected values. This component can be further divided into two parts: The first part collects the monitored signals at specified intervals, and the second component provides the expected behavior.

Collecting signals in most cases is simply accessing the signals by using hierarchical paths. If comparison is written as a task that takes in monitored signals as arguments, the signals need to be packed as an array before calling the comparison task. As an example,

```
// pack monitored signals into an array
signal_array[0] = top.design.fpu.mm1.q;
signal_array[1] = top.design.mmu.tbl.q;
...
signal_array[N] = top.design.xmt.regf.q;

check_signals(signal_array); // comparison task
```

Checking a memory's contents often requires preprocessing: Search and package the desired entries into an array before calling the checking routine. In the following example, memory I_mem contains instructions, and the instructions retired at the end of the simulation have the MSB set. Before checking, the retired instructions are searched for in I_mem and are copied to an array that is later passed to a checker routine that can check for the correct number of retired instructions, the correct sequence of retired instructions, and other properties:

```
for(i = 0; i <= N; i = i + 1)
begin
   ri = I_mem[i];
   if(ri[63] == 1'b1)
   begin
      retired_instr [j] = ri;
      j = j + 1;
   end // end of if
end // end of for loop
end

$check_retired_instruction(retired_instr);
```

Generating the expected behavior can be done off-line or on the fly. In off-line mode, the generator runs the stimuli to be applied to the design and creates the expected behavior in a file. During a simulation, the file is read and searched for the expected values of the monitored variables. In on-the-fly mode, a model of the design runs with the simulation in lock step. At the end of each step, the model generates and sends the expected behavior to the comparison routine. This model can be in RTL or C/C++, and can communicate via PLIs. More details about on-the-fly mode are provided later in the chapter.

If the monitored signals do not match the expected behavior, errors are detected and are classified into various levels of severity. Depending on the severity of the error, a self-checking test bench may continue simulation after issuing error messages or may exit the simulation all together. Furthermore, signals traces may be dumped. The self-checking test bench structure is illustrated in Figure 4.12.

An example of a self-checking test bench for a multiplier is shown next. The first part of the code is a gate-level implementation of a multiplier; the second part compares the product with the result from an algorithmic behavior of the multiplier. If the results are not equal,

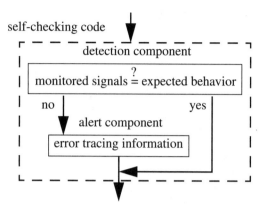

**Figure 4.12** Self-checking test bench structure

the simulation stops immediately, and the values of the multiplicands and the product from the design are printed:

```
// the design: a multiplier
// RTL code of a multiplier
multiplier inst(.in1(mult1),.in2(mult2),.prod(product));

// behavior model of a multiplier
expected = mult1 * mult2;

// compare results
if (expected != product)
begin // alert component
   $display("ERROR: incorrect product, result = %d,
multiplicant1 = %d, multiplicant2 = %d", product, mult1,
mult2); // the alert component
   $finish; // exit
end
```

It is a good practice to separate checking code from design code. For one reason, the checking code will not be part of the final design and the verification engineer should make it straightforward to remove the checking code if needed. One way to accomplish this is to put the checking code inside an `ifdef` pragma so that the checking code is included only when the pragma is explicitly defined:

```
// the design: a multiplier
// RTL code of a multiplier
multiplier inst(.in1(mult1),.in2(mult2),.prod(product));

`ifdef VERIFY
// check result against behavior code
// behavior model of a multiplier
expected = mult1 * mult2;

// compare results
if (expected != product)
begin // alert component
    $display("ERROR: incorrect product, result = %d,
multiplicant1 = %d, multiplicant2 = %d", product, mult1,
mult2); // the alert component
    $finish; // exit
end
`else
`endif
```

A further improvement is to encapsulate the checking code in a task that resides along with other verification utility routines in a separate file, and invoke the task inside the `ifdef`. In file `design.v`,

```
... // design

`ifdef VERIFY
check_multiplication (product, mult1, mult2, status);
`else
`endif
```

In file `verification_utility.v`,

```
task check_multiplication;
input product;
input multiplicant1;
input multiplicant2;
output status;
begin
    ...
end
endtask
```

Another way to derive expected behavior is to use a C/C++ routine that can be executed on the fly or off-line. Using a C/C++ routine to compute the expected behavior increases the confidence of the correctness of the design, because, in general, the more different the two methods are that give the same result, the higher the confidence with the correctness of the result. Two models of the same design written in the same language are biased toward some aspects imposed by the characteristics of the language.

Using the previous multiplier example, a sample self-checking test bench using a PLI user task to compute the product on the fly is as follows:

```
`ifdef VERIFY
// check result against behavior code
// behavior model of a multiplier
$multiplication(mult1, mult2, expected);
if (expected != product)
begin
    $display("ERROR: incorrect product, result = %d,
multiplicant1 = %d, multiplicant2 = %d", product, mult1,
mult2);
    $finish;
end
`else
`endif
```

The C code corresponding to user task $multiplication is as follows:

```
void multiplication()
{
...
m1 = tf_getp(1); // get the value of the first argument of
                       the user task
m2 = tf_getp(2); // get the value of the second argument
ans = m1 * m2; // multiply
tf_putp(3, ans); // return the answer to the third argument
                     of the user task
}
```

When user task $multiplication in the RTL is encountered during the simulation, the simulation pauses and the corresponding C routine is executed. The C routine retrieves the two multiplicands using PLI function tf_getp(), computes the product, and returns the product in the third argument of the user task, which is register expected. When the C routine finishes, the simulation resumes and compares with the product from the RTL code.

Using a user task is a very versatile modeling technique because the C/C++ language can easily compute many complex behaviors, such as cosine transformation, Fourier transformation, data compression, encryption, and so on, for which a significantly different Verilog model is difficult to write as a reference. Furthermore, RTL and C/C++ implementations are sufficiently different to have an added degree of confidence. The major disadvantage is the overhead in communication between the languages, and hence a significant performance penalty. Every time a simulation is run, the simulation must pause for PLI user task executions and must wait for a considerable amount of time spent on data transfer from Verilog to C/C++ and vice versa.

In situations when the expected behavior takes a relatively long time to compute, the on-the-fly mode is not suitable. In these cases, the program computing the expected behavior is run off-line to generate and store the expected responses in a table or a database. The table or database, if small enough, can reside in RTL code; otherwise, a PLI user task must be called from the RTL code to access the table or database. Hard coding expected responses in RTL sacrifices portability but gains performance. The opposite is true for external tables and databases.

When there are too many input stimuli and expected responses that access time to the database severely impedes performance, a prerun method can be used. With a prerun method, as opposed to running the stimuli once and storing the response permanently in an accumulative database, the expected behavior program is run each time prior to a simulation run and the response is stored. This way, the database is smaller and offers faster access. The prerun scheme is most appropriate for moderate runtime cost for the expected behavior program. An example of prerun response generation is compiling a C-test program to run on a CPU design. In the prerun stage, the C-test program is compiled into the CPU's machine codes, then the binary image is loaded into the CPU's instruction and data memory and the CPU is started to run off the memory. During the run, all registers are saved as expected responses.

**Cosimulation with reference model.** Verification using a reference model is a well known self-checking technique. The reference model, usually written in a high level language, captures the design's specifications and compares its responses with those of the RTL model. The RTL model runs on a simulator which in turn runs on the host, while the reference model runs directly on the host. The reference model can be run along with or independent of the RTL model. If running independently, the reference model takes in the set of stimuli to be simulated on the RTL model and writes the responses to a file. The RTL model is then simulated on the same set of stimuli and its responses are compared with those of the reference model. Running the reference model concurrently with the RTL model is also called cosimulation. Cosimulation with a reference model requires a more sophisticated architecture.

Cosimulation models have two levels of accuracy: instruction and cycle-accurate level. A model of instruction-level accuracy guarantees that the correct state at the end of the instruction cycle but not each end of the cycle. A cycle-accurate model guarantees the exact behavior of the design at the cycle boundaries. Thus, a model of instruction-level accuracy is of a higher level of abstraction than that of a cycle-accurate model, and hence is significantly faster. A cycle-accurate model, in turn, is significantly faster than an RTL model.

Here we examine the major components of a cosimulation verification architecture. A block diagram showing the major components of a cosimulation system is presented in Figure 4.13. The arrows among the blocks also indicate the flow of execution. To make the following discussion more concrete, we need to make several assumptions. Bear in mind that the basic principles remain the same without these assumptions. First, the RTL model is in Verilog and the reference model is in C/C++. So, the communication channel, which (in a general setting) is a set of application programming interfaces, now becomes PLIs. Second, we need to assume that the initializer, synchronizer, and comparator are in the reference model, even though they can very well reside in the RTL model. In practice, they are

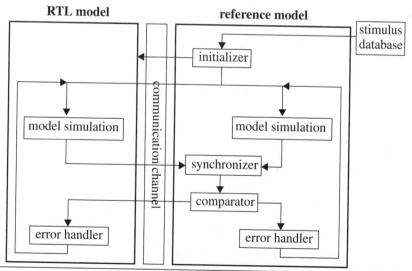

**Figure 4.13** Cosimulation with a reference model

usually in the reference model because they are easier to implement in C/C++. Finally, we need to assume that the design under verification is a microprocessor, and the reference model is instruction-level accurate. In the context of processor verification, the reference model simulates instructions and thus is often called an *instruction set simulator* (ISS).

At the beginning of a simulation, the initializer takes in a stimulus from the stimulus database and initializes both the RTL model and the reference model. An example of a stimulus is a binary image of a program that runs on the processor. The initializer copies this image to the data and instruction memory of both models and registers. Loading an image to the RTL model is done by first getting hold of the RTL memory using `tf_nodeinfo` and writing the image to the array. Refer to "Memory Loading and Dumping Mechanism" on page 186 for more details. Initializing registers in the RTL model is done with the `tf_put` routine.

After both models are initialized, the RTL model is started by toggling a reset signal. At the same time, the reference model starts simulating off its memory. The two models can be resynchronized after each instruction or after the entire image in memory has been simulated. Let's say that they resynchronize after each instruction. The synchronization mechanism uses tokens to indicate both models have completed simulating instructions. In the RTL model, the token can be a signal (say, `instruction_retired`) that goes high after an instruction has been retired. In the reference model, the token can be the program counter. When `instruction_retired` rises, the RTL model blocks and passes the registers' values and memory to the reference model for comparison. Sample RTL code is as follows:

```
always @(instruction_retired)
// check result at each instruction completion
begin
   if(instruction_retired == 1'b1)
   begin
      halt_simulation; // task to pause RTL model
      $pass_states_for_compare(instruction_retired,
            registers,..., memory,...);
      resume_simulation; // task to resume RTL model
   end
end
```

Task `halt_simulation` can stop the simulation by disabling the system clock. If user task `$pass_states_for_compare` is blocking, then `halt_simulation` and `resume_simulation` are not required. In the reference model, after the current instruction has been simulated and before the program counter is incremented, the registers' contents are compared with those from the RTL model.

An implementation creates three threads. The reference model thread executes the instruction pointed to by the current program counter and waits for the comparison of the states of the two models before continuing to the next instruction. The RTL model thread executes the C/C++ routine of `$pass_states_for_compare` which retrieves the states from the RTL model and waits for state comparison. The comparator thread waits for the previous two threads to reach the comparison state and then compare their states. After the comparison is done, the first two threads are signaled to continue. This process continues until an error exit condition is encountered or the end of the simulation is reached.

A sample implementation using semaphores for this part of the reference model is illustrated next. A semaphore is a device used to synchronize threads. A thread can increment or decrement a semaphore, using `sem_wait` or `sem_post` respectively. A thread uses `sem_wait` to wait for a semaphore. If the semaphore's value is positive, the thread continues and the semaphore's value is decremented by one. If the value is zero, the thread continues to wait until the value is incremented by other threads.

All semaphores are initialized to zero. The reference model thread increments semaphore `reference_compare` when it is ready to compare state, and then waits on semaphore `reference_resume` to continue. This is similar for the RTL model thread. The comparator thread waits for both `reference_compare` and `RTL_compare`. When both are ready, it compares the states. If the result warrants simulation continuation, it increments both `reference_resume` and `RTL_resume`:

```
sem_t reference_compare;
sem_t RTL_compare;
sem_t reference_resume;
sem_t RTL_resume;

// reference model thread
void execute_instructions()
{
done = 0;
while(!done){
   next_pc(&pc);
   execute_instruction(pc);
   sem_post(&reference_compare); // ready to compare
   sem_wait(&reference_resume); // wait for continue signal
   } // end of while(!done)
}
```

```
// RTL model thread
void pass_states_for_compare()
{
   ...
   // get states from RTL model
   instruction_retired = tf_get(1);
   gr1 = tf_get(2);
   ...
   sem_post(&RTL_compare); // ready to compare
   sem_wait(&RTL_resume); // for for continue signal for RTL
}

// comparator thread: compare states and resume both models
void compare_state()
{
   ...
   sem_wait(&RTL_compare);
   sem_wait(&reference_compare);
   // now both models are ready to compare
   // compare states
   ...
   if(!errors) {
   // continue simulation
      sem_post(&RTL_resume);
      sem_post(&reference_resume);
   }
   else  // error handler here
}
```

Finally, if errors occur, the simulation may exit with error messages or continue after issuing error messages. The decision here is a function of error severity and user requirements.

### 4.5.3   Checking Temporal Specifications

So far, we have dealt with functional correctness. After functional correctness is checked, timing correctness emerges. Timing specifications can be of synchronous or asynchronous type. With the synchronous type, transitions are synchronized to a reference clock, and timing requirements are expressed in terms of clock cycles. With the asynchronous type, there is not a reference clock among the involved signals; hence, timing requirements are imposed as absolute time intervals among transitions of the signals. Asynchronous timing requirements can arise from a system with asynchronously interacting components or as a result of a design in its early stage. Timing specifications in a system can have both synchronous and asynchronous types.

An example of a timing requirement is that a transition, triggered by other transitions, must occur within a time interval. In Figure 4.14, in_a, in_b, and out are the inputs and output of a design that requires that signal out, triggered by both in_a and in_b, which are synchronized to clock, rise when both inputs fall. A timing specification for this rising transition is that the transition must occur between the second and the third clock rising edge. This is synchronous specification, because the timing requirement references a clock. An example of an asynchronous timing specification is that signal out falls when both inputs have risen, and must do so within the interval [2ns, 6ns].

The timing specifications among in_a, in_b, and out can be verified in the test bench. These timing requirements are examples of the generic specification form that mandates that a transition occur within an interval [$t_1$, $t_2$]. Checking transitions occurring in an interval can be done in two steps. The first step checks the lower limit, and the second step checks the upper limit—$t_1 \leq t$ and $t \leq t_2$. If the transition is caused by a condition on a set of inputs (for example, both inputs are high), time is measured from when the condition is met to the time the transition occurs. The measured time is compared with $t_1$ or $t_2$. A condition bit is used to denote that a triggering condition is met. Let's use the previous timing specifications for illustration and assume the time unit is 1ns. The first specification is

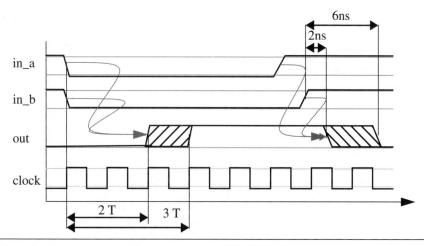

**Figure 4.14** Synchronous and asynchronous timing specifications

that signal out rises in [2T, 3T], when both in_a and in_b become low. The second specification is that out falls in [2ns, 6ns], after both in_a and in_b become high.

The first specification is a synchronous specification and it is broken into two specifications: $2T \leq t$ and $t \leq 3T$. Because it is a synchronous timing specification, the clock is used to strobe the checking code. For example, the following code checks for early arrival of signal out, making sure that signal out will arrive at least two cycles after both inputs are low ($2T \leq t$):

```
assign condition = in_a | in_b;

@(negedge condition)
begin
   if ( out == 1'b0 )
   begin // now start measuring time
      repeat (1)
         @ (posedge clock)
         if (out == 1'b1) $error ("violate lower time limit.");
   end
   else // out == 1'b1
      $error("violate lower time limit.");
end
```

Because signal *condition* is triggered by clock, the clock transition seen by the block is the second transition after the one that changes *condition*—in other words, repeat (1) instead of repeat (2).

Now, check the upper time limit, $t \leq 3T$:

```
@(negedge condition)
begin
   arrived = 1'b0;
   if ( out == 1'b0 )
   begin // start measuring time
      repeat (3)
         @ (posedge clock)
         if (out == 1'b1) arrived = 1'b1;
   end
   else
      // anomaly, out did not start with 0, error.

   if(arrived == 1'b0) error ("violate upper time limit");
end
```

This code checks for the arrival of signal out for every clock transition up to three. If it arrives, flag arrived is set. Here we assume that in_a and in_b cannot cause *condition* to have another rising transition before three cycles have passed. If this is violated, the block checking the upper limit would miss the second rising transition of *condition.*

For asynchronous timing specifications, there is no reference clock; therefore, real time is used. Checking the lower limit can be done by creating two processes when the condition bit rises: One watches for time and the other watches for the out transition. When the lower limit expires, it checks output for the required transition. If it has already arrived, the lower limit is violated. Sample Verilog code is as follows.

```
condition = in_a & in_b; // both inputs high => trigger
...
@ (posedge condition)
begin // start checking
   arrived = 1'b0;
   fork: check_lower_limit
      // delay T1 (2ns), the lower time limit

      #2 disable check_lower_limit;
      // record arrival of signal out
      @ (negedge out) arrived = 1b'1;
   join

   // error if out comes before lower time limit expires.
   if(arrived == 1b'1) error ("lower time limit is violated");
end
```

Checking the upper time limit is done similarly, as shown here:

```
condition = in_a & in_b; // both inputs low => trigger
...
@ (posedge condition) begin // start checking
   arrived = 1'b0;
   fork: check_upper_limit
      // delay T1 (6ns), the upper time limit

      #6 disable check_upper_limit;

      // record arrival of result signal
      @ (negedge out)
      begin
```

```
        arrived = 1b'1;
        disable check_upper_limit;
   end
join

// error if result is late.
if(arrived != 1b'1) error ("upper time limit is violated");
end
```

Note that the `fork` process should be disabled on the arrival of signal `out` or when the upper time limit expires. In the check lower limit case, the `fork` process is disabled only when the lower time limit expires.

The previous discussion focuses on verifying timely transitions. The other side of the coin is to verify the absence of transitions in a time interval (in other words, stability of signals). An example of such a specification is that the address line of memory be stable until the ongoing write is completed. The idea is to use the event operator @ to detect transitions on the signals required to be stable until the enforcement period is over. This can be done by forking two processes, one detecting transitions and the other disabling the first process when the enforcement period expires. Using the memory example, assume that CS (chip select) is lowered when memory is accessed. Therefore, the enforcement period is the interval when CS is low (see Figure 4.15). Sample code verifying that the address is stable during an access cycle is as follows:

```
@(negedge CS) // CS is active low
begin
   fork: stable_address
      @(address[0] or address[1] or ... address[31])
         $error("address changes while accessing.");
      @(posedge CS) disable stable_address;
   join
end
```

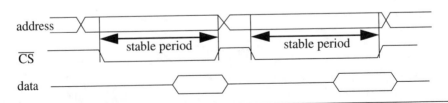

**Figure 4.15** Ensuring a stable address line during an access

Note that to detect changes in address bits, the bits must be spliced in the event operator, because @(address) will only detect changes in the LSB of the address, per IEEE standard 1364-1995.

## 4.6 Verification Utility

Every test bench uses utility routines common to other test benches. Therefore, for productivity, it is best to create a library of common facilities shared by most, if not all, test benches. This utility library can be included as a header file in a test bench. These utilities can take on the form of a module, a task, or a function. If it is designed as a module, it has to be instantiated when it is used. If it is designed as a task or a function, it must reside inside a generic module and must be accessed through a hierarchical path. The following code illustrates this idea. File utility.v is a library of utility routines and should be included when the utility routines are used. In this example, the test bench file uses two utility routines. The first is a module, util_dump_reg, and the second is a task, util_check_cache. To use the module routine, it instantiates the module. To use the utility task, the generic module containing it is instantiated. The task is then accessed through a hierarchical path:

```
// this file, utility.v, contains utility routines

// this routine takes the form of a module
module util_dump_reg (register, clock);
input register;
input clock;

always @(clock) $display("...", register);
...
endmodule

// this routine contains all utility tasks and functions
module util_generic_routines;
...
task util_check_cache(...);
begin
...
endtask
...
endmodule
```

```
// test bench file
`include utility.v

module top;
...
util_dump_reg util_mod (...); // call utility module by
   instantiation
...
util_generic_routines util_gen (...); // instantiate generic
   routines
...
top.util_gen.util_check_cache(...); // access utility task
   via hierarchical path
...
endmodule
```

Some commonly used verification utility routines are a random error generator and an injector, error and warning display routines, memory loading and dumping mechanisms, assertions of various kinds, routines that search patterns in memory, and other miscellaneous types. The list varies from project to project. Here we will study only a subset in detail.

## 4.6.1    Error Injector

An error injector generates pseudorandom errors and forces specified signals to take on the erroneous values. An error injector can be turned on or off. When it is off, it has no effect on the design. The purpose of an error injector is to verify that the design can gracefully detect errors, handle errors, recover from errors, or accomplish a combination thereof.

A block depicting a general-purpose error injector is shown in Figure 4.16. The error injector has a switch to turn error injection on or off. If turned off, its output is directly connected to the input without any error injected. If turned on, it modulates the incoming signal with a random or a hard-coded error. The resulting output can have a varying degree of impairment. One extreme is an output that is independent of the original signal and is simply the output of a random number generator. This models the situation when the original signal is totally disconnected and the receiver picks up noise. The other side of the impairment spectrum consists of modulation of the original signal by an error of varying degree of severity. Error modulation can be implemented in many ways. One way is to add a random number to the input signal; another is to invert certain bits of the input.

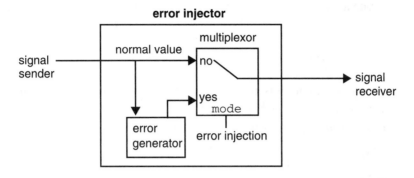

**Figure 4.16** Components of an error injector

The following module is an implementation of an error injector of additive noise:

```
module error_injector (in, mode, out);
input [31:0] in;
input mode;
output [31:0] out;

reg [31:0] error;

   always @(in[31] or ... or in[0])
      error = $random;

   assign out = (mode == 1'b1) ? out = in + error : out = in ;

endmodule
```

## 4.6.2 Error and Warning Alert Mechanism

When an error or a warning is encountered, two decisions need to be made. The first one decides the amount of information to be displayed (or dumped) about the error or warning. The second decision determines the course of the simulation: continue, stop, or pause for user direction. Both decisions are a function of the severity or priority of the error or warning. Therefore, a well-developed error and warning alert mechanism shows the type of error or warning it encounters, the scope in which it occurs, and has the capability to ignore certain types of errors or warnings and to halt the simulation if a serious error is encountered. Such a mechanism can be implemented as a user PLI task or a Verilog task. A sample user PLI task interface is

```
$alert(type, severity, action, message);
```

where `type` indicates whether it is an `ERROR` or a `WARNING`; `severity` classifies the type of error or warning as `HIGH`, `MEDIAN`, or `LOW`; action can be `HALT` the simulation or `IGNORE` the error or warning; and `message` is any user-specified text output. A sample use is as follows:

```
if ( toggle == 32'hDEAD )
    $alert (ERROR, HIGH, HALT, "unexpected dead signal. Inputs
are: sig_1=%h, sig_2=%h,...", sig_1, sig_2, ...);
```

And the result is halting the simulation with an output message like

```
At time 23292: ERROR (high priority, simulation stopped):
    In module top.design...arb, unexpected dead signal.
    Inputs are: sig_1=32'h024ad, ...
```

The time stamp and the scope are built into the task and show the time the error or warning was encountered and the scope in which the erroneous signal resides. Other arguments in an error and warning task are possible, such as arguments to specify dump filename and dump format if the error or warning warrants signal tracing.

### 4.6.3  Memory Loading and Dumping Mechanism

Loading memory involves opening a file, reading the data, and writing the data to memory. A memory loader is used frequently during verification. For example, during initialization, a compiled image is loaded into the design's instruction memory, or at the end of a simulation, data memory is dumped out for examination. Thus, a memory loading and dumping mechanism should be included in a verification utility library.

Such a memory mechanism can be implemented using Verilog constructs or C/C++ construction. The former is simpler and its basic operation is the Verilog system task `$readmemh` and `$readmemb`, which take a filename and memory, and store the data in the file to memory. For example, the following statement transfers data in file `image` to memory `mem`:

```
// declaration of a 1K 32-bit wide memory
reg [31:0] mem [1023:0];

// transfer data in file image to memory mem
$readmemh ("image", mem, START_ADDR, END_ADDR);
```

The first argument is the filename, the second is memory, the third is the starting address, and the fourth is the last address to be written. System task $readmemh assumes the data are in hex, and $readmemb assumes the data are binary.

A better equipped memory loading and dumping mechanism issues warnings when anomalies are encountered, such as cannot open a file, memory argument not declared as memory, amount of data overflows the specified address range, or starting and ending addresses fall out of the range declared. To have this feature, the memory loader is usually implemented as a user PLI task instead of a Verilog task, because memory cannot be passed to a Verilog task. However, memory can be passed to a user PLI task, with the PLI task argument being a "word select" of the memory. However, the entire memory can be retrieved using PLI function tf_nodeinfo(), regardless of the word select. Once the memory is on hand inside the C/C++ code, it can be read or modified. The following call to a user PLI task illustrates passing memory mem to user PLI task $util_memory_loader, where word select 0, mem[0], is arbitrary as long as it is within the declared range:

```
// passing memory to user PLI task
$util_memory_loader ("file_name", mem[0],...);
```

Even though only the first word of mem was passed to $util_memory_loader, all memory can be accessed in the C/C++ code through the PLI function tf_nodeinfo(). As an example, suppose the following C code implements $util_memory_loader():

```
void util_memory_loader()
{
    s_tfnodeinfo nodeInfo;
    char *first_address;

    tf_nodeinfo(2, &nodeInfo);
    first_address = nodeInfo.node_value.memoryval_p;
    ...
}
```

The Verilog memory mem is accessed using tf_nodeinfo(2, &nodeInfo), the first argument of which is the argument number in the user PLI task call. In this case, it is 2 and it points to mem[0]. Data in memory mem are then stored in nodeInfo, which points to a data structure:

```
struct t_tfnodeinfo
{
    ...
```

```
    union {
      ...
      char *memoryval_p;
      ...
    } node_value;
    int node_ngroups;
    int node_vec_size;
    int node_mem_size;
  } s_tfnodeinfo;
```

where `node_value.memoryval_p` points to the start of the memory array, `node_mem_size` contains the number of words in the Verilog memory, `node_vec_size` is the width of a word, and `node_ngroups` is the number of groups in a word. A group is an 8-bit byte of a Verilog memory word; thus, a 32-bit word has four groups. The concept of a group becomes clearer when the mapping from Verilog memory array to C array is illustrated.

A Verilog memory array is mapped to a C array in PLI. Each word is represented by a C array of characters. A word, which can be of any length, is partitioned into groups of 8 bits. Each group is represented by a C character. For example, a word of 32 bits is represented by four groups. The groups for a word form an element in the C array. Therefore, the number of words in Verilog memory is equal to the number of elements in the C array.

In Verilog, a bit range can be declared in decreasing or increasing order, such as `reg [31:0] mem [...]` or `reg [0:31] mem [...]` respectively. In decreasing order, bites $7, 6, \ldots, 0$ form the LSB, whereas in increasing order bytes $24, 25, \ldots, 31$ form the LSB. Bytes in a Verilog word are represented by groups in a C element. The mapping between the bytes and the groups is that the LSB is always mapped to group 0. Figure 4.17 shows the mapping. Note that the bit order within a byte and a group is preserved.

Similarly, Verilog memory's word range can be declared in decreasing or increasing order, such as `reg [...] mem [1023:0]` and `reg [...] mem [0:1023]` respectively. However, address 0 is always mapped to `C_array[0]`, and so on, as shown in Figure 4.17.

In four-state representation, two bits `(a,b)`, are used to represent the four possible values 0, 1, X, and Z. The 2-bit patterns are `(0,0)` for 0, `(1,0)` for 1, `(0,1)` for Z, and `(1,1)` for X. The two bits are called `aval` and `bval`. So, for every byte in Verilog, two groups are required in C. A 32-bit word thus requires eight groups in C to be represented. The mapping of the bytes and the groups is shown in Figure 4.18.

**word-to-element mapping**

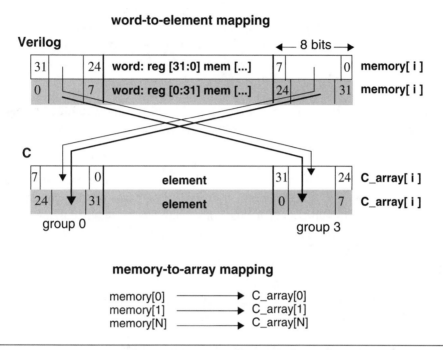

**Figure 4.17** Mapping between Verilog memory array and C char array in PLI

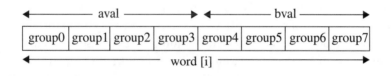

**Figure 4.18** Word-to-group mapping in a four-state representation

With mapping understood, a location at address N in Verilog memory can be accessed in C using the following indices. aval and bval are obtained separately, and their bits are combined to get the four-state value:

```
char *avalp; // pointer to aval
char *bvalp; // pointer to bval
char *base_addr; // the starting address of the memory
int word_size; // word size in C
int ngroups; // number of groups
```

```
base_addr = nodeInfo.node_value.memoryval_p;
ngroups = nodeInfo.node_ngroups;
word_size = ngroups * 2; // 2 comes from aval and bval
avalp = base_addr + N * word_size;
bvalp = avalp + ngroups;
```

For example, if the first bytes of `aval` and `bval` are 10100000 and 11000000 respectively, then the four-state value of the first byte is XZ100000. Using the previous indices, Verilog memory can be read or written to any value.

A memory dumping mechanism can be similarly constructed. The basic Verilog system tasks are `$dumpmemh` and `$dumpmemb`. A user system task can provide checks for valid address range, file permission, and a format conversion facility. Again, passing in memory can be achieved through a PLI argument using `tf_nodeinfo()`. An example memory dumping facility interface is as follows:

```
if(dump_flag) // a typical usage
   $util_memory_dump("file_name", memory[0], start_address,
end_address, format);
```

### 4.6.4   Sparse Memory and Content Addressable Memory (CAM)

When a design has a lot of on-chip memory or is simulated with large off-chip memory, this memory is called *design memory* and often it is not fully used during verification. For instance, a CPU may have hundreds of megabytes of memory, but data and instructions in a simulation occupy only a small portion. To save memory, a simulator needs to allocate only simulation memory space equal to the amount actually used, and to use a dynamic table to map the occupied design memory locations to the simulation memory. This technique is called *sparse memory modeling*. The dynamic table mapping from design memory address space to simulator memory address space can be implemented with CAM, with an address that is a simulation address and with data in a design address (see Figure 4.19).

When a location in design memory is accessed, the upper bits of the address, [ASIZE:LB], are sent to the CAM address map to look up the corresponding CAM address. That part of the design memory address is compared with the contents of CAM. If it is found, the CAM address is returned and concatenated with the remaining lower bits of the design address, [LB-1:0], to access simulation memory. If it is not found, an entry is added to the CAM address map if it is not yet full. The new entry in the map contains the upper bits of the design memory address as the data and the corresponding CAM address as the address. If the CAM address map is full, it can be augmented or fewer upper bits of

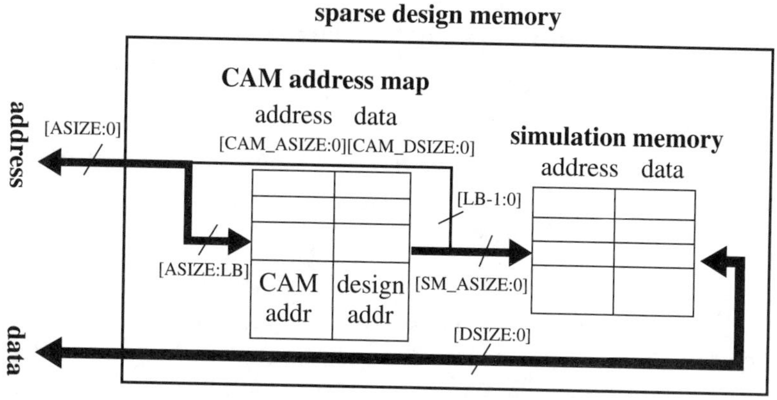

**Figure 4.19** Sparse memory implementation using CAM

the design address are used in the map or the simulation memory is expanded. Using this sparse memory technique, the size of the simulation memory shadows the amount of design memory actually used, which is often a small percentage of the entire design memory. The following code illustrates a sparse memory implementation:

```
// ASIZE := design address size
// DSIZE := design data width
// SM_ASIZE := address of simulation memory
// LB := lower bits for CAM
// CAM_ASIZE := CAM address size, CAM_ASIZE = SM_ASIZE - LB
// CAM_DSIZE := CAM data size, CAM_DSIZE = ASIZE - LB

module sparse_memory (address, data, mode, cs);
input ['ASIZE:0] address;
inout ['DSIZE:0] data;
input mode;
input cs;

reg ['DSIZE:0] simulation_memory ['SM_ASIZE:0];
reg ['SM_ASIZE:0] sim_address;
reg ['DSIZE:0] mem_out; // simulation memory output

reg ['CAM_DSIZE:0] CA_memory ['CAM_ASIZE:0];
reg ['CAM_ASIZE:0] cam_address;
reg ['CAM_ASIZE:0] last_CAM_address; // pointer to last CAM
                                     // address
```

```
reg found; // 1 = found entry in look-up table
reg add_ok;// 1 = successfully added an entry to look-up table

initial
begin
   last_CAM_address = (`CAM_ASIZE+1)'b0;
end

always @(negedge cs)
begin
   // get simulation address for design address
   CAM_lookup(address[`ASIZE:`LB], cam_address, found);
   // task modeling a CAM
   if( found != 1'b1)
   begin // add entry to look-up table
      CAM_add(address[`ASIZE:`LB], cam_address, add_ok);
      if(add_ok != 1'b1) $error("CAM full");
   end

   // form address to access sparse memory address
   sim_address = {cam_address,address[`LB-1:0]};
   if (mode == `WRITE)
      simulation_memory[sim_address] = data;
   else if (mode == `READ)
      mem_out = simulation_memory[sim_address];
   else
      $error("unknown mode");

   end // end of always

assign data = (mode == `READ) ? mem_out : {(`DSIZE+1){1'bz}};

// task modeling CAM
task CAM_lookup;
input [`CAM_DSIZE:0] address_s; // shortened design address
output [`CAM_ASIZE:0] cam_address;
output found;

begin: cam_block
reg [`CAM_ASIZE:0] i;

found = 1'b0;
for (i=0; i<=last_CAM_address; i=i+1)
```

```
   begin
     if(address_s == CA_memory[i])
     begin
        found = 1'b1;
        cam_address = i;
        disable cam_block;
     end
end // end of for
end // end of named block
endtask

// add an entry to the look-up table
task CAM_add;
input ['CAM_DSIZE:0] address_s;
output ['CAM_ASIZE:0] cam_address;
output status; // 1 = ok

begin: cam_add_block
   // check whether CAM is full
   if(last_CAM_address == {('CAM_ASIZE+1){1'b1}}) // CAM full
   begin
      status = 1'b0;
      disable cam_add_block;
   end
   else // CAM not full
      // add the address to the look-up table
      // return the next available simulation address
      status = 1'b1;
      cam_address = last_CAM_address;
      last_CAM_address = last_CAM_address + 'CAM_ASIZE'b1;
      CA_memory[cam_address] = address_s; // add entry
end
endtask

endmodule
```

Task CAM_lookup takes in an address and compares it with every stored entry in CAM—from 0 to last_CAM_address. If found, the corresponding CAM address is returned and the task is disabled. Task CAM_add checks whether the CAM address map is full by comparing the size of CAM with last_CAM_address. If it is not full, last_CAM_address becomes the CAM address of the incoming address, which is then stored as data in the new entry.

The value of LB has the following effects on the CAM address map and simulation memory. If it is too small, the address map becomes large. In the extreme when LB equals zero, every design memory address is stored in the address map; hence, the size of the address map is the same as the simulation memory. If it is too big, the simulation memory becomes large, approaching the size of design memory, and thus has many unused locations. In the extreme when LB equals the design memory address, all design address bits become the simulation memory address bits; meaning, the simulation memory is identical to the design memory. An example of storing data in sparse memory is shown in Figure 4.20, in which the value of LB was chosen to be 2. The upper 3 bits of address 101001, 101, are mapped to 0, which combines with the lower 2 bits 01 to form an address for the simulation memory: 001. The data, 11010, are then stored at 001 in simulation memory. Design memory uses a 5-bit address, but simulation memory uses only a 3-bit address.

### 4.6.5  Assertion Routines

Verification utility can also include assertions. Assertions compare signals with their expected values on the fly and issue warnings or errors when they do not match. There are some common assertions that should be standardized within a project and should be grouped in a verification utility library. The set of such assertions varies from project to project. These assertions can be implemented either as Verilog tasks or user PLI tasks. For an in-depth study on assertion, please refer to "Assertions" on page 232.

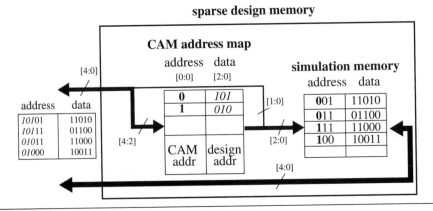

**Figure 4.20** Sample data in sparse memory

## 4.7   Test Bench-to-Design Interface

The interface mechanism between the test bench and the design under verification is responsible for exchanging data between them and can be a combination of the following methods: design I/O ports, hierarchical access, file I/O, and PLI. Using the design's I/O ports to pass data between the design and the test bench is usually reserved for signals to which the design would physically connect in an application—for example, data and address lines of a microprocessor. Observation signals for verification purposes (such as internal registers or nodes) are accessed through hierarchical paths, because to read an internal register through the design's I/O ports could require an elaborate and long control sequence. In contrast, hierarchical accesses can read or write to the nodes or registers immediately. However, hierarchical accesses must not be used among nodes inside the design, because a hierarchical access cannot be implemented physically.

The design and its test bench can also exchange data through file I/O. The design prints signal values to a file, which is then read by the test bench for analysis, or the test bench writes stimuli to a file to be read by the design. File I/O is slow and is usually used for one-time I/O (for example, loading stimuli at the start of a simulation or dumping out node values at the end of a simulation). An advantage of file I/O is that analysis can be done after the simulation is finished, so it is used when simulator time is scarce (such as with a hardware simulator) or when the analysis time is long.

The design can also have an interface with a test bench through PLI. In this case, the interface on the Verilog side is a user-defined system task. Whenever the user task is encountered during simulation, the control is transferred to the C program implementing the task. Using PLI can be slow, especially for a hardware simulator, in which the simulator stops and transfers control to the host machine to execute the C program. On the other hand, when test bench computation is extensive and difficult to implement in Verilog, PLI is the preferred interface. Examples of a test bench using PLI are reference model verification and memory initialization. With reference model verification, a model of the design is implemented in C and is run in lock step with the Verilog code of the design. At the end of each cycle, registers in Verilog are compared with the corresponding ones in the C model. Memory initialization loads the object codes for the design into the Verilog model of the data and instruction memory of the design. The loader is implemented using PLI. Figure 4.21 illustrates and summarizes various interfaces.

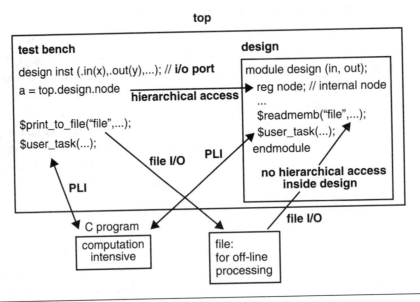

**Figure 4.21** Various design and test bench interfaces and their applications

## 4.8   Common Practical Techniques and Methodologies

Making a test bench or a verification environment work for a design is the first step toward the ultimate goal: building on the working environment and creating a verification infrastructure that is easily adaptable for future designs. In this section, we look at several common techniques and methodologies that further this goal. The general principle is to hide implementation details and raise the verification infrastructure to a level where people can operate and maintain it with simple and meaningful procedures. Some such practices include centralizing configuration parameters and encapsulating low-level operations with high-level procedures.

### 4.8.1   Verification Environment Configuration

A configuration is a particular combination of certain representations and/or components in the design and test bench. For example, a design can have a gate-level as well as an RTL representation, a design can have a representation in which all scan chains are connected or disconnected, or a verification setup can have all assertions turned on or off. An example of a configuration, a combination of these representations, is a circuit in RTL with all scan chains disconnected and all assertions turned on. A configuration can be regarded as a view of the design. An RTL configuration is a view of the design at the RTL, whereas a gate-level configuration is another view of the same design. Similarly, the same design can also have a synthesis view and a verification view, among many other possibilities. Multiple configurations, or views, can be embedded in a single set of design files using `ifdef` directives. A pictorial illustration of embedding multiple views in a single set of files is shown in Figure 4.22. In the figure, there are three views: a gate-level view, an RTL view, and a verification view. The gate-level and the RTL views describe the same functionality—an adder. The verification view has the verification codes for the adder. All these views can be embedded in the same file.

Directives `ifdef`, `else` and `endif` form the delimiters to include or exclude code to be compiled. An example use is

```
`ifdef X
   code_A
`else
   code_B
`endif
```

If variable X is defined, using `define X` in Verilog or `+define+X` on the command line, `code_A` is included and compiled, and `code_B` is omitted. If variable X is not defined,

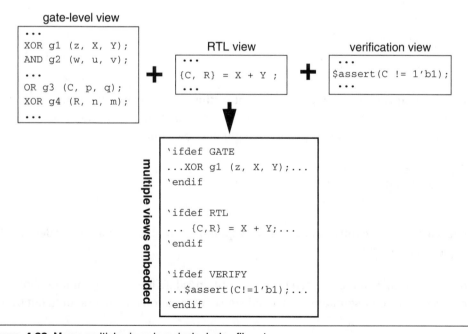

**Figure 4.22** Merge multiple views in a single design file set

`code_B` is included instead. Directive `` `else `` is optional. Therefore, a configuration is just a collection of variable definitions. For instance, to include the RTL and the verification views of the adder in Figure 4.22, simply do the following in Verilog:

```
`define RTL
`define VERIFY
```

or, on a command line of a Verilog compiler, do this:

```
verilogCompile -f filelist +define+RTL+define+VERIFY ...
```

The following example shows two `ifdef`'s embedding four possible configurations: SYNTHESIS and VERIFY, NOT SYNTHESIS and VERIFY, SYNTHESIS and NOT VERIFY, and NOT SYNTHESIS and NOT VERIFY:

```
`ifdef SYNTHESIS
  RTL codes A ...
`else
  RTL codes B ...
  `ifdef VERIFY
    RTL codes C ...
  `else
    RTL codes D ...
  `endif // VERIFY
  RTL codes E
`endif // SYNTHESIS
```

A configuration can also change component settings by using macro-defined directives. For example, in the following code, macro PERIOD takes on any value in the range 1, ..., 8, and thus PERIOD alone generates eight possible configurations:

```
// define macro PERIOD to have value 7
`define PERIOD 7
$Generate_clock(PERIOD, ...);
```

To select a particular configuration, the corresponding variables or macros are defined (for example, `` `define `` PERIOD 7).

To make the design and test bench easy to maintain, all defines and definitions should be centralized to the configuration file. Outside the configuration file, there should not be any

variable defines or macro definitions. The configuration file is included as a header whenever necessary. A sample configuration file, `config.v`, may look like

```
// system configurations
// defines
`define VERIFY
`define SCAN
...
// macro definitions
`define SEED 7
`define ITER 100
...
```

Besides defining variables and macros in a header file, they can also be defined on the
command line. For example,

```
compile filelist +define+VERIFY +define+SCAN +define+SEED=7
+define+ITER=100 ...
```

In this case, the configuration file can be an input to a `make` utility, which generates the
compile command with the correct configuration. Again, the purpose of having a centralized configuration file is to eliminate passing to a verification environment multiple conflicting configurations or missing configurations.

### 4.8.2 Bus Functional Model

As the name implies, a bus functional model encapsulates detailed operations among the
test bench and devices under verification as high-level procedures that resemble commands a device receives from or sends to the bus. To exercise a device, high-level bus
instructions, instead of bit patterns, are issued. The instructions are translated into low-
level bits and are applied to the device by predefined procedures. So, interactions among
the test bench and the devices are done at the transaction level. To have a bus functional
model for a device is to create a wrapper that enables the device to receive and send bus
commands, and the wrapper disassembles and assembles the commands into bits. The
wrapper is sometimes called an *interpreter* or a *transactor*. The main advantage of using a
bus functional model or a transaction-based model is that people only deal with high-
level meaningful instructions, making the verification environment easy to maintain and
adapt, and less prone to errors. Figure 4.23 shows the components and structure of a bus
functional model in a verification environment. The figure has two devices under verification: a CPU and memory. Each device has its own command set. In this example, for simplicity, each device is assumed to have three commands: read, write, and status. For each

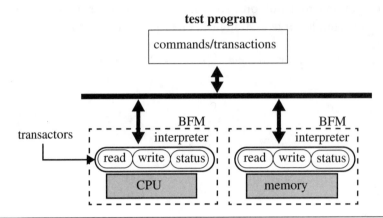

**Figure 4.23** Components and structure of a bus functional model

command, there is a corresponding transactor that translates the command into bit vectors required for the device to perform the command. For example, a READ command calls the read transactor, which generates bit patterns for read and passes to the device. The set of transactors for a device is the device's interpreter or bus functional model wrapper. The test program consists of high-level commands, also called *transactions*, and sends them to the bus functional model of the devices.

To make the concept more concrete, let's construct a bus functional model for memory. Let memory have the simplified read and write timing diagrams shown in Figure 4.24.

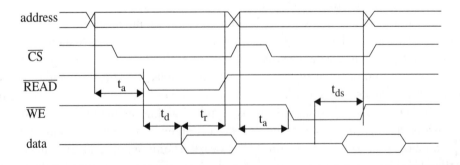

**Figure 4.24** Memory read and write timing diagrams

To read, signal READ must be $t_a$ after the address is stable, and READ must persists for $t_r$ so that the output data are stable. Data will be available for reading after $t_d$. To write, signal WE must wait for $t_a$ after the address is stable, and WE must hold for $t_{ds}$ after the data are stable.

The following code illustrates a test bench that instantiates memory, defines two tasks (one for read and the other for write), and uses the tasks to test the memory. Let's use parameters to represent the timing constraints. This code structure is a standard verification setup but is not a bus functional model:

```verilog
module testbench;
//memory instance
memory mem (.CS(CS), .read(READ), .WE(WE), .address(address),
.data(data));

// read from memory
task read_memory;
input [31:0] in_address;
output [63:0] out_data;
begin
   address <= in_address;
   CS <=1'b0;
   #'ta READ <= 1'b0;
   #'td out_data <= data;
   # 'tr READ <= 1'b1;
   CS <= 1'b1;
end
endtask


// write to memory
task write_memory;
input [31:0] in_address;
input [63:0] in_data;
begin
   address <= in_address;
   CS <=1'b0;
   #'ta WE <= 1'b0;
   data <= in_data;
   #'tds WE <= 1'b1;
   CS <= 1'b1;
end
endtask
```

```verilog
// test bench code testing the memory
always @(posedge clock)
begin
   write_memory(addr1, data1);
   @(posedge clock)
   write_memory(addr2, data2);
   ...
   @(posedge clock)
   read_memory(addr1, result1);
   ...
end

endmodule // test bench
```

A bus functional model groups the design and its associated tasks as a module so that operations on the resulting module, BFM, are performed by calling the module's tasks. Translating this code into memory BFM, we have

```verilog
module memory_BFM (address, data);
input [31:0] address;
inout [31:0] data;

   memory mem (.CS(CS),.read(READ),.WE(WE),
.address(address),.data(data)); // memory instance

   task read_memory;
      ...
   endtask

   task write_memory;
      ...
   endtask
endmodule // memory_BFM

module testbench;
   memory_BFM mem(.address(addr),.data(data));

   always @(posedge clock)
   begin
      mem.write_memory(addr, data);
      @(posedge clock)
      mem.write_memory(addr, data);
      ...
```

```
        @(posedge clock)
        mem.read(addr, data);
    end
endmodule // test bench
```

Note that the bus functional model now has only two arguments: `address` and `data`. Signals `CS` and `WE` are no longer visible. This is reasonable because address and data are transaction-level entities whereas `CS` and `WE` are signaling entities that the user should not be concerned with. Of course, if the user intends to verify the signaling correctness of all interface signals, `CS` and `WE` need to be specified in the model.

At first glance, the difference between this bus functional model and the standard test bench seems to be quite trivial. This difference becomes significant when there is more than one component under verification. In this case, if the bus functional model is used, each component has its own encapsulated operational tasks. All operations in the test bench are done at the transaction level by simply calling the device's task. An example test bench with multiple devices may look like

```
module testbench;
// device instances
device1 dev1 (...);
device2 dev2 (...);
device3 dev3 (...);

// transactions among devices
@(posedge clock)
dev1.op1(...);
dev2.op1(...);
dev3.op1(...);
...
@(posedge clock)
dev1.op3(...);
dev2.op5(...);
dev3.op2(..);
...
endmodule // test bench
```

Try to use the standard test bench structure to set up the previous multiple-device environment. You will notice the confusion among the operations for the devices. If two devices have the same operation (for example, READ), you will have to use different names for the tasks. Furthermore, the tasks are not grouped with their design and hence are more difficult to maintain.

## 4.9  Summary

In this chapter we studied the major components of a test bench: initialization, clock generation and synchronization, stimulus generation, response assessment, test bench-to-
design interface, and verification utility. For initialization we looked at initialization using
RTL code and PLI routines, with predetermined and random values. For clock generation
and synchronization we presented several methods to produce clock waveforms. We also
discussed the effect of time scale and resolution. We then examined the synchronization of
different clock domains. For stimulus generation we presented synchronous vector application, asynchronous stimulus application, and instruction code stimuli. For response
assessment we considered assessment methods using waveform tracing, monitors, golden
files, and self-checking routines. Besides functional assessment, we presented methods to
monitor timing constraints. We examined four types of test bench and design interface
mechanisms: I/O ports, hierarchical paths, PLIs, and file I/O. As noted, verification utility
routines are often-used routines and are created to increase productivity. The set of such
routines varies from project to project. In this chapter we discussed methods for error
injection, memory loading and dumping, sparse memory modeling, and some widely
used assertions.

We examined some common methodologies and practices. In particular, we discussed
test bench configuration and the bus functional model. Test bench configuration embeds
several views of the same RTL code in the same file. A view can be invoked by defining a
particular variable. The bus functional model encapsulates detailed operations into high-
level transactions. We discussed how a design and its operational commands are packaged to form a bus functional model, and how the bus functional model can be used in a
test bench.

## 4.10  Problems

1. Write a test bench that initializes the following FF output to value VALUE. Can you initialize the input of the FF instead of its output?

```
module flip=flop (Q, D, clock);
output Q;
input D;
input clock;

always @(posedge clock)
  Q <= D;
endmodule // end flip-flop
```

```
module testbench;

// initialization codes here
...

flip_flop ff(.Q(q), .D(d), .clock(clk));

endmodule // test bench
```

**2.** Initialize the following latch, driven by the clock, to VALUE.

```
module latch (Q, D, clock);
...
always @(clock or D)
   if(clock == 1'b1)
      Q <= D;
endmodule // end of latch

module testbench;
initial
clock = 1'b1;

always #1 clock = ~clock;

latch gate(.Q(q), .D(d), .clock(clk)); //
endmodule // end of test bench
```

**3.** Initialize memory data_mem, 32 bits wide and 4K deep, with the data from file file_mem and then dump out the memory contents. Simulate your code to check whether the memory was loaded correctly.

   a. The starting address is 0. Load the entire file.
   b. The starting address is 12'h12. Load only 1KB.

**4.** Initialize the memory in the previous problem to random values.

**5.** Write a PLI routine to initialize memory with data from a file. (This problem is for those who know how to write PLI functions.)

**6.** Write Verilog code to generate the waveforms presented in Figure 4.25.

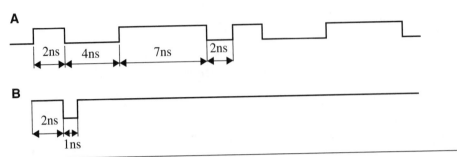

**Figure 4.25** Waveform generation (A) periodic waveform (B) aperiodic waveform

7. The following clock starts at high. The transition to high at time zero may be considered a legitimate transition for some simulators but may be ignored by others. Write Verilog code to determine whether your simulator ignores time zero transition.

8. What clock periods do you expect to see when the following code is simulated?

    a. Clock generation code:

    ```
    'timescale 10ns/1ns
    initial clock = 1'b0;
    parameter C = 10;
    always #(C/3) clock = ~clock;
    ```

    b. Same as part a, except for the following line:

    ```
    always #(C/3.0) clock = ~clock;
    ```

    c. Same as part b, except for the following line:

    ```
    'timescale 10ns/100ps
    ```

9. Write Verilog code to generate the clock waveform shown in Figure 4.26, where the falling and rising transitions have 300ps and 200ps jitter respectively. In other words, the edges transit randomly with an uncertainty of 300ps on rising and 200ps on falling.

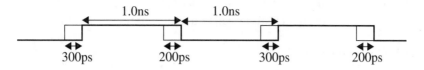

**Figure 4.26** Waveform having 300ps jitter

**10.** Compile the following C code. Call the resulting binary file `test.o`. Load `test.o` into 32-bit, 8-MB memory, `test_mem`. Write RTL code to process the loaded memory to determine whether the content has odd or even parity:

```
main () {
   int x, y, z;
   x = 12;
   y = 21;
   z = x + y;
}
```

**11.** Write RTL code to dump out variables within a module of your choice using `$dumpvars`.

a. Simulate your code and determine what format `$dumpvars` produces.
b. Dump out variables at all levels from the current module scope.
c. Dump out only the variables at the current module scope.

**12.** In this problem, calculate the cost in maintaining golden files. Assume that 10% of RTL changes require updating the golden files and there are ten RTL changes per day. For each golden file update, the design is simulated over 100 diagnostics, each of which takes ten minutes.

a. What percentage of simulation time is spent on updating golden files each month if only one simulator is used, assuming a month has 30 days?
b. How many simulators must be used to reduce the percentage to 5% assuming linear speedup in parallel simulation?

**13.** Self-checking code can be executed through RTL or a PLI. For the following design, write a self checking test bench. You need to decide whether it should be in RTL or PLI.

a. A 32-bit Booth multiplier

```
module Booth_multiplier (product, in1, in2, clock);
   output [63:0] product;
   input [31:0] in1, in2;
   input clock;
   ... //
endmodule
```

b. A 64-bit circuit implementing an N-sample fast fourier transform

```
module FFT_coefficient (i, H, h0, ..., hN);
   output [63:0] H; // ith coefficient
   input [31:0] i; // index
   input [63:0] h0, ..., hN; // time samples
   ...
endmodule
```

**14.** Write tasks to detect the timing violations in the waveforms shown in Figure 4.27.

**15.** A communication device receives an incoming clock up to $M$ Hz. Write a task to enforce this maximum frequency requirement. It should issue an error if the input frequency exceeds $M$ Hz.

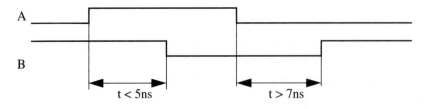

**Figure 4.27** Timing constraints

**16.** Write an error injector for the following memory model. The injector selects a random address and writes a random data to the address at rising transition of a global signal add_error:

```
module memory (ReadWrite, CS, address, data);
input ReadWrite; // 1=read, 0=write
input CS; // 0 = select
input [31:0] address;
inout [31:0] data;
...
endmodule
```

## CHAPTER 5

# Test Scenarios, Assertions, and Coverage

## Chapter Highlights

- Hierarchical verification

- Test plan

- Pseudorandom test generator

- Assertions

- SystemVerilog assertions

- Verification coverage

In the previous chapters we discussed the static removal of design errors, structure and use of simulators, and test bench creation. With these components in place, we are ready to simulate designs. The challenge in simulation is to uncover as many bugs as possible within a realistic time frame. To uncover all bugs, one can exhaustively enumerate all possible input patterns and internal states. However, this is seldom feasible in practice. In the following discussion, we will compute an upper bound on the number of input patterns to test exhaustively a sequential circuit having S states. A sequential circuit is completely tested or verified if all of its reachable states are visited from an initial state, and all transitions from that state are verified. So, the first step is to derive an input that drives the circuit from an initial state to any of the S states. There are S such inputs, one for each state. Next, all transitions from that state need to be tested, and there are, at most, R transitions.

*R* is the number of possible inputs. Therefore, an upper bound for the number of inputs is *SR*. If the circuit has *N* FFs and *M* inputs, then $S = 2^N$ and $R = 2^M$, and the number of input patterns becomes $2^{(M+N)}$, which prohibits any practical attempt to verify exhaustively any useful design.

In this chapter we discuss how a design should be simulated over a well-selected subset of all inputs in a systematic manner to achieve a reasonably high level of confidence that the design is correct. With regard to deriving a well-selected subset of input, we discuss test plans and test scenarios, which aid verification engineers in determining corner cases and in enumerating functionality. Because only a subset of all possible inputs is simulated, a measure of quality of the input subset, or of the confidence level of the simulations, is required. We address the issue of simulation quality by discussing several types of coverage measures. Test plans, test scenarios, and coverage can be regarded as methods of search or exploring the input space of the design. After input space exploration is considered, we focus on output space—namely, detecting errors. Detecting errors by comparing a design or a module's primary outputs with the desired responses may not be the most efficient, because any internal error will have to be propagated to a primary output to be detected, and there can be many cycles after its occurrence. As an example, consider the architecture in Figure 5.1 of a simplified CPU design. If only the I/O port is monitored,

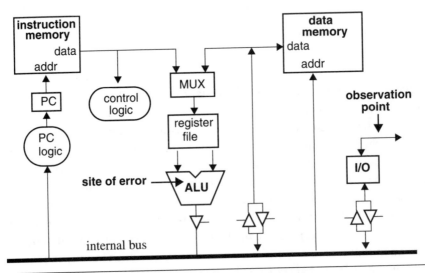

**Figure 5.1**   An error at the ALU is observed many cycles after it is stimulated

marked as the observation point, and we assume that an error in the ALU occurs, then the error is detected only when the I/O block is activated to output data from the internal bus, and the ALU is driving the bus. It is likely that the error in the ALU is stored in data memory or the register file, gets used in later operations, and manifests, if at all, in a totally different form at the I/O port many cycles later. This latency makes debugging difficult, because the designer needs to dump out signals in a time window covering the moment the error originates and trace back through the many cycles. It is also possible that the error will never make it to the I/O port and thus will go undetected.

One approach to reducing bug incubation is to enhance the design's observability to detect errors as soon as they surface. For example, observation points can be added to the output of the ALU, instruction and data memory data and address lines, control logic, program counter (PC) logic, and the register files. Then, when the error occurs in the ALU, it will be caught immediately. In this chapter we will look at a methodology to enhance systematically a design's observability by using assertion technology.

We will address the bug-hunting process in three stages: enumerating input possibilities with test plans, enhancing output observability with assertions, and bridging the gap between exhaustive search and practical execution with coverage measures, as illustrated in Figure 5.2.

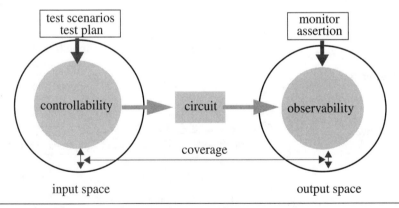

**Figure 5.2**    The relationships among test scenarios, assertions, and coverage

## 5.1  Hierarchical Verification

Directly verifying a large system can be difficult for a number of reasons: verification tool capacity limitation, long runtime, and limited controllability and observability. Therefore, a larger system needs to be decomposed into smaller subsystems, which may be decomposed recursively until their sizes are manageable by verification tools. Decomposition can be based on the organization of the design architecture. For instance, a CPU can be decomposed into an instruction prefetcher, an instruction dispatcher, an integer unit, an FPU, an MMU, a cache, a testability structure, and an I/O unit. Furthermore, decomposition can be accomplished at multiple levels. A typical hierarchy includes the system level, unit level, and module level. That is, a system is decomposed into an interconnection of units, each of which is further decomposed into an interconnection of modules. To verify lower level components, their specifications or functionality are derived from those of higher level components. At the top level, the functionality of the overall system is extracted from specifications independent of the implementation architecture. However, the functionality for lower level components is dependent on both the overall specifications and the implementation architecture. The functionality of lower level components working together as determined by the implementation architecture, must accomplish the higher level functionality. Therefore, to verify the overall system specifications, two steps are required. The first step verifies that the units or modules meet their specifications. The second step verifies that the architecture, comprised of the units and modules, as a whole meets the overall specification. The concept of hierarchical verification—structural and functional decomposition—is illustrated in Figure 5.3.

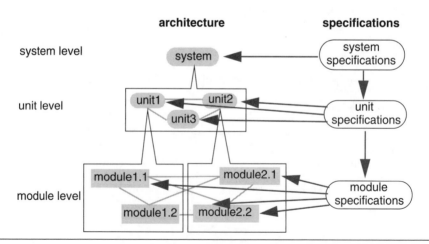

**Figure 5.3**   Hierarchical verification: structural and functional decomposition

### 5.1.1   System Level

The design under verification at the system level can be the entire chip, a connection of multiple chips, or a board. Because of the complexity at this level, hardware simulators are often used. If software simulators are used, multiple threads or multiple parallel simulators are run concurrently over a network of computers (in other words, a computing grid or computer farm), with each thread or simulator verifying a part of the system. A diagram of a computing grid is shown in Figure 5.4. The two-way processor system with look-through L2 caches is mapped to the computing grid, where each computing unit simulates a unit of the two-way processor system. The dashed arrows denote the mapping from the design units to the computing units. The computing units communicate with each other through TCP/IP sockets.

Two areas of verification are addressed at the system level: (1) verifying interconnections of the subsystems that make up the system and (2) verifying the complete system at the level of implementation. For the first area, the units are assumed to have been verified, and thus they are often modeled at a high level of abstraction (for example, functional models). For example, the CPUs can be replaced by C programs called ISSs. The goal is to demonstrate that the units were constructed to achieve the design objectives of the system. For the second area of verification, all units of the system retain their implementation

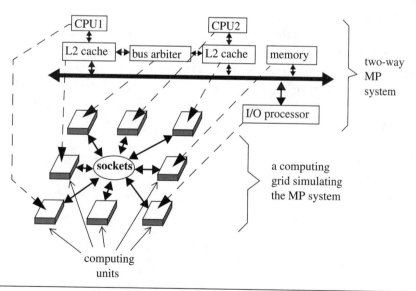

**Figure 5.4**   Simulating a system on a computing grid

details (such as RTL), and the entire system is simulated using a computing grid or a hardware simulator. The advantage of this approach is the added confidence that comes from not relying on the functional model, which comes at the expense of high complexity and slow runtime.

With system-level simulation, because the entire system is self-contained, stimuli can be application programs loaded in the system's memory or can come from the I/O ports when the system is embedded in an application environment. For instance, if the system is a microprocessor, the stimulus can be a compiled program loaded into instruction memory.

To generate test scenarios for system verification, a typical approach is to "benchmark" representative applications and run the benchmarks on the system. Choose benchmarks that represent normal as well as abnormal operations, just to test design robustness. For instance, to test error handling and recovery, select an application that produces illegal operations. To garner better coverage for normal operations, select applications that exercise all functional units in the system. For example, choose an application that uses integer operations, as well as floating point operations, or an application that uses two processors and causes frequent cache misses and snoop push-backs.

### 5.1.2   Unit Level

A unit can be a large functional unit consisting of dozens of files. An example is an integer unit in a CPU. Specifications for the unit are derived from the overall system specifications. For example, if the system specification is a 64-bit SPARC processor with a minimum clock frequency requirement, then a specification derived for the integer unit would mandate the ability to handle 64-bit operands, the set of predefined SPARC integer operations, at a speed bound by the timing constraints for an integer unit to meet the overall clock frequency.

The derived specifications can be used to create test cases. Additionally, one can apply other functional categories to generate test scenarios. For instance, using the integer unit as an example, one can create tests with operands that cause overflow, underflow, or error traps.

### 5.1.3   Module Level

Verification at the module level is similar to that at the unit level, except that the functionality is at a lower level.

## 5.2  Test Plan

Although it is practically impossible to stimulate all inputs, two principles make verification in practice useful without exhaustively stimulating all inputs. First, it is seldom necessary to stimulate all possible input vectors in order to find all bugs because a bug can be stimulated by several "equivalent" inputs, thus stimulating only the representative inputs is sufficient to achieve our goal. Second, bugs have priorities. The definition of a bug may even become blurred in the lower priority categories. High-priority bugs affect a design's essential functions. Such a bug renders the design useless, whereas low-priority bugs or features may be a nice-to-have option or an unspecified behavior in the architecture. Therefore, a practical and effective verification prioritizes input stimulation accordingly.

A test plan is the foundation for a successful verification effort; it plays the same role that an architectural blueprint does for designers. A test plan starts with the architectural specifications of the design, derives and prioritizes functionality to be verified, creates test scenarios or test cases for each of the prioritized functionalities, and tracks the verification progress. These four steps can be regarded as a top-down process that refines high-level specifications to lower level ones, and hence to more concrete requirements that eventually yield test cases. Let's discuss each of these four steps:

1. Extract functionality from architectural specifications
2. Prioritize functionality
3. Create test cases
4. Track progress

### 5.2.1  Extract Functionality from Architectural Specifications

An architectural specification of a system is a high-level description that mandates the I/O behavior of the system, without spelling out the details of how the behavior can be accomplished. An example architectural specification is a processor with a 32-bit instruction and address space, running SPARC instructions, supporting IEEE 754 floating point format, and using memory mapping for I/O. A design implementation constructs an interconnection of modules and components to accomplish the required I/O behavior. To verify that the design implementation meets the specification, functionality must be extracted from the specification to enable the verification engineer to confirm concretely that all functionality works as specified. An example of functionality of the aforementioned specification is a specific SPARC instruction, such as LDSTUB. A set of specifications often leads to a large amount of functionality; thus, to verify all functionality, a systematic way is required to enumerate functionality from a specification. An ideal enumerating methodology extracts functionality from the specifications as well as that implied by the specifications. Getting a complete list of functionality or features often depends on one's experience with the design and its applications. Let's discuss a *state machine-based approach*, which will serve as a guide in our exploration of features and functionality verification. This method partitions complete design behavior and its interaction with its environment into six categories. For each category, functionality and features are extracted to generate test scenarios. If all categories are verified completely, the design is completely verified.

Any digital system can be described as a finite-state machine with behavior that is completely determined by its initial state, its transition function for the next state, the input it accepts, and its output. An input or state is valid if it is valid according to the specification. Denote all valid initial states, valid states, valid inputs, and valid outputs by $Q_0$, $S$, $I$, and $O$ respectively. The complements of these sets are invalid and are denoted with a bar over the set (for example, $\bar{I}$). Thus, if the design receives an unexpected input $F$, then $F \in \bar{I}$. Furthermore, we use $X \rightarrow Y|_I$ to denote a transition from state $X$ to state $Y$ under input $I$, and $X \Rightarrow Y|_I$ to denote a sequence of transitions that starts at state $X$ and ends at state $Y$ under an application of consecutive inputs (a sequence of input). (The input in this case should be a Cartesian product of the input set I: $I \times \ldots \times I = I^*$. For convenience, we abuse the notation a little bit.) Finally, $\Omega$ is a don't-care or wild card, which can be a don't-care state or an input. As an example, $\Omega \Rightarrow Q_0|_I$ means the circuit starting at any state, under the application of a sequence of valid input, enters a valid initial state.

Based on this notation, $\Omega \rightarrow \Omega|_\Omega$ is the set universe describing all possible input and behavior of the design. We partition this universe into the following categories of functionality or features to facilitate generating test scenarios:

1. $\Omega \rightarrow \Omega|_{\bar{I}}$. In this category, the design receives an input not specified in the specifications. This can happen if the design is used in an unintended environment. The functionality to be verified for this category could be that the design detects the invalid input, protects itself from drifting to an invalid state, and issues appropriate warnings.

2. $\bar{S} \rightarrow \Omega|_\Omega$. The design is an invalid state, which can be caused by component failures or a design error that drove the design to this state. The functionality to be verified is how the design responds and recovers from it.

3. $S \rightarrow \bar{S}|_I$. The design starts at a valid state, receives a valid input, but ends up at an invalid state. This is a design bug. Test scenarios belong to this category only after they expose bugs. Then these test scenarios are preserved for regression to ensure the bugs do not occur in future versions of the design.

4. $S \rightarrow S|_I$. This behavior, starting at a valid state and transiting to another valid state under a valid input, is the correct behavior. This category states what the

system specifications are. Therefore, the functionality in the specifications is the test scenario.

5. $\Omega \Rightarrow Q_0$. The design on power-on enters into a valid initial state from any state. Tests in this category are sometimes called *power-on tests*. When the design is first powered on, it can be in any state. The functionality to verify is that the design can steer itself to a valid initial state.

6. $\Omega \Rightarrow \overline{Q_0}$. The design on power-on is at an invalid state. If the design enters an invalid state, the design has a bug. Test scenarios in this category try to drive the design to an invalid initial state (for example, through various combinations of inputs during the power-on stage).

To understand that these six categories form the complete input and state space, we reverse the partitioning as follows. Categories 5 and 6 together verify $\Omega \Rightarrow \Omega_0$, which is the complete initialization from any state to any initial state. Categories 3 and 4, $S \rightarrow \overline{S}|_I$ and $S \rightarrow S|_I$, together verify $S \rightarrow \Omega|_I$, which covers transition from a valid state to any state under a valid input. Combining $S \rightarrow \Omega|_I$ with category 2, $\overline{S} \rightarrow \Omega|_\Omega$, gives $\Omega \rightarrow \Omega|_I$, which forms the complete transition from any state to any state under a valid input. Finally, by combining $\Omega \rightarrow \Omega|_I$ with category 1, $\Omega \rightarrow \Omega|_{\overline{I}}$, the entire transition space is covered: $\Omega \rightarrow \Omega|_\Omega$. Therefore, the six categories cover the entire state and input space: $\Omega \Rightarrow \Omega_0$ and $\Omega \rightarrow \Omega|_\Omega$. Hence, if each of the six categories is thoroughly verified, the design is thoroughly verified.

Note that these categories explore situations when the design is in both expected and unexpected environments (for example, the design starts in an invalid state or the design receives an invalid input). Verifying under an unexpected or even illegal environment is required, especially for reusable designs with applications that are not known to the designers beforehand.

---

**Example 5.1**

Generate test scenarios using a state space approach. To demonstrate the use of the six categories, generate test cases for the generic disk controller shown in Figure 5.5. A disk is partitioned into concentric circular strips called *tracks* and each track is further partitioned into blocks called *sectors*. The atomic storage unit is a sector. There are six types of registers: command, track, sector, data-out, data-in, and status. The command register contains the command being performed on the disk. The track and sector registers record the track and sector numbers. The data-in and data-out registers have the data to be written and read. The status register shows the state of the disk and the status of the operation. Figure 5.5 shows the status bits of the status register.

*continues*

**Example 5.1    (Continued)**

The commands can be RESTORE, SEEK, STEP, STEP-IN, STEP-OUT, READ SECTOR, WRITE SECTOR, READ ADDRESS, READ TRACK, WRITE TRACK, and FORCE INTERRUPT. The RESTORE command determines the head position when the disk drive is first turned on. To move the head to a track, the motor is pulsed to step in or out. To read or write a number of bytes, the head is first stepped to the right track and is then positioned over the sector. The track and sector IDs are stored in the track register and sector register respectively. The data in the data-in register are then written to the disk in write mode or the read data are placed in the data-out register in read mode.

Now let's create test scenarios for this disk controller using the state space approach. Each of the six categories are considered in turn. For category 1, test cases send illegal input to the controller. An illegal input can be an address in a forbidden range, consecutive data requests before the grant signal occurs, or an active interrupt signal while in scan mode. The goal in these test cases is to determine how the disk controller responds to an illegal input.

For category 2, test cases set the controller in an illegal state, which can be an illegal opcode in the command register, or out-of-range data in the track and sector registers, and observe whether the controller detects the problem and recovers from it.

For category 3, test cases assume that the controller is in a valid state and attempt to drive it to invalid states with valid input. For example, are there any track and sector values that cause the controller to produce incorrect pulses for signal step and a wrong turn-on time for signal motor on? Given the head position, are there track values that would produce a wrong value from signal direction?

For category 4, test cases assume that the controller is in a valid state, drive it with valid inputs, and determine whether it will end up in a valid state. For example, with a certain pattern in the data-out register, does the CRC bit in the status register flag or not? For given values in the track and sector registers, are the outputs from step and direction correct?

For category 5, sample test cases include the following: What are the values of the registers when the controller is first powered on? Are these values correct or expected? Immediately after power-on, do registers track and sector contain the correct track and sector number of the head position if the RESTORE command is executed?

| Example 5.1   (Continued) |
|---|
| Finally, in category 6, the questions to ask include the following: Under what conditions will the controller enter an illegal state on power-on? On power-on, is it possible that the interrupt signal is active, that output `motor on` is continuously on, or that output `step` is out of range? What input pattern could drive the controller to these abnormal outputs on power-on? |

Table 5.1 summarizes the six categories. In each category, a name or a brief description of the state space test generation method is given.

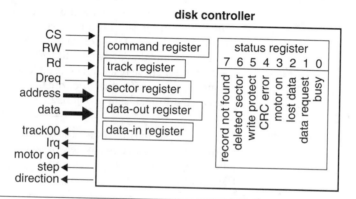

**Figure 5.5**   A disk controller model

**Table 5.1**   Functional Categories for Test Scenario Enumeration

| Category | Name | Description | State Space Coverage |
|---|---|---|---|
| 1 | Input guarding | Create test scenarios that have illegal inputs and determine the design behaves correctly. | $\Omega \rightarrow \Omega|_{\bar{I}}$ |
| 2 | Error handling and recovery | Create test scenarios that preload the design in illegal states and drive it with both legal and illegal inputs. | $\bar{S} \rightarrow \Omega|_{\Omega}$ |

*continues*

**Table 5.1**    Functional Categories for Test Scenario Enumeration (Continued)

| Category | Name | Description | State Space Coverage |
|---|---|---|---|
| 3 | Operational anomaly | Create test scenarios from the functionality specified in the specifications. Set the design in a legal state and drive it with a legal input. If the design transits to an illegal state, a bug is found. These tests are preserved for regressions to ensure that the bugs are fixed in future design revisions. | $S \rightarrow \bar{S}\vert_I$ |
| 4 | Normal operation | Create test scenarios from the functionality specified in the specifications. Set the design in a legal state and drive it with a legal input. If the design transits to a legal state, that particular functionality passes. These tests can be used for regression to ensure this functionality passes in future design revisions. | $S \rightarrow S\vert_I$ |
| 5 | Power-on test | Create test scenarios to start the design from any state and determine whether the design arrives at a correct initial state. If yes, the power-on test passes. | $\Omega \Rightarrow Q_0$ |
| 6 | Initialization anomaly | If a test scenario makes the design arrive at an incorrect initial state, a bug is found in the power-on test. This test is preserved for regression. | $\Omega \Rightarrow \overline{Q_0}$ |

## 5.2.2   Prioritize Functionality

When functions to be verified are determined, they need to be prioritized, because there is always not enough time to verify all functionality. Furthermore, as a project progresses, specifications and architectures change, resulting in new verification tasks. When the project deadline gets close, some lower priority verification tasks have to be sacrificed.

The priorities are determined by the application priorities of the design. The mission-critical applications attain the highest priority. Thus, the highest priority functionality consists of the basic operational requirements and minimum features the design must have to be usable for the mission-critical application. A basic operational requirement is that the design will power up and enter a correct initial state. Another basic operational requirement is that the design, once powered on, executes basic commands correctly so that the minimum feature set is operational. Yet another basic operational requirement is that the debugging facility be operational. The next-priority functionality consists of the features that expand the design's applicability for a large market share (for instance, supporting a newly adopted standard or being compatible with formats from other regions in the world).

### 5.2.3    Create Test Cases

Once the set of functionality is determined and prioritized, it is translated into test cases. The translation process involves two stages. During the first stage we need to ask specific questions about what constitutes the passing of a functionality, and how to tell from signal values that a functionality has passed or failed. The list of functionality in a test plan is also called *test items*. The next stage is about how to create stimuli to exercise the signals to determine whether the functionality passes.

### 5.2.4    Track Progress

Finally, progress is tracked by listing all functions to be verified in order of decreasing priority, documenting which function's test bench has been written, and checking off the functions that have been tested and passed. This process is also known as *scoreboarding*.

In summary, the following four steps are used to create a test plan:

1. Enumerate all functions and features to be verified. This can be done systematically using the state space approach. These functions and features are then prioritized according to their importance to the design's applications.
2. For each function or feature, determine the signal or variable values that constitute a pass or fail.
3. Create a procedure or a test bench to stimulate the design and monitor the signals or variables.
4. Collect all testcases and document their progress.

**Example 5.2**

Create a test plan for a cache and testability unit.

In the following discussion we delve further into the application of the functional categories in conjunction with hierarchical verification using a cache unit and a testability unit in a processor. Two specifications at system level are (1) the system is in a testable state on power-on (category 5, power-on test) and (2) the system's performance is not hampered by the slower off-chip memory (category 4, normal operation). These two specifications, when translated to the unit level, become (1) the testability structure (which consists of a scan chain, a Joint Test Action Group (JTAG) boundary scan, and a Built-In Self Test (BIST) structure) is able to reset and test the system to ensure that it is in a correct state and (2) the cache is operational. The next question is what constitutes an operational testability structure and a cache unit?

The testability unit, a scan chain, is shown in Figure 5.6. In test mode, pin mode is 0, the FFs are stitched together as a shift register that can shift in value for the FFs. When mode is normal mode, 1, the FFs operate in normal mode. For the testability unit, a list of objectives includes the following:

1. The scan test multiplexors switch correctly between normal mode and test mode.
2. The scan chain shifts in a predetermined pattern and loads it into the FFs.
3. The scan chain shifts out the states through the scan-out pin.
4. The JTAG instruction register loads correctly.
5. The Test Access Port (TAP) controller functions correctly.
6. The JTAG instructions are executed correctly.
7. The BIST structure loads patterns correctly.
8. The BIST functions correctly during power-on.
9. The BIST detects injected errors and exits as expected.
10. The power up sequence is executed correctly.
11. The error registers log in the correct error type.

For the cache unit, a list of objectives includes the following:

1. When there is a miss, the miss flag raises (similarly for a hit).
2. The block fill operation correctly loads the data.
3. In a multiple-processor configuration, the shared bit is set when more than one processor has a copy of the block.
4. If processor A is requesting a READ and processor B has a modified copy, a back-off message is issued to stop processor B from reading. The modified copy is transferred to memory so that processor A can now gain access.
5. In a multiple-processor configuration, the snoop mechanism generates the correct flags.

**Example 5.2    (Continued)**

When a list of functionality is at hand, the functionalities are prioritized. In this example, all except possibly the last entry of the testability unit belong to the first priority.

Then stimuli are designed to create the scenarios to test the functionality. For example, to test functionalities 1, 2, and 3 of the testability unit—proper switching of the scan multiplexor and shifting and loading of the scan chain—a stimulus can be set up as follows. First we need to determine what procedure constitutes pass. Instead of testing each functionality separately, the following procedure will verify them together. The mode pin is toggled to `test` mode, a bit pattern is shifted in and is loaded to the FFs. Then the output of the FFs is shifted out. If the shifted-out pattern is identical to the shifted-in pattern, the load and shifting features pass. The scan multiplexor switch is only tested for the `test` mode. To complete the test for `normal` mode, the mode pin is toggled to `normal` mode after a pre-determined pattern is loaded, then the clock is toggled once. The mode pin is again toggled to `test` mode and the outputs of the FFs are shifted out for examination. If the shifted-out pattern is equal to the state of the machine running one cycle starting from the loaded state, the multiplexor passes. If it fails, it could be caused by the multiplexor or the next-state function of the machine. Further tests are required to isolate the culprit of the problem. This example shows that sometimes it is not possible to isolate a single feature or function for verification.

To test functionality 4 of the cache unit—if a processor is requesting a `READ` and another processor has a modified copy, a back-off message is issued, followed by a write-back—a stimulus loads a pattern to a particular address in the cache of a processor and has that processor modify it. Then the stimulus causes another processor to read that address. At the end of the read, the content at the address in memory and the flag of the cache are examined for correct operation. The flag for back-off should have been raised when another processor attempted to read the modified data. If the back-off was indeed raised, that feature passes. Furthermore, if the memory content at the address is the same as the modified data, then the write-back feature also passes.

Table 5.2 is a sample test plan for verifying the cache and testability units. It contains specific, prioritized functionalities; verification status for the functionality; and the results. There are six test items in the plan. This test plan keeps track of what has been verified and how it progressed.

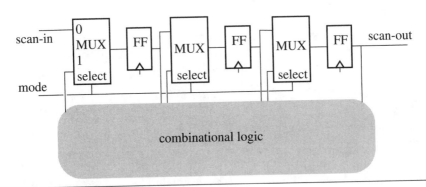

**Figure 5.6**   Testability structure: a scan chain

**Table 5.2**   Example Test Plan for a Cache Unit and a Testability Unit

| Priority | Unit/Owner | Function (test case name) | Status | Result |
|----------|-----------|---------------------------|--------|--------|
| 1 | Cache/<br>J. Doe | Invalidate signal should be active before modifying a shared cache block (INVD) | Written: 10/11/03, revised: 11/11/03 | Passed |
| 1 | | Snoop push-back transfers modified cache data to memory before another processor reads (SNP_PB) | Written: 12/01/03, revised: 12/11/03 | Passed |
| 1 | | Back-off flag should raise if a processor is to read another processor's modified cache block (BKOFF) | Written on 12/02/03 | Failed; flag is delayed by one cycle |
| 1 | | Cache state changes to shared when more than one processor has the same cache block (SHARED) | Work in progress | N/A |
| 1 | Testability unit/<br>D. Smith | Shift and load operation (SHFT_LD) | Work in progress | N/A |
| 2 | | Error register value (ER_RG) | Work in progress | N/A |

## 5.3  Pseudorandom Test Generator

Tests derived from specifications are called *directed tests*, and do not cover all possible scenarios. Oftentimes, directed tests are biased and are only as good as the specifications. If a specification detail is missing, that detail will not be tested. To fill in the gaps, pseudorandom tests are used. Another reason to use pseudorandom tests is that they are generated; hence, the cost of creating the tests is much lower compared with directed tests. It is important to know that random tests are not arrows shot in the dark. On the contrary, random tests are only random within a predetermined scope. That is, the pseudorandomly generated tests are not entirely random. They emphasize certain aspects of the target component while randomize on other aspects. The emphasized aspects are determined by the user and are usually centered at the directed tests. Hence, random tests can be regarded as dots neighboring directed tests. As an example, to test a 32-bit multiplier, the emphasized aspects can be that the multiplicands are at least 30 bits and that 50% of time they have opposite signs. The random aspect is the specific values of the multiplicands. The emphasized aspects are dictated to the test generator through the use of seeds. A seed can be a signal number or a set of parameters. In the following discussion, we look at the key components of a random test generator with a pseudorandom CPU instruction generator.

There are several requirements and objectives that a pseudorandom test generator should meet. First, the generated tests must be legal operations, because not all combinations of instructions are legal. For instance, the target address of a branching instruction must be within the code address range; it cannot be in the privileged address space. Second, the tests generated must be reproducible. This requirement ensures that the same tests can be used in reproducing and hence debugging the problems found by the tests. Thus, using date, time, or a computer's internal state as a seed is prohibited. Third, the tests should have a target objective as opposed to a collection of haphazard instructions. A target objective can be testing a particular unit of a system or the intercommunication among functional blocks. Fourth, the distribution of random tests should be controllable so that, although random, it can be directed to explore one area more than another. Finally, random tests should be complete probabilistically; that is, all input patterns should have a nonzero probability of being generated.

In this section we will look at the architecture of a generic pseudorandom test generator through the design of a pseudorandom processor code generator. A pseudorandom processor code generator creates programs run on a processor so that the programs test certain emphasized instructions and architectural features. The major components of a pseudorandom code generator are shown in Figure 5.7.

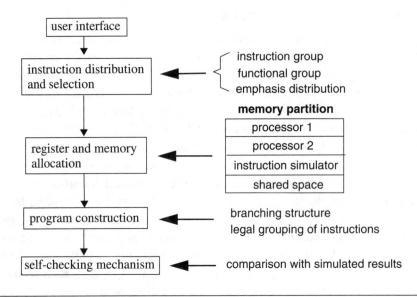

**Figure 5.7**   Major components of a random code generator

## 5.3.1   User Interface

The user interface allows the user to specify the instructions and system features to be emphasized in generating random code. For instance, one may stress storing and loading instructions and test MMU functionality. Another example is to test cache coherency.

The user selectable parameter set usually consists of instruction groups, architectural function, selection probability distribution, degree of data dependency, and methods of result checking. Instructions are grouped according their functionality and nature. Typical groups are floating point instructions, arithmetic and logical operations, branching instructions, address instructions, memory instructions, exception and interrupt instructions, multiple-processor communication instructions, and data movement instructions.

Architectural function stresses individual functional units and interactions among functional units. Example architectural function groups are cache coherency, interrupt and exception handling, instruction prefetcher, instruction decoding, cache line-fill, Multiple-Processor (MP) snooping operation, multistrand operation, privileged memory access, atomic operations, Direct Memory Access (DMA) operation, pipeline stall and flush, and many others. When a feature is selected, the generator assembles the necessary instructions to create the scenario.

The selection probability distribution forces some instructions or functions that are more likely or less likely to be included in the generated programs. The code generator should have the capability of automatically generating default distributions and should be able to vary the distribution from run to run to explore fully the entire space. For example, a Gaussian distribution on architectural function groups selects the function groups according to a Gaussian distribution, with the highest emphasized group located within one sigma of the mean, the second highest emphasized group within two sigmas, and so on.

Data dependency is accentuated by restricting instructions to use only a specific subset of registers and memory locations. Address space restriction creates data dependency because an instruction can be executed only after all its operands are available. For example, an instruction attempting to store to a register must wait until the existing content of the register is no longer needed and can be safely overwritten. The same principle applies to memory addresses. Therefore, the smaller the size of the address space, the stronger the dependency among the instructions.

Randomly generated code is useful only if there are ways to check the results. Therefore, it is imperative that a code generator create self-checking routines for the random code. Self-checking routines require expected results for comparison. Result checking can be done at the end of each instruction or at the end of a program run. To produce the expected results, the generated instructions are simulated by the generator. This simulation can be done at the time the generated program is executed or at the time the program is generated. Expected results usually consist of register values and values at some memory locations. Then the expected results and comparison operations are inserted in the generated program. The advantage of simulating instructions concurrently with running the generated code is the added testing capacity, at the expense of a longer execution time.

## 5.3.2   Register and Memory Allocation

In addition to the restriction on register and memory location use imposed by data dependency, registers and memory space are partitioned for different use. Some registers are reserved for storing address offsets, some as link registers for storing the base address on returning from a subroutine call, some as general-purpose registers, and some as data registers for store and load instructions. The memory space is partitioned so that each processor has address ranges for data and stack. All branching instructions should have target addresses within this assigned range. If instructions are simulated concurrently with code execution, the simulator will have a code and data memory segment to carry out the simulation. Finally, there is shared memory space among processors and the instruction simulator.

### 5.3.3   Program Construction

During this step, instructions are selected according to the user-specified distribution and to form a program to stress the selected system features. The instructions are assembled in a legal order. For branching instructions, target addresses are resolved after the instructions are put in order. For loop constructs, the loop index is initialized and the exit condition is added. After the main program is created, an initialization subroutine is inserted at the beginning and an exit subroutine is appended. The initialization routine saves the registers' values and then fills the registers and data memory space with random numbers constrained by the user-set distribution. The exit subroutine ensures that all processes have finished or an error is recorded. Then it restores the registers' values saved by the initialization routine. An example program structure and program layout are shown in Figure 5.8. In this example (Figure 5.8A) shows a template of the program flow. The flow contains a loop in a shaded area that has two exit points to label A and label B. Many different flows can be selected with this generator. Once the flow of the program is set, the code for the blocks is randomly selected according to the user-chosen distribution. Figure 5.8B shows a sample random code segment. The initializer and exit routines are library routines that are linked with the generated program. The loop is implemented with an index initializer and an index decrementer, and two jump instructions to exit the loop (see the upper shaded area of Figure 5.8B).

### 5.3.4   Self-checking Mechanism

A randomly generated program runs on a processor and, at the end of the run, the result (values of registers and some memory locations) is compared with the expected result. The expected values are computed by the program generator. For instance, if a random program is a multiplication operation and the multiplicands are 7 and 5 respectively, the result will be in register R1. The generator computes the product and stores 35 as the expected value in a register, say R31. Then the generator appends a comparison routine to the random program that checks the value of R1 with that of R31. If they are equal, a passing message is issued; otherwise, a fail message is issued. Figure 5.9 shows a code segment of the self-checking mechanism.

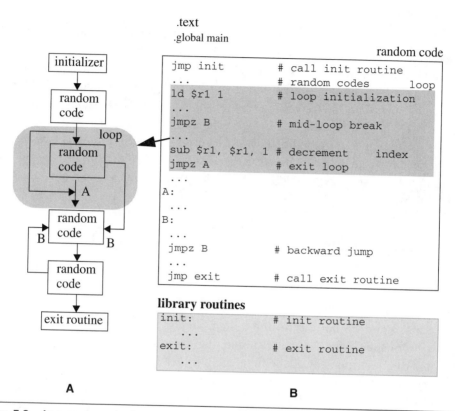

**Figure 5.8** An example randomly generated program. (A) A program template (B) A sample random code generated from the template

```
    ...
    add $r0 0 7       # store 7 in R0
    add $r1 0 5       # store 5 in R1
    mult $r1 $r1 $r0  # calculate 7x5, result in R1
    jmp CHK # call check routine at the end of program
CHK:
    add $r31 0 35     # store 35 in R31
    sub $r30 $r1 $r31 # R31 has expected value 35
    jmpz EQ           # if equal, call pass routine
        jmp error
    EQ: jmp pass
```

random program

self-checking
routine

**Figure 5.9** Self-checking mechanism in a random program

## 5.4   Assertions

Assertions are constructs, in the form of a group of statements, a task, or a function, that check for certain conditions on signals or variables in a design over a span of time. If the condition is violated, an error message is issued identifying the location of the occurrence. Assertions have been used in software and hardware design for quite some time, but only until recently have there been coordinated efforts to standardize assertions. An example of an assertion is that chip select, CS, is high whenever active low signal WR is low. If this requirement is not met, the assertion fails and an error message is issued. If it passes, there may not be a message issued. However, it is essential to keep track of the assertions that have passed, because silence could also mean that an assertion was not encountered. In Verilog, the assertion is

```
// assertion
`ifdef ASSERTION
if ( (WR == 1'b0 && CS == 1'b1) == 1'b0 )
    $display ("ERROR: !WE && CS failed. WR=%b,CS=%b",WR,CS);
`endif
```

Because assertions are not part of the design, they should be able to be conveniently excluded from the design. In this example, the assertion is enclosed within directive `ifdef` and can be stripped by not defining `ASSERTION`.

Assertions can be inserted throughout a design to detect any problems immediately. Without using assertions, one must rely on primary outputs to detect errors that often manifest the errors many cycles after their occurrence, hence making debugging difficult. Worse still are the situations when errors never propagate to a primary output. Therefore, a major advantage of using assertions is enhancing error observability and debugability. Furthermore, by standardizing assertions, they can be reused for new projects. In addition, assertions, now mainly used in simulation-based verification, can be used as specifications of design properties during formal method-based verification. For instance, the previous code example in a standard form may look like `assert(!WR && CS)`, which can be a simple runtime check on `(!WR && CS)` during simulation verification, or it could be a property that `(!WR && CS)` must be proved to hold in formal verification.

### 5.4.1   Defining What to Assert

There are two approaches to defining assertions. We can start with specifications and conclude that certain behaviors or properties of the design must hold, which we call *affirmative assertion*, or we can start with design faults or some features or behaviors not specified, derive conditions that when met indicate that anomalies have occurred, and assert that these conditions must not be satisfied, which we call *repudiative assertions*. Both affirmative and repudiative assertions originate from and codify the items in a test plan. Test items state the functions or features to verify, and assertions translate these objectives into measurable signals and variables. Therefore, a comprehensive test plan should precede a robust assertion plan.

To illustrate deriving assertions from test items, consider a test item in category 1 (input guarding) of the state space approach for test plan creation. The item specifies which input patterns are illegal. Hence, an assertion can check for the specified input pattern. If an illegal pattern is encountered, a warning or error is issued. This application of assertions is especially informative when embedded in IPs or reusable designs for which the designers assumed a particular input configuration and have no control of the environment in which the design would reside. By embedding the input-guarding assertions, users can be warned of application environments unsuitable for the design. As another example, consider a category 5 (power-on test) test item that states that after power-on, all stacks are empty. To codify this condition, an assertion is activated by a power-on signal and, once activated, it compares all stack pointers with their respective top pointer. If they are unequal, an error is issued.

Like specifications, assertions exist in a hierarchy. There are assertions at the system level, unit level, and module level. As an example of a module-level assertion, consider a First-In-First-Out (FIFO) and a category 3 (operational anomaly) property. An assertion is that an error occurs if an empty FIFO is popped or if a full FIFO is pushed. During hierarchical verification, assertions can be classified into those monitoring internal operations of the building blocks and those ensuring correctness of the interactions among the building blocks. An example of the latter is the following: In a system consisting of a processor and external memory connected through a Peripheral Component Interconnect (PCI) bus, an assertion that a `store` command issued by a processor will cause the PCI bus to execute the corresponding transactions and after a number of cycles, memory has the data stored at the specified address.

## 5.4.2   Assertion Components

Assertions have four major components: signal archive, triggering condition, assertion, and action. The first component is a signal archive that stores past values of signals and variables. The archive is just a memory or a shift register storing past values of signals and variables. The second component is a triggering expression that activates the assertion if the expression evaluates to true. This expression sets the conditions the assertion should be effected. An example of a trigger expression is `if (init_done)`, which turns off all assertions during the power-on period in which unknown signal or variable values may falsely fire assertions and turns on assertions when initialization is done. The third component is an expression that checks a property or condition on signals and variables, and calls the fourth component if the check fails or passes. The fourth component is an action, called by the assertion, that prints a message if the assertion fails or may do nothing if it passes. Figure 5.10 depicts the four components of a sample assertion.

An expression contains variables or signals over a range of cycles. An expression is combinational only if it uses the current value of its variables or signals; otherwise, the expression is sequential. Denote the value of signal $S$, $n$ cycles in the past, by $S^{-n}$ and $m$ cycles in the future by $S^{+m}$. Then the value at last cycle is $S^{-1}$, the value at next cycle is $S^{+1}$, and the current value is simply $S$. The following Boolean expression is combinational,

```
(A != 0 ) && ( A & (A-1))
```

whereas the following expression is sequential. It states that if signal A changes value from last to current cycle, then it should hold the value for the next cyle. -> is the implication operator.

$$(A^{-1} \mathrel{!=} A) \rightarrow (A == A^{+1})$$

An assertion is combinational if all of its expressions are combinational; otherwise, it is sequential. To code combinational assertions, only combinational logic is needed. To code sequential assertions, a finite state machine is required. If an expression has both past and future variables, the time interval from the most past and to the most future is the time window of the expression.

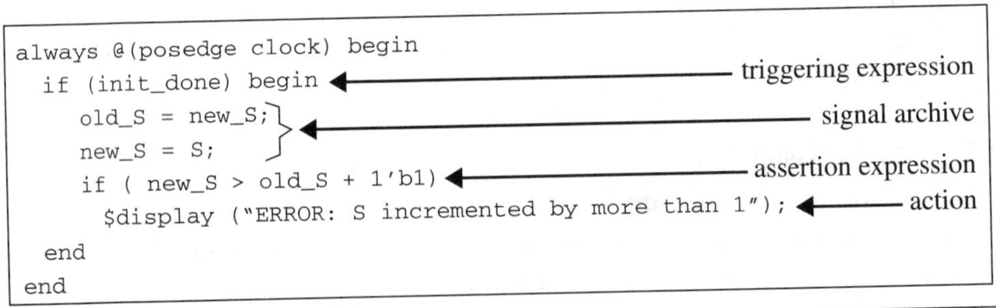

```
always @(posedge clock) begin
   if (init_done) begin  ◄─────────────────── triggering expression
      old_S = new_S;⎫ ◄───────────────────────── signal archive
      new_S = S;     ⎭
      if ( new_S > old_S + 1'b1) ◄──────────── assertion expression
         $display ("ERROR: S incremented by more than 1"); ◄──── action
   end
end
```

**Figure 5.10** Illustrating the four components of an assertion

### 5.4.3 Writing Assertions

In the following section, we discuss common techniques of checking signal properties. First we discuss combinational assertions and then sequential assertions.

**Signal range.** The simplest kind of assertion is to check whether a signal's value is within an expected range. This is done by comparing the signal's value with a minimum bound and a maximum bound. If the value lies outside the bound, an error has occurred. Sample code that checks whether signal S is out of bound [LOWER, UPPER] is

```
if ( `LOWER > S || S > `UPPER )
    $display ("signal S = %b is out of bound");
```

An assertion may not need to be active all the time. For instance, during the power-on period, S may not have been initialized and the assertion should be turned off. To activate the assertion only when S is ready, a triggering expression is added:

```
always @(posedge clock)
begin
if ( (ready_to_check == 1'b1) && (`LOWER > S || S > `UPPER ))
    $display ("signal S = %b is out of bound");
end
```

Many assertions used in practice fall into this type of assertion, sometimes with a minor change. Table 5.3 lists some variants of signal range assertion.

**Table 5.3**  Variants of Signal Range Assertion

| Assertion Name | Description | Code Variation |
|---|---|---|
| **Constant** | Signal's value remains constant | UPPER = LOWER; or, simply use event control to detect change (e.g., @(signal)) |
| **Overflow and underflow** | Signal's value does not overflow or underflow the limit of operation | No code modification is needed |
| **Maximum, minimum** | Signal's value does not exceed a lower or upper limit | No code modification is needed |

**Unknown value.** Checking whether signal A is equal to an unknown value cannot be done by simply comparing A with x, because A can have multiple bits. The assertion is to ensure that no bit of A has an unknown value. This assertion is needed in situations when data patterns are expected on A (for instance, data bus to memory during a write cycle). A brute-force check is to go through each bit of A and compare it with 1'bx. A more clever method is to use a reduction XOR operator, ^. A reduction XOR performs XOR on all the bits of A, one by one. By definition of XOR, if at any time an operand has an unknown value x, the result will be X regardless of other operand values. Thus, the reduction XOR will produce an X result if and only if at least 1 bit of A is X. The Verilog code to check for an unknown value on signal A is

```
if ( ^A == 1'bx) $display ("ERROR: unknown value in signal A");
```

**Parity.** A signal is of even parity if the number of 1s is even, such as ^A gives 0. Similarly, A is odd parity if ^A is 1. To check a signal's parity, reduction XOR can be used, which XORs all bits of the signal together. If the result is 1, the parity is odd; otherwise, it is even. The code for checking even and odd parity is as follows:

```
if ( ^A == 1'b0) $display ("info: signal A is even parity");
if ( ^A == 1'b1) $display ("info: signal A is odd parity");
```

**Signal membership.** This type of assertion ensures that a set of signals is equal to expected values or differs from unwanted values. The expected values can be valid opcodes and the unwanted values can be unknown values. As an example, consider checking an instruction variable, instr, against a list of legal opcodes: BEQ, JMP, NOOP, SHIFT, and CLEAR. The comparison can be done by using either the comparison operator == or a case statement, as shown here. The macros (for example, BEQ and JMP) are defined to be the corresponding opcode binaries:

```
`define BEQ 1010
`define JMP 1100
`define NOOP 0000
`define SHIFT 0110
`define CLEAR 1000

if( (instr != `BEQ)   && \
    (instr != `JMP)   && \
    (instr != `NOOP)  && \
    (instr != `SHIFT) && \
    (instr != `CLEAR))
$display ("ERROR: illegal instruction instr = %b", instr);
```

Or, embed a case statement within the body of an instruction execution unit:

```
case (instr):
    'BEQ: execute_beq();
    'JMP: execute_jmp();
    'NOOP: execute_noop();
    'SHIFT: execute_shift();
    'CLEAR: execute_clear();
    default:
    $display ("ERROR: illegal instruction instr = %b", instr);
endcase
```

Several signals can be concatenated and compared against a set of legal values. Grouping signals gives a more concise description and opens the opportunity for using don't-cares in the expressions, as illustrated next. For example, we restrict single-bit signals reset, load, cs, hold, and flush to the values {10001, 11001, 00110, 01110}, which can be simplified using don't-care notation to {1?001, 0?110}. To take advantage of don't-cares, use the casez statement for comparison:

```
sigSet = {reset, load, cs, hold, flush}
casez (sigSet) :
    1?001: // do something
    0?110: // do other thing
    default:
        $display ("illegal control signals ...);
endcase
```

**One-hot/cold signals.** A signal is one-hot if exactly 1 bit of the signal is 1 at any time. To assert the one-hot property, instead of comparing the bits of the signal with 1, 1 bit at a time, the following simple code accomplishes the goal:

```
if ( |(B & (B - 'WIDTH'b1)) != 1'b0 )
    $display ("Bus B is not one hot, B = %b", B);
```

The macro WIDTH is the width of signal B. To see why this expression detects the one-hot property, try sample value B equals 8'b00010000.

```
              B : 8'b00010000
        B - 8'b1 : 8'b00001111    // subtract 1
  B & (B - 8'b1) : 8'b00000000    // bitwise AND
| (B & B - 8'b1 ) : 1'b0          // reduction OR
```

If B is not one-hot, say, B equals 8′01001000, then

```
              B  :  8′b01001000
        B-8′b1  :  8′b01000111
   B & (B-8′b1) :  8′b01000000
 | (B & B-8′b1 ) :  1′b1
```

Sometimes, a one-hot signal has a 0 value when it's not active. A 0 value is not one-hot, but the previous algorithm does not distinguish 0 from one-hot. To eliminate 0 from being one-hot, a comparison with 0 is added to the previous one-hot-checking code, as follows:

```
if ((B != ‘WIDTH′b0) &&                          \
    ( |(B & (B - ‘WIDTH′b1)) != 1′b0 ))
```

In addition, some bits of a one-hot signal can have unknown values when inactive. To accommodate this situation, a check for value X is added to the one-hot assertion, as follows:

```
if ((^B != 1′bx) &&                              \
    (B != ‘WIDTH′b0) &&                          \
    (|(B & (B - ‘WIDTH′b1)) != 1′b0 ))
```

OR

```
if ((^B != 1′bx) &&                              \
    (|B != 1′b0) &&                              \
    (|(B & (B - ‘WIDTH′b1)) != 1′b0 ))
```

A one-cold signal, on the other hand, has exactly 1 bit equal to 0. To assert on one-cold, invert the signal bit by bit and use the previous one-hot assertion:

```
B = ‘WIDTH{1′b1} ^ C;
```

where C is a one-cold signal. ‘WIDTH{1′b1} creates a mask of 1s with a width of ‘WIDTH (for example, 8{1′b1} is 8′b11111111).

**Sequential assertions.** Sequential assertions involve past and future values of variables. The key is to create a signal archive that saves the past values of the variables. A simple archiving mechanism is a shift register file for each variable such that at each clock transition, the current value of the signal is shifted in. The length is the register file equal to the history required for the assertion. The shift register file is clocked by the variable's reference clock. To evaluate an assertion, the past values of the variables are retrieved from the shift register files. For example, consider writing an assertion to ensure that a Gray-coded signal changes exactly 1 bit at a time. The assertion first determines the bits that change by bitwise XORing the current value with that of the previous cycle ($A = S^{-1} \wedge S$). If a bit changes, the corresponding bit position in A is 1. Then we determine how many bits have changed by checking whether A is one-hot. If A is one-hot or 0, the signal did not violate the Gray coding requirement. The code is as follows:

```
A = prev_S ^ S; // prev_S = S⁻¹
if (~((A == 0) || |(A & (A-1) ) ) ) // one-hot assertion
    $display ("ERROR: Gray code violation");
```

where $S^{-1}$ or prev_S can be obtained as follows:

```
initial
    curr_S = 1'bx;
    // assume S (curr_S) is one-bit; or use (`WIDTH-1){1'bx}

always @(posedge clock)
begin
    prev_S = curr_S;
    curr_S = S;
end
```

In general, to save the previous $N$ values of a signal S, a circular queue of length $N + 1$ is used, $N$ slots for the $N$ past values plus one slot for the current value. Figure 5.11A illustrates the movement of the queue over time. The queue rotates clockwise one slot every time the clock clicks, but the external labels (S, $S^{-1}$, . . ., $S^{-N}$) remain stationary. At every clock tick, the current value of the signal is stored in the slot facing label S. Thus, after $N$ clock ticks, the queue is fully filled, and now all $N$ past values are available. A piece of Verilog code implementing the circular queue is as follows:

```
reg [`WIDTH-1:0] CQ [N:0]; // WIDTH is width of S
integer i; // index of the current cycle
integer j; // index of the jᵗʰ past value

initial i = N;

always @(posedge clock)
```

```
begin
    i = i +1 ;
    i = i % (N+1); // circular index
    CQ[i] = S; // present value
    // retrieving kth past value
    j = (i-k>=0) ? i-k : i-k+N+1 ; // calculate circular index
    prev_k_S = CQ[j];
end
```

Signal $S^{-k}$, $1 \leq k \leq N$, is CQ[j], where $j = i - k$, if $i - k$ is nonnegative; otherwise, $j = i - k + N + 1$.

---

**Example 5.3**

Storing and retrieving past signal values in a circular queue.

After the first clock transition, i takes the value of 0, after the modulo operation. So the value at clock 1 is stored in CQ[0]. On the next clock, the queue rotates one slot and stores the new value in slot CQ[1]. After yet another clock transition, the queue rotates again and stores the current cycle value in CQ[2]. Now, index *i* is 2, meaning CQ[2] has the current cycle value. To get the value two cycles before (in other words, $k = 2$), we compute index *j*, which gives 0, meaning CQ[0] contains the value two cycles ago. Figure 5.11B confirms that these slots have the correct values.

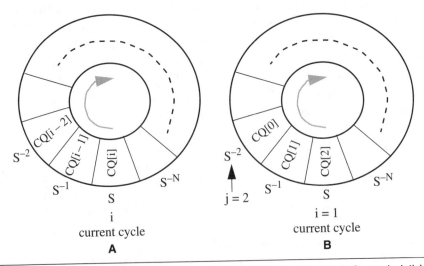

**Figure 5.11** Circular queue for archiving past signal values. (A) A circular queue before a clock tick. (B) A circular queue rotates after a clock tick.

Once past values of signals are computed, assertions over the history of signals can be written. An example assertion over signal history is as follows: If `Req` is 1, then `Ack` should turn 1 three cycles later, which causes `Req` to return to 0 seven cycles later. Whether this assertion fails will not be known until ten cycles after `Req` becomes 1. After `Req` has become 1, values of `Ack` and `Req` are archived for the next ten cycles. By then, passage of the assertion can be determined. Let `prev_k_Req` and `prev_k_Ack` denote the past $k^{th}$ value of `Req` and `Ack` respectively:

```
if (prev_10_Req == 1'b1)
    if ( !(prev_7_Ack == 1'b1 && Req == 1'b0) )
        $display ("ERROR: assertion failed");
```

If we use circular queues `CQ_Ack` and `CQ_Req` for `Ack` and `Req` respectively, and choose N to be ten, then the assertion takes the form

```
always @(posedge clock) begin
    i = i + 1;
    i = i % 11;
    j1 = (i-7>=0) ? i-7 : i + 4; // index for Ack
    j2 = (i-10>=0) ? i-10 : i + 1 ; // index for Req
    if (CQ_Req[j2] == 1'b1)
        if (!(CQ_Ack[j1] ==1'b1 && Req == 1'b0) )
            $display ("ERROR: assertion failed");
end
```

Note that this code should be activated after at least ten cycles have passed.

**Signal change pattern.** Once the history of signals is available, and together with combinational assertions, we can obtain assertions on many common signal properties. Table 5.4 lists some common assertions that can be implemented by applying combinational assertions to the signals' history.

**Table 5.4** Assertions on Signal Change Pattern

| Assertion | Description | Coding Algorithm |
|---|---|---|
| **Gray code** | At most, 1 bit changes. | $S \wedge S^{-1}$ is one-hot. |
| **Hamming code, H** | Number of bit changes is $H$. | $S \wedge S^{-1}$ has $H$ 1s. |

*continues*

**Table 5.4**   Assertions on Signal Change Pattern (Continued)

| Assertion | Description | Coding Algorithm |
|---|---|---|
| **Increment, decrement** | A signal increments or decrements by a step size, $N$. | $(S - S^{-1})$ or $(S^{-1} - S)$ is $N$. |
| **Change delta** | A signal's change is with a range, $R$. | $|(S - S^{-1})|$ is within range $R$. It is a signal range check. |
| **Transition sequence** | A signal's transition follows a certain value pattern. | $\{S, S^{-i}, S^{-j}, S^{-k},....\}$ has the pattern value. It is a signal membership check. |
| **Cause and effect** | Signal A asserting at $T$ causes signal B to assert at $T + N$. | $A^{-N}$ -> B asserts. |

**Time window.** These assertions determine whether a signal does or does not change over a period of time or within a period specified by a pair of events. In the case when the time period is delimited by a number of cycles, a repeat statement can be used to perform the check. For instance, assume that a signal must remain unchanged for N cycles, checking is done by embedding $S^{-1}$ == S in a repeat loop of N iterations. In the code, signal S is sampled at every positive transition of the clock, and thus it is assumed that signal S changes only at the clock's transitions:

```
repeat (N)
    @(posedge clock)
        if (prev_S != S) $display ("ERROR...");
```

In the other case, when the interval is specified by a pair of events, signal S must be monitored for changes all the time, not just at some clock transitions. This situation is related to an asynchronous timing specification and it is discussed in "Unclocked Timing Assertions" on page 243.

**Interval constraint.** The previous discussion assumes that the cycle numbers of the past values are specified, such as signal value seven cycles ago. On the other hand, if a range, instead of the exact cycle number, is specified (such as, S will rise some time between cycle 1 and cycle 10 after signal Ready is set), then the assertion can be implemented by

embedding the check inside a loop with an iteration count equal to the range. The following code asserts that S will rise some time from cycle 1 to cycle 10 after signal Ready is
set. When S rises, it immediately disables the block. If S does not rise after ten cycles, an
error will be displayed:

```
always @(posedge clock)
begin: check // named block
   if ( Ready == 1'b1 ) begin
      repeat (10) @(posedge clock)
         if ( (prev_S != S) && (S == 1'b1) ) disable check;
         // assertion passes, exit
   end
   $display ("ERROR: assertion failed");
end // end of named block, check
```

Similar code can be written for assertions about an event that should not occur throughout an interval. If the event occurs within the interval, an error message is issued and the
block is disabled. If not, it silently passes. For example, the following code asserts that
address [31:0] remain steady for ten cycles after WR becomes one:

```
always @ (posedge clock)
begin: check
   if (WR == 1'b1)
      repeat (10) @(posedge clock)
         if ( prev_Address != Address ) begin
            $display ("ERROR: Address changed");
            disable check;
         end
end // end of named block, check
```

**Unclocked timing assertions.** Thus far, all timing constraints use a clock cycle as a reference.
Hence, this way of constraining is not applicable for assertions that do not have explicit
clocks. Situations with no reference clocks occur in high-level specifications that constrain
the order of event occurrences without specific delays. For instance, signal R must toggle
after signal S becomes high, with no requirement on how soon or how late R must toggle.
Here we need to study two cases. The first case concerns an expression that should remain
steady within an interval. The second case involves an expression that should attain a certain value within an interval. The interval is defined by two transitions—one signaling the
beginning and the other signaling the end. An example of the first case is that the WR line

must not change during a read or write cycle. An example of the second case is that signal Grant must attain a value and remain steady until signal Req is lowered. Unlike the situations with reference clocks, the expressions must be constantly monitored throughout the interval, as opposed to monitoring just at clock transitions.

To construct an assertion for the first case, the delimiting events and the expressions or signals are monitored using the event control operator @. The following example code ensures that S remains steady throughout the interval marked by events E1 and E2. The assertion is active only when guarding signal Start becomes 1. Two named blocks are created: one block waiting on E1 and the other, E2. If the assertion fails before E2 arrives, an error is displayed and the E2 block is disabled. If E2 arrives before any failure, the E1 block is disabled:

```
always @(posedge Start) // guarding statement
begin: check
   @ (E1);
   @ (S) $display ("ERROR: signal S changes");
      disable stop;
end

always @(posedge Start) // guarding statement
begin: stop
   @ (E2) disable check;
end
```

To prevent a hanging situation, when E2 never arrives, another block can be created to impose an upper time limit on the interval:

```
always @(E1)
begin: time_limit
   @(posedge clock) begin
      count = count + 1;
      if (count > LIMIT) begin
         $display ("ERROR: interval too long");
         disable check
         disable stop; // disable the ending block
      end
   end
end
```

When E2 arrives, block stop has to disable block time_limit because the stopping event has already arrived, and thus checking for its arrival time becomes irrelevant. A modified block stop is as follows:

```
always @(posedge Start) // guarding statement
begin: stop
   @ (E2) begin
      disable check;
      disable time_limit;
   end
end
```

For the second case that asserts an expression to attain a certain value during an interval, the expression is first monitored for changes. When a change occurs, the value of the expression is checked. The interval is instrumented with two named blocks: one block waiting on the first event and the second block waiting on the second event. We can illustrate this with the assertion that signal Grant cannot change until signal Req lowers, and this sequence of transitions must happen within an interval:

```
always @(posedge Start) // guarding statement
begin: check
   @ (E1); // beginning event
   @ (posedge Grant)
   if (Req != 0) $display ("ERROR: Grant changed at Req = 1");
   disable stop;
end
```

```
always @(posedge Start) // guarding statement
begin: stop
   @ (E2) $display ("ERROR: Grant never changed");
   disable check;
end
```

If signal Grant changes within the interval, the value of Req is checked. If it's not 0, an error will be issued. In either case, block stop will be disabled when Grant arrives. Therefore, if block stop has not been disabled when event E2 arrives, it means that signal Grant never changed in the interval and an error will be issued.

**Container assertions.** Assertions can also check the integrity of the data without knowing the exact data values. These assertions ensure that design entities remain intact after they have gone through processing. An example is that the content of a cache block remains the same from the time it is filled to the time it is first accessed. Another example is that a packet received is the same packet sent, even though the packet has undergone encoding, transmission, and decoding. This type of assertion contains local variables or containers that save the data before processing, and use the saved data to compare with the data after processing. Let's illustrate container assertion with following packet transmission example.

In this example, packets are encoded, transmitted through a network, decoded, and received. We want to write a container assertion to check for data integrity of the packets— namely, the received packets were not corrupted during the process. We assume each packet is identified by an ID and packets arrive in the same order they are sent. When a packet is sent, signal SEND toggles. The container assertion has two FIFOs: one for the packet ID and the other for the packet data. When SEND toggles, the packet's ID is pushed into id_FIFO, and the packet data are saved in packet_FIFO. The packet also carries its ID. On the receiving end, when a packet arrives, signal RECEIVE toggles, and the ID is extracted from the packet. At the same time, an ID is also retrieved from id_FIFO for com- parison. If there is no match, an error is issued. If there is a match, the saved data are retrieved from packet_FIFO and are compared with those of the received packet. If there is a difference, an error has occurred. The transmission process and its interactions with the assertion are shown in Figure 5.12.

The following abridged code implements the container assertion:

```
always @(SENT) begin
    push_idFIFO(ID);
    push_packetFIFO( packet);
end

always @(RECEIVE) begin
    ID1 = get_FIFO();
    ID2 = get_ID_from_packet (packet);
    if (ID1 != ID2) $display ("ERROR: packet out of order");
    else begin
        org_packet = get_original_packet (packet_FIFO);
        ok = compare_packet (packet, org_packet);
        if (!ok) $display ("ERROR: packet corrupted");
    end
end
```

If we relax the requirement that the arrival order has to be the same as the sending order, id_FIFO would have to be replaced by another data structure that is more amicable to matching, such as a hash table. Then, when a packet arrives, its ID is extracted and is sent to the hash table for matching. If it is found, the packet content is then compared; other- wise, an error has occurred.

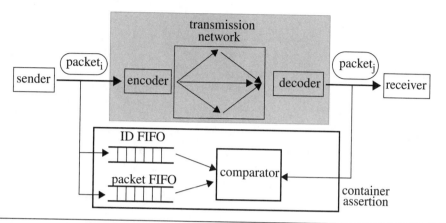

**Figure 5.12** Assertion to check for packet integrity

### 5.4.4  Built-in Commercial Assertions

An alternative to writing assertions from the ground up is to use commercial built-in assertions. Most of these assertions are called as functions and they offer checks in a wide range of abstraction, from signal level to protocol level. Assertions can be categorized according to the functions they check. Some common categories are signal or variable property, temporal behavior, control and data checks, interface, and protocol.

Signal property may consist of checks for "one-hot-ness," Hamming distance, parity, overflow, and change range. Temporal behavior may include checks for expected transitions or no transitions within a time interval, sequences of signal transitions at specified times, and order of signal changes without timing constraints. Control and data checks ensure proper operations for functional units such as FIFOs, stacks, memory, encoders, finite-state machines, and buses. Example checks include: no "enqueue" command is issued when the FIFO is full, data popped are identical to those pushed, memory address lines are steady when the read or write line is active, no unknown values are written to the memory, read locations have been initialized, state transitions are within the legal range, and, at most, one driver is active on a bus at any time. Interface checks ensure that interactions among functional units are correct. Some example checks include signal sequencing is correct during a DMA transaction, commands to the Universal Serial Bus (USB) port are legal, and data sent from the processor are received intact by main memory. Protocol checks are high-level assertions that ensure the specific protocols are followed at all times. Example checks include: PCI protocol is followed, bus arbitration is fair to all units, and critical shared resources are "mutexed" (in other words, only one client can access at a time).

These built-in assertions can be invoked as stylized comments, inline functions, or PLI system tasks:

```
always @(posedge clock)
    case (PS):
        NS = ...; // assertion check_state_transition(PS, NS,
                       clock,...), stylized comment assertion


    grant= (idle ? ... ) ;
    check_arbitor (grant, request, ...); // inline assertion


    if( command == 'ENQUEUE) ...
    $check_fifo(command, ...); // PLI system task assertion
```

Stylized comment assertions are recognized only by the assertion tool that processes the assertions and performs the checks during simulation. Because these assertions are comments, they are not part of the design. Inline assertions are implemented as modules, functions, tasks, or keywords specific to the vendor's simulator. They should be enclosed within a pair of Verilog pragmas for easy removal from the design code, such as `ifdef ASSERTION ... `endif. PLI system task assertions perform the checks in C/C++ code. Like inline assertions, they should be embedded within a pair of `ifdef`s.

## 5.5  SystemVerilog Assertions

Although most assertions can be written using Verilog constructs, the assertion code can be lengthy and complicated, making maintenance and reuse difficult. In SystemVerilog language, built-in constructs are proposed to express succinctly the intent of commonly encountered verification assertions. These assertion constructs can serve as assertions for simulation-based verification as well as property specifications for formal method-based verification. Although at the time of this writing the exact syntax of SystemVerilog assertions has not been standardized, the concepts conveyed should persist into the future. Let's discuss the key constructs of SystemVerilog assertions. For more details and updates about the syntax, please refer to IEEE 1364 SystemVerilog documents.

### 5.5.1  Immediate Assertions

Two types of assertions exist in SystemVerilog: immediate assertions and concurrent assertions. Immediate assertions follow simulation event semantics for their execution and are executed like any procedural statement. Assertions of this nature are also called *procedural assertions*. They are just Verilog procedural statements preceded with the SystemVerilog keyword `assert`, which behaves like an if–then–else statement. If the assertion passes, a set of statements is executed; if it fails, another set of statements is executed. For example, to write to memory, we first ensure $\overline{CS}$ (chip select) is low and signal WR is high. If the condition is not met, an error will be printed; otherwise, a confirmation message is printed. The checking part and the associated actions are succinctly encapsulated by the immediate assertion construct. The following is an immediate assertion inside an always block:

```
always @(posedge write_to_memory)
begin // a Verilog procedural block
   data = ...;
   address = ...;
   ...
   // here is an immediate assertion
   assert (WR == 1'b1 && CS == 1'b0)
      $display ("INFO: memory ready"); // pass clause
      else $display ("ERROR: cannot write"); // fail clause
...
end // end of always
```

Either or both of the pass and fail clauses may be omitted. Immediate assertions check on Verilog expressions only and do not describe temporal behavior. Immediate assertions are used during simulation-based verification.

### 5.5.2  Concurrent Assertions

Concurrent assertions, on the other hand, describe temporal behavior, and, analogous to Verilog's module, they are declared as stand-alone properties and can be invoked anywhere, just as a module is defined and instantiated. Assertions of this nature are also called *declarative assertions*. Concurrent assertion is to Verilog module as immediate assertion is to procedural statement. Concurrent assertions use sequence constructs to describe temporal behavior. The syntax of a concurrent assertion is

```
assert property (property_definition_or_instance) action

action ::= [statement] | [statement] else statement
```

If the property holds, the first statement is executed; otherwise, the statement after else is executed. The keyword property distinguishes a concurrent assertion from an immediate assertion. An example of a concurrent assertion is

```
assert property (hit ##1 read_cancel) $display("ok"); \\
   else $display ("ERROR: memory read did not get \\
   cancelled in time");
```

which asserts on the property `hit ##1 read_cancel` that if there is a hit on cache, the read to main memory should be canceled during the next cycle. If the property fails, an error is issued.

The main ingredient in modeling temporal behavior in property definition is sequence. A sequence is a list of Boolean expressions ordered in time—for example, sequence = $\{(B_i, T_i), i \in V\}$, where $B_i$ is the Boolean expression at time $T_i$, and $V$ is a set of integers denoting time (such as a cycle number). For example, $\text{sequence}_1 = \{ (x_1 \,\&\, \overline{x_2} + x_3, 1), (x_1 + x_2, 2), (x_2 \,\&\, x_3, 3) \}$. For a sequence to be true, $B_i$ must be true at $T_i$ for all i in V. Therefore, for $\text{sequence}_1$ to be true, $(x_1 \,\&\, \overline{x_2} + x_3)$ must be true at time 1, $(x_1 + x_2)$ must be true at time 2, and $(x_2 \,\&\, x_3)$ must be true at time 3. The property in this example is a sequence that translates into $\{(\text{hit}, N), (\text{read\_cancel}, N+1)\}$, where $N$ is the current cycle. Sequences can model waveforms. A waveform with a 50% duty cycle and a period of 2 can be represented by sequence

$$\{(S_i, i), S_i = 1, i \text{ is even. } S_i = 0, i \text{ is odd. } i \in Z\}.$$

A sequence can represent several waveforms. If a waveform satisfies the sequence, then the sequence is said to have resulted in a match. For instance, sequence

$$\{(S_i, i), S_i = a, i \text{ is even. } S_i = \overline{a}, i \text{ is odd. } i \in Z\}$$

represents two waveforms, one of which is the previous waveform of the 50% duty cycle and period 2. The previous waveform has resulted a match in this sequence. The other waveform is the complement of the first waveform.

Sequence, therefore, is an extension of Boolean expression over time. Furthermore, although a Boolean expression describes the behavior of variables at an instance in time, a sequence describes the behavior of variables over time. A waveform is to a sequence as a vertex is to a Boolean function. For example, a Boolean function is true for the set of vertices in the on set of the function, and a sequence is true for the set of waveforms that result in matches for the sequence. Figure 5.13 contrasts Boolean domain with temporal (sequence) domain. Let's divide our discussion on SystemVerilog assertions into two parts: the first part on sequence constructors and the second part on sequence connectives or operators.

**Sequence constructors.** The key to construct a sequence is to specify time. The first constructor is cycle delay specifier ##. ## N means N cycles of delay, and N can be any nonnegative integer. As an example, $x_1$ ##3 $x_2$ is equivalent to $(x_1, t_1), (x_2, t_1 + 3)$. In other words, $x_2$ is delayed by three cycles relative to the previous time. The argument of ## can be a range (##[n,m]), which means the delay is any number between $n$ and $m$. For example, $x_1$ ##[1,3] $x_2$ is equivalent to the following:

```
x₁ ##1 x₂
or
x₁ ##2 x₂
or
x₁ ##3 x₂
```

Therefore, for $x_1$ ##[1,3] $x_2$ to be true, at least one of the previous sequences must be true. Consider the waveforms in Figure 5.14. $x_1$ ##1 $x_2$ is satisfied at time 7 if it starts at time 6, because at time 6, $x_1$ is true and one cycle later $x_2$ becomes true. However, $x_1$ ##2 $x_2$ and $x_1$ ##3 $x_2$ are not satisfied with the waveforms. Therefore, sequence $x_1$ ##[1,3] $x_2$ is satisfied

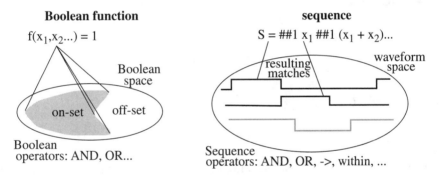

**Figure 5.13** A comparison between Boolean domain and sequence domain

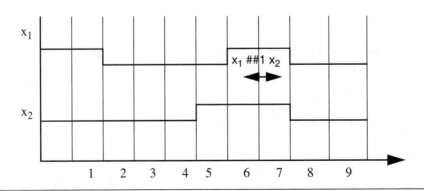

**Figure 5.14** Example waveforms for a sequence

by the waveforms. To specify an unbound upper limit, $ denotes infinity (for example, ##[1,$] means any number of cycles greater than or equal to one).

As an example of constructing a sequence, consider the order of events that signal req is high, followed by ack going high three cycles later, and ending with req going low seven cycles later. The sequence is req ## 3 ack ## 7 !req. This sequence is true if and only if the previously described pattern occurs.

Just like a module, a sequence can be defined, with or without arguments, and can be instantiated elsewhere. For example, to define sequence mySeq, enclose the sequence within keywords sequence and endsequence:

```
sequence mySeq ;
    ##1 x_1 (x_2) [*2,8] ##3 x_1
endsequence
```

To instantiate, just call mySeq,

```
##3 y mySeq #1 x
```

which is equivalent to

```
##3 y ##1 x_1 (x_2) [*2,8] ##3 x_1 #1 x
```

To define a sequence with parameters, enclose the parameters inside a pair of parentheses. The following is an example of defining sequence seqParam with parameters a and b:

```
sequence seqParam (a,b);
    ##1 a x_1 ##3 (x_1 == b)
endsequence
```

To instantiate the sequence, just call it by name and substitute values for the parameters. For example, the following instantiation

```
seqParam (y,z)
```

is equivalent to

```
##1 y x_1 ##3 (x_1 == z)
```

Larger sequences can be formed from sequences using connective operators, so that complex assertions can be built from simple ones. To duplicate a sequence over time, use the repeat operator [*N], which duplicates the preceding expression $N$ times with one cycle delay inserted between duplicates. For example, $(x_1 + x_2)$ [*3] is equivalent to the following:

```
(x₁ + x₂) ##1 (x₁ + x₂) ##1 (x₁ + x₂) // duplicate 3 times
```

Note the one cycle delay was inserted between the duplicates. The argument of the repeat operator, N, can be a range (for example, [*n,m]) in which case it repeats for a number of times $j$, where $n \le j \le m$. For example, $(x_1 + x_2)$ [*1,3] is equivalent to the following:

```
(x₁ + x₂)
or
(x₁ + x₂) ##1 (x₁ + x₂) // duplicate 2 times
or
(x₁ + x₂) ##1 (x₁ + x₂) ##1 (x₁ + x₂) // duplicate 3 times
```

Instead of separating duplicates by exactly one clock cycle, as in repeat operator [*n,m], two variants of the repeat operation allow arbitrary intervals between duplicates. To create patterns with arbitrary intervals between duplicates, use nonconsecutive repeat operators [*->n,m] or [*=n,m]. Nonconsecutive repeat operators allow other patterns to be inserted between duplicates. For example, $x_1$ ##1 $x_2$ **[*->1,3]** ##1 $x_3$ is equivalent to

```
x₁ ##1 (y ##1 x₂) [*1,3] ##1 x₃
```

where $y$ is any sequence, including an empty sequence, that does not have $x_2$ as a subsequence. Therefore, $x_1$ ##1 $( x_4$ ##1 $x_2$ $)$ $( x_5$ ##1 $x_2$ $)$ ##1 $x_3$ is an instance described by the previous construct. Formally, $x_1$ ##1 $x_2$ [*->1,3] ##1 $x_3$ is equivalent to

```
x₁ ##1 (!x₂ [*0,$] ##1 x₂) [*1,3] ##1 x₃
```

Repeat operator [*=n,m] is a slight variation of [*->n,m] in that a ##1 is appended to the end of the result from the operator. That is, $x_1$ ##1 $x_2$[*=n,m] ##1 $x_3$ is equivalent to

```
x₁ ##1 (!x₂ [*0,$] ##1 x₂) [*n,m] ##1 !x2 [*0,$] ##1 x₃
```

**Sequence connectives.** Sequence connectives build new sequences from existing ones. To build new sequences, some connectives place constraints on the operand sequences, such as start times and lengths, to make the resulting sequences meaningful. The idea behind sequence connectives is twofold. First, it makes creating complex sequences easy; second, it extends operations on variables to sequences so that temporal reasoning can be carried out in a manner similar to Boolean reasoning.

**Operator AND.** Sequences can be logically ANDed if the operand sequences start at the same time. Operation (sequence$_1$ AND sequence$_2$) is true at time t if and only if both sequences are true at time t, or one sequence is true at time t and the other has already become true at time t. Let's use Figure 5.15 to determine whether S$_1$ AND S$_2$ is true on the given signal waveforms of $x_1$ and $x_2$, where S$_1$ = $x_1$ ##2 $(\overline{x_1}x_2)$ and S$_2$ = ##1 $x_2$ ##2 $\overline{x_1}$ ##1 ($x_1$ + $\overline{x_2}$). The horizontal double–arrowhead lines denote the time interval a sequence is evaluating, from the start time to the end time. Assume the current time is cycle 1. Sequence S$_1$ starts at time 1 and ends at time 3. In cycle 1, the expression is $x_1$, and in cycle 3, the expression is $\overline{x_1}x_2$. From the waveforms, we see $x_1$ is 1 at cycle 1, and at cycle 3, $x_1$ is 0 and $x_2$ is 1. Therefore, S$_1$ becomes true at cycle 3. The same goes for sequence S$_2$, which becomes true at cycle 5. Hence, S$_1$ AND S$_2$ is true at cycle 5.

**Operator intersect.** For intersection operation, sequence$_1$ intersect sequence$_2$ is true at time t if and only if both sequences are true at time t. This operator requires that both operand sequences start and end at the same time. Therefore, we cannot perform intersection on the sequences in Figure 5.15, because S$_1$ and S$_2$ do not end at the same time. If we change S1 to S$_1$ = $x_1$ ##2 $(\overline{x_1}x_2)$ ##2 $(x_1x_2)$, then S1 intersect S2 becomes true at cycle 5, because both sequences become true at cycle 5, as shown in Figure 5.16. If either or both sequences have multiple matches, only the pairs with the same start time and end time are considered.

**Operator OR.** sequence$_1$ OR sequence$_2$ is true at time *t* if and only if at least one sequence is true at time *t*. Both operand sequences start at the same time. For the sequences in Figure 5.15, S$_1$ OR S$_2$ is true at cycles 3, 4, and 5, because S1 is true at these times.

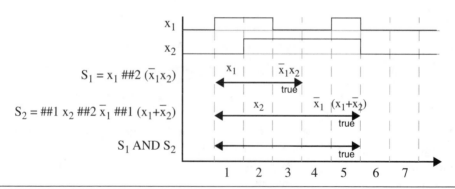

**Figure 5.15** Signal traces satisfy S$_1$ AND S$_2$

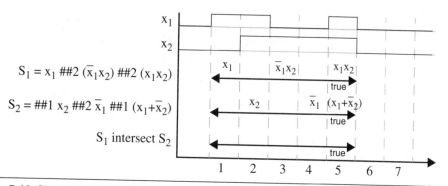

**Figure 5.16** Signal waveforms satisfy $S_1$ intersect $S_2$

**Operator first_match.** Operator first_match takes in a sequence and returns true when the sequence first becomes true. If the sequence expands to multiple sequences, such as ##[2,5]$(x_1x_2)$ expanding to four sequences, then first_match returns true when any of the expanded sequences first becomes true. The end time of first_match is the time it turns true. Figure 5.17 shows the operation of first_match. ##[2,5]$(x_1x_2)$ expands to ##2 $(x_1x_2)$ or ##3 $(x_1x_2)$ or ##4 $(x_1x_2)$ or ##5 $(x_1x_2)$. Of these, ##2 $(x_1x_2)$ and ##4 $(x_1x_2)$ are true for the given waveforms. Operator first_match detects the first time the sequence becomes true (at cycle 3, ##2 $(x_1x_2)$). Thus, it returns true and ends at cycle 3.

**Operator implication.** This operation is the same as the Boolean implication operator, $A \rightarrow B$, which is $\bar{A} + A \cdot B$, except that it applies to sequences (for example, sequence$_1$| $\rightarrow$ sequence$_2$). If sequence$_1$ is true at time t, then sequence$_2$ starts at time $t$ and its valuation

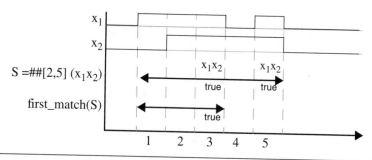

**Figure 5.17** Example of first_match operation

determines that of the implication. If sequence₁ is false, then the implication is true. A variation is sequence₁| ⇒ sequence₂ for which sequence₂ starts one cycle after sequence₁ becomes true. An example use of sequence implications is when an interrupt signal arrives, the processor will wait up to three cycles. If all in-flight tasks are finished by then, the control is transferred to an interrupt handler in the next cycle. Once control is in the interrupt handler, it should finish within 20 cycles. Let INT, TASK, HDLR, and DONE denote the interrupt signal, in-flight task done signal, signal to start the interrupt handler, and handler finish signal respectively. This behavior can be described as (INT ##[0,3] TASK) |=> HDLR ##[1,20] DONE. This assertion is true at cycle 7 for the waveforms shown in Figure 5.18.

**Operator throughout.** This operator, E throughout S, where E is an expression and S is a sequence, imposes expression E on sequence S. Recall that a sequence is {(B_i,T_i),i ∈ V}. Then E throughout S is {(E · B_i,T_i),i ∈ V}. An example is $(x_1 + x_2)$ throughout (##1 $x_3$ ##2 $x_4$), which is equivalent to (##1 $(x_1 + x_2)x_3$ ##2 $(x_1 + x_2)x_4$). An applicable situation is that S is an assertion packaged with an IP module and the expression E is a condition describing the environment in which the IP is instantiated (for example, the peripheral signals to the IP must be within the correct ranges required by the IP). For the IP instance to work properly, the environment expression E must be true throughout the interval that the IP assertion holds. In other words, E is a predicate guarding the application of S.

**Operator within.** This operator, sequence₁ within sequence₂, tests whether sequence₁ lies within sequence₂ and whether both sequences succeed. This operation requires that sequence₁ start and end after sequence₂ starts and before sequence₂ ends. The operation

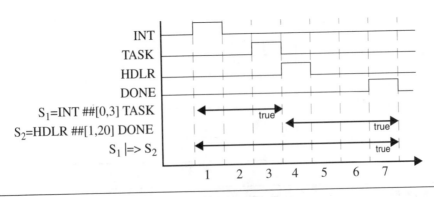

**Figure 5.18** Implication assertion for interrupt and interrupt handler

becomes true if sequence₁ has become true and is true when sequence₂ is true. $S_1$ `within` $S_2$ is equivalent to

```
(1 ##1)[*0,$] S₁ (##1 1)[*0,$] intersect S₂
```

An applicable scenario is that throughout an exclusive critical-section access, the ID of the accessing process should not change. Let $S_1$ = `(ID == $past(ID))` `[*6]`, $S_2$ = `(ACCESS ##5 !ACCESS)`, where `$past` is a system function that returns the previous cycle value of the operand and thus `(ID == $past(ID))` is true if ID has not changed since the last cycle. Sequence $S_1$ asserts that the ID remain the same for six cycles, whereas sequence $S_2$ asserts that access be done within five cycles. If variable ID does not change while the critical section is being accessed, ($S_2$ `within` $S_1$) is true. Figure 5.19 shows signal waveforms satisfying ($S_2$ `within` $S_1$).

**Operator container.** A sequence can have local variables to hold data that are to be used in the future. An application checks whether the data that popped out of a stack are the same as the data pushed. The assertion uses a local variable to save the data when it is pushed into the stack. When the data are popped out, they are compared with the saved version. The following assertion does this check, saving data when push is high and comparing the popped data seven cycles later when it is popped. The container operator is the comma that separates push and the assignment in the antecedent:

```
int orig;
(push, orig = data |-> ##7 data = orig)
```

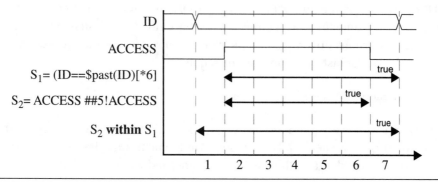

**Figure 5.19** Asserting critical-section access using the `within` operator

**Operator ended.** This operator, `sequence.ended`, returns true at the time when the sequence becomes true—that is, the time the sequence ends. An example application is to assert that signal `grant` lowers two cycles after sequence `transaction` finishes:

```
( transaction.ended ##2 !grant )
```

**System functions.** Built-in system functions in SystemVerilog simplify code writing. Some system functions are `$onehot`, `$onehot0`, `$inset`, `$insetz`, `$isunknown`, `$past`, and `$countones`. A system function takes in expressions as arguments, such as `$onehot` `(bus)` and returns the advertised value. The following list explains some common system functions.

1. `$onehot (expression)` returns true if only 1 bit of `expression` is 1. `$onehot0()` returns true if at most 1 bit of `expression` is 1.
2. `$inset (e1, e2, ...)` returns true if `e1` is equal to one of the subsequent expressions. For example, `$inset (cmd, `WR, `RD, `CLR)` is true when `cmd` is either `` `WR ``, or `` `RD ``, or `` `CLR ``. This function replaces the longer and more complex case statement. `$insetz` is similar to `$inset` except it treats `z` and `?` as don't-cares, as in `casez`.
3. `$isunknown (expression)` returns true if any bit of `expression` is of x value.
4. `$past (expression, i)` returns the value of `expression` i cycles prior to the current cycle. If `i` is omitted, it is defaulted to 1. `$past(e, i)` is equivalent to $e^{-i}$. With this function, asserting Gray code is simply `$onehot(S ^ $past(S))`.
5. `$countones (expression)` returns the number of `1s` in `expression`. An application is checking the parity of a signal.

**Multiple clocks.** So far, the number of clock cycles has referred to an inferred clock, which is the clock of the domain in which the assertion is embedded. SystemVerilog provides a way to write assertions driven by multiple clocks and references the cycle delays to their respective clocks. For instance, consider the sequence

```
@(posedge clk₁) x₁ ## (posedge clk₂) x₂
```

In this sequence, at every `posedge` of $clk_1$, $x_1$ is checked. If $x_1$ is true, then on the next `posedge` of $clk_2$, $x_2$ is checked. Thus, ## serves as a synchronizer between the two clocks. If $clk_1$ and $clk_2$ are identical, the previous sequence reduces to

```
@(posedge clk₁) x₁ ##1 x₂
```

Sequences can be defined and instantiated, making sequences reusable. The following code defines a sequence, called `testSequence`, to be ##1 A ##2 B ##3 (x == y):

```
sequence testSequence (x,y);
   ##1 A ##2 B ##3 (x == y)
endsequence
```

To instantiate `testSequence`, just invoke it by name, as shown here in another sequence definition:

```
sequence callSequence;
   @(posedge clock) testSequence(p,q)
endsequence
```

This sequence, `callSequence`, is equivalent to

```
@(posedge clock) ##1 A ##2 B ##3 (p == q)
```

The purpose of defining sequences is to facilitate assertion construction on temporal behavior. Expressions and sequences form properties that are asserted by assertions. SystemVerilog provides a way to define properties and instantiate them in different assertions. To define a property:

```
property SteadyID (x)
   @(posedge clock) (x == $past(x))[*6]
endproperty
```

To assert property `SteadyID`:

```
assert property SteadyID(procID)
```

## 5.6   Verification Coverage

Because it is impossible to verify exhaustively all possible inputs and states of a design, the confidence level regarding the quality of the design must be quantified to control the verification effort. The fundamental question is: How do I know if I have verified or simulated enough? Verification coverage is a measure of confidence and it is expressed as a percentage of items verified out of all possible items. An "item" can have various forms: a line of code in an implementation, a functionality in a specification, or application scenarios of the design. Different definitions of item give rise to different coverage measures. As you may have anticipated, even defining item can be challenging, such as defining an item to be a functionality. Furthermore, it is even more challenging to determine all possible items, such as all functionalities in a design. In this section we will look at three coverage measures: code coverage, parameter coverage, and functional coverage. In our discussion we will assume that the underlying verification mechanism is simulation, because it is applicable to all three measures and because it lends itself to a clear explanation.

Code coverage provides insight into how thoroughly the code of a design is exercised by a suite of simulations (for example, the percentage of RTL code of a FIFO stimulated by a simulation suite). Parameter coverage (also referred to by some as *functional coverage*) reveals the extent that dimensions and parameters in functional units of a design are stressed (for example, the range of depth that a FIFO encounters during a simulation). Finally, functional coverage accounts for the amount of operational features or functions in a design that are exercised (for example, operations and conditions of a FIFO simulated: push and pop operation, full and empty condition). Note that code coverage is based on a given implementation and it computes how well the implementation is stimulated, whereas functional coverage is based on specifications only, and it computes the amount of features and functions that are verified. Parameter coverage can depend on implementation, specification, or both. Therefore, sometimes code and parameter coverage are referred to as implementation coverage, and functional coverage is referred to as specification coverage. Coverage measures based on implementation are only as good as the implementation, but are more concrete and hence easier to compute, whereas those based on specification are more objective and direct to what the design intended, but are more difficult to compute.

Each of the three coverage metrics has its shortcomings, but together they complement each other. Because code coverage assumes an implementation, it only measures the quality of a test suite in stimulating the implementation and does not necessarily prove the implementation's correctness with respect to its specifications. As an extreme example, consider a 64-bit adder that is mistakenly implemented as a 60-bit adder. It is possible that every line of the 60-bit adder has been stimulated, giving 100% code coverage, but the implementation is wrong. This problem would be detected if parameter coverage is used. To detect it, parameter coverage has to include the size range of operands. Even with parameter coverage, problems with the carry-in bit may not be detected (for example, the carry-in bit does not propagate correctly). Functional coverage includes all specified functions (for example, in this case, addition of numbers, carry-in, and overflow operations). In this case, functional coverage analysis will detect the carry-in error. By defining operand size to be parameters, better verification coverage is achieved by supplementing functional coverage with parameter coverage which explores different sizes of operands.

Because it is easy to leave out parameter and functional items, code coverage can provide hints on missing functionality. Therefore, in practice, a combination of code, parameter, and functional coverage is used.

With regard to ease of use, the code coverage metric ranks first, followed by parameter then functional coverage. Code coverage is easiest because it involves accounting for code exercised. The difficulty of using parameter coverage stems from finding most, if not all, parameters in a design and determining their ranges. Once the parameters are found, an apparatus needs to be set up to track them. Functional coverage is the most difficult to construct and use, because all functions and features have to be derived from the specifications, and there is no known method of systematically enumerating all of them. Often, enumerating all functionality is not possible because specifications can be ambiguous or incomplete.

Coverage metrics are used to measure the quality of a design and can also be used to direct test generation to increase coverage. It should be noted that coverage numbers do not necessarily project a linear correlation with verification effort. It is conceivable that the last 10% of coverage takes more than 90% of the overall verification effort.

## 5.6.1   Code Coverage

Code coverage depicts how code entities are stimulated by simulations. These code entities are statement, block, path, expression, state, transition, sequence, and toggle. Each entity has its own percentage of coverage; for example, code coverage may consist of 99% statement coverage, 87% path coverage, 90% expression coverage, and 95% state coverage.

**Statement coverage.**   Statement coverage collects statistics about statements that are executed in a simulation. For example, consider the code in Figure 5.20. Excluding the `begin`, `end`, and `else` lines, there are ten statements. If a simulation run is such that `a > x` at line 5 and `x == y` at line 12 are true, then all statements except those on lines 10, 11, 14, and 15 are executed in this run, as indicated by the arrows in Figure 5.20. Again excluding the `begin`, `end`, and `else` lines, eight lines are executed, giving 80% statement coverage.

**Block coverage.**   Note that in Figure 5.20, once line 3 is reached, line 4 will be executed; that is, coverage of some statements automatically implies coverage of other statements. Therefore, the statements can be grouped into blocks such that a statement in a block is executed if and only if all other statements are executed. The dividing lines for blocks are flow control statements such as if–then–else, `case`, `wait`, `@`, `#`, and `loop` statements. The code in Figure 5.20 can be grouped as shown in Figure 5.21. During the simulation run, four of six blocks are executed, giving 66.67% block coverage. Because of the similar nature

```
➤ line 1: always @(posedge clock)
   line 2: begin
➤ line 3:    a = b + c;
➤ line 4:    x = (y << 4);
➤ line 5:    if (a > x)
   line 6:    begin
➤ line 7:       y = b & a;
➤ line 8:       x = (x >> 2);
   line 9:    end
   line 10:   else
   line 11:      b = b ^ c;
➤ line 12:  if ( x == y)
➤ line 13:      c = y;
   line 14:   else
   line 15:      y = a;
   line 16: end // end of always
```

**Figure 5.20** Sample code for statement coverage

```
line 1: [always @(posedge clock)]   block 1
line 2: begin
line 3:    [a = b + c;
line 4:     x = (y << 4);]           block 2
line 5:    if (a > x)
line 6:    begin
line 7:       [y = b & a;
line 8:        x = (x >> 2);]         block 3
line 9:    end
line 10:   else
line 11:      [b = b ^ c;]            block 4
line 12:   if ( x == y)
line 13:      [c = y;]                block 5
line 14:   else
line 15:      [y = a;]                block 6
line 16: end // end of always
```

**Figure 5.21** Sample code for block coverage

of statement coverage and block coverage, and the fact that blocks better reflect the code structure, block coverage is often used in lieu of statement coverage. Near 100% statement and block coverage are easy to achieve and should be the goal for all designs. There are types of code that prevent 100% statement or block coverage. The first type is dead code, which cannot be reached at all and is a design error. The other type of statement is executed only under particular environment settings or should not be executed under correct operation of the design. Examples of this type are gate-level model statements that are not activated in a behavioral simulation, or error detection statements or assertions that do not fire if there are no errors. Thus, these types of statements should be excluded from statement and block coverage computation.

Coverage tools should have a pragma to delimit statements that are to be excluded. For example, when a coverage tool encounters the pragmas `coverage_tool off` and `coverage_tool on`, it skips the statements between them and excludes the statements from coverage calculation. The exact pragma name and syntax depend on specific coverage tools:

```
// coverage_tool off
if( state == 'BAD )
    $display ("ERROR: reach bad state");
// coverage_tool on
```

**Path coverage.** Conditional statements create different paths of execution. Each conditional statement is a forking point. Path coverage metrics measure the percentage of paths exercised. For the code in Figure 5.20, a graph of paths of statement execution is shown in Figure 5.22. There are four possible paths (in other words, four combinations of if–else branches). Two paths are shown in Figure 5.22. Path P1 is where both `if` statements are true, whereas path P2 is where both `if` statements are false. For the simulation run under discussion, P1 is executed, giving 25% path coverage. The number of paths grows exponentially with the number of conditional statements. Therefore, it is difficult to achieve 100% path coverage in practice.

**Expression coverage.** How a statement is executed can be further analyzed. For example, when the statement containing several variables or expressions is executed, a detailed analysis is: which of the variables or expressions did get exercised. For example, that the expression $(x_1x_2 + x_3x_4)$ evaluates to 1 can be broken down into whether $x_1x_2$ or $x_3x_4$ or both evaluate to 1. Such detailed information, revealing the parts of the expression executed, can be used to guide test generation to exercise as many parts of the expression as possible.

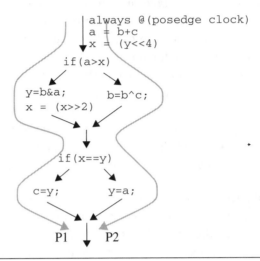

```
                          always @(posedge clock)
                          a = b+c
                          x = (y<<4)
                      if(a>x)

             y=b&a;              b=b^c;
             x = (x>>2)

                      if(x==y)

             c=y;                y=a;

                  P1      |      P2
```

**Figure 5.22** Sample code for path coverage

To compute expression coverage, operators in an expression are layered. For instance, $E = (x_1x_2 + x_3x_4)$ is considered to have two layers:

```
layer 1: E = y₁ + y₂
layer 2: y₁ = x₁x₂, y₂ = x₃x₄
```

The second layer consists of two subexpressions: $y_1$ and $y_2$. Each expression or subexpression is a simple operator (such as AND or OR) and its expression coverage can be calculated from the minimum input tables shown in Tables 5.5 through 5.7. A minimum input determining an expression is the fewest number of input variables that must be assigned a value in order for the expression to have a Boolean value. For simple operators, all minimum inputs determining the value of the expression can be tabulated. The number of entries in the table consists of all cases for the expression to be exercised completely. Expression coverage is the ratio of the cases exercised to the total number of cases.

As examples, Tables 5.5 through 5.7 show the minimum inputs for $f = x_1 \& x_2$, $f = ( y ? x_1 : x_2)$, and $f = (x > y)$.

This table indicates three cases for the AND function. For AND to evaluate to 0, only one input needs to be assigned 0. Because there are two inputs, two minimum inputs exist for the AND expression to be 0. For AND to evaluate to 1, both inputs must be 1. Therefore, there are a total of three minimum input entries. A line in the table represents a don't-care for the variable. If, during a simulation run, $x_1$ and $x_2$ take on values $x_1 = 1$ and $x_2 = 0$ at one time and $x_1 = 1$ and $x_2 = 1$ at another time, then two cases are exercised, giving two-thirds or 66.67% expression coverage for this operator.

For the quest operator $y\ ?\ x_1 : x_2$, if y evaluates to true, the expression's value is then determined by that of $x_1$. If y evaluates to false, $x_2$ then determines the expression's value. There are four cases, as shown in Table 5.6.

For the comparison operator, there are two cases: true or false.

For composite operators, expression coverage can be computed for either the top layer of the operator or for all layers. Deciding on the top or all layers is a matter of how precisely one wants to test an expression. As an example, consider $f = ((x > y) \parallel (a\ \&\ b))$. Input $(x,y,a,b)$ takes on values $(3,4,1,1)$ and $(2,5,1,0)$ during a simulation. The top layer operator is OR $(\parallel)$. In this simulation, the operands of OR evaluate to $(x > y) =$ false, $(a\ \&\ b) = 1$, and $(x > y) =$ false, $(a\ \&\ b) = 0$, for $(3,4,1,1)$ and $(2,5,1,0)$ respectively.

**Table 5.5**   Minimum Input Table for $f = x_1\ \&\ x_2$

| $f = x_1\ \&\ x2$ | $x_1$ | $x_2$ |
|:---:|:---:|:---:|
| 0 | 0 | — |
| 0 | — | 0 |
| 1 | 1 | 1 |

**Table 5.6**   Minimum Input Table for $f = (y\ ?\ x_1 : x_2)$

| $f = (y\ ?\ x_1 : x_2)$ | y | $x_1$ | $x_2$ |
|:---:|:---:|:---:|:---:|
| 1 | 1 | 1 | — |
| 0 | 1 | 0 | — |
| 1 | 0 | — | 1 |
| 0 | 0 | — | 0 |

**Table 5.7**    Minimum Input Table for f = (x > y)

| f = (x > y) | x > y | meaning |
|:-----------:|:-----:|---------|
| **0**       | 0     | $x <= y$ |
| **1**       | 1     | $x > y$  |

Because the OR operator has three cases, this simulation has a top-layer expression coverage of two-thirds, or 66.67%. For the next layer, (x > y) evaluates to false, giving 50% expression coverage; (a & b) has two-thirds, or 66.67%, coverage.

**State coverage.** State coverage calculates the number of states visited over the total number of states in a finite-state machine. The states are extracted from RTL code, and visited states are marked in simulation. Consider the following three-state finite-state machine (depicted by the state diagram shown in Figure 5.23), where state $S_1$ is the initial state and $S_4$ is not defined in the RTL code.

```
initial present_state = 'S1;
always @(posedge clock)
    case ({present_state,in})
        {'S1,a} : next_state = 'S3;
        {'S2,a} : next_state = 'S1;
        {'S3,a} : next_state = 'S2;
        {'S1,b} : next_state = 'S2;
        {'S3,b} : next_state = 'S3;
    endcase
```

If a simulation causes input in to be b,a,b,a,b,a,..., then the state machine will only transit between $S_1$ and $S_2$, giving two-thirds, or 66.67%, state coverage. Because at

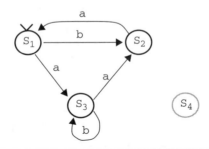

**Figure 5.23** State diagram extracted from RTL for state coverage

least 2 bits are required to encode the three states, $S_4$ is an implicit state and the state machine may end up at $S_4$ for some error combinations on state bits. A more reliable design should consider how to handle $S_4$. Therefore, for state coverage purposes, S4 should be counted and the previous input sequence would give only 50% state coverage. Because S4 can never be reached under normal operation, state coverage can be, at best, 75%. This should alert the designer that some states cannot be reached.

**Transition coverage.** Transition or arc coverage records the percentage of transitions traversed, and, in some cases, the frequency they are traversed. For the state machine in Figure 5.23 and the input sequence b,a,b,a,....,, only two of five transitions are traversed, giving 40% transition coverage. Note that the transition from S2 under input b is not defined. Including this undefined transition in the coverage calculation lowers the coverage to 33.33% and alerts us to potential problems. Therefore, the total number of transitions can be the actual number of transitions in the design or all possible transitions, which is equal to the number of states multiplied by the number of all possible inputs.

**Sequence coverage.** Sequence coverage calculates the percentage of user-defined state sequences that are traversed. A state sequence is a sequence of states that a finite-state machine traverses under an input sequence. For example, under input sequence a,b, a,a,a,b,b, the previous state machine traverses state sequence $S_1, S_3, S_3, S_2, S_1, S_3, S_3, S_3$. A user may specify that certain state sequences be traversed in a simulation run or not be traversed (in other words, illegal sequences). These state sequences could represent crucial functions or corner cases of the design. Sequence coverage provides information about which legal sequences are traversed and which illegal sequences are encountered.

**Toggle coverage.** Toggle coverage measures whether and how many times nets, nodes, signals, variables, ports, and buses toggle. Toggle coverage could be as simple as the ratio of nodes toggled to the total number of nodes. For buses, bits are measured individually. For instance, a change from 0000 to 1101 for a 4-bit bus gives three-fourths, or 75%, toggle coverage, instead of 100%, as if the entire bus were considered as a whole.

**Code instrumentation.** To compute code coverage, the RTL code must first be instrumented to have a mechanism installed to detect statement execution and to collect coverage data. There are two parts in instrumenting RTL code. The first part adds user-defined PLI tasks to the RTL code so that whenever a statement is executed, a PLI task is called to record the activity. Using the example in Figure 5.20, the code can be instrumented as shown in Figure 5.24. The user PLI task $C1(), ... , $C5() is added to the RTL code. When the statements in block 1 are executed, $C1 is called. If $C2 and $C4 are called at the same

```
line  1: always @(posedge clock)
line  2: begin
line  3:   $C1(); // instrumented PLI
line  4:   a = b + c;
line  5:   x = (y << 4);                              block1
line  6:   if (a > x)
line  7:   begin
line  8:     $C2(); // instrumented PLI
line  9:     $C_exp(y,b,a); // y = b & a;    block2
line 10:     x = (x >> 2);
line 11:   end
line 12:   else begin
line 13:     $C3(); // instrumented PLI
line 14:     b = b ^ c;                               block3
line 15:   end
line 16:   if (x == y) begin
line 17:     $C4(); // instrumented PLI
line 18:     c = y;                                   block4
line 19:   end
line 20:   else begin
line 21:     $C5(); // instrumented PLI
line 22:     y = a;                                   block5
line 23:   end
line 24: end // end of always
```

**Figure 5.24** Instrumented RTL code to collect coverage data

simulation time, then path $P_1$ of Figure 5.22 is traversed. The C/C++ code of PLI tasks $C1, $C2, and so on, record the simulation times when they are called and write out the results to a file for postprocessing. A fragment of code is shown here:

```
void C1()
{
    ...
    time = $tf_gettime(); // get simulation time
    C1_call_frequency[time].frequency++;
    // C1_call_frequency is a look-up table taking in
    // simulation time T and returns the number of times C1 is
    // called at time T.
    ...
}
```

At the end of the simulation, table C1_call_frequency is dumped out to a file. The file's contents may look like

```
routine:      time called:     frequency:
C1            1002             1
C1            1201             2
...
C2            1002             2
C2            1201             1
...
C3            1002             1
C3            2001             1
...
```

The second part of instrumenting creates a map that correlates the dumped output file to the filenames and line numbers of the RTL code. In Verilog, there are no means to get the filename and line number at which a user-defined PLI task is invoked; thus, a map is created to accomplish this goal. During the first part of instrumentation, when the PLI tasks are inserted, the filename and the line number of the insertion point are recorded in the map. Thus, a map contains data similar to

```
routine:      file name:      line number:
C1            cpu.v           3
C2            cpu.v           7
...
```

With this map, the calling frequencies of statements can be computed, from which block, path, state, transition, and other coverage metrics are derived. To evaluate expression coverage, the expression is passed to a PLI task and it is evaluated in C, through which the expression's coverage is computed. For example, the expression on line 9, $y = (b \text{ \& } a)$ is replaced by PLI task $C\_exp(y,b,a)$, which takes on the values of variables $a$ and $b$, computes the expression coverage, and returns the value of $y$. A postprocessor combines the dumped output and the map to create a coverage report.

**Performance and methodology.** After RTL code is instrumented, the instrumented code, instead of the original code, is simulated. Because PLI calls are involved in measuring code coverage, simulation performance is significantly decreased. It is not uncommon to see a 100% decrease in simulation speed once coverage measurement is turned on. Let's look at several techniques to mitigate this problem.

For the code that the user does not want to measure coverage, or code that has been measured already, or error detection code that normal operations should not invoke, turn off the code with a pragma. Although the syntax of the pragma varies from tool to tool, a general form is shown here. The pragma is in the form of a stylized comment and it has no effect on other tools. Some coverage tools automatically ignore code marked by pragmas of synthesis tools, such as `synthesize_off`.

```
// coverage_tool off
// the following code is to be ignored
always @(posedge clock)
    begin
        ...
    end
// coverage_tool on
```

Ordering tests according to the parts of the circuit they cover improves performance. Coverage areas by each test can be represented using a Vann diagram. Tests with coverage that is subsumed by other tests should be eliminated. One way of ordering tests is to start with the test having the largest coverage area, followed by the test that has the most incremental change to coverage. When running a set of tests, the covered portions of circuit can be turned off progressively using pragmas so that a new test only measures the code that has not been covered by previous tests. When the code of a design is changed incrementally, we would identify the part of the design that is affected and update the coverage only for that part of the design.

A well-designed coverage tool should exhibit progressive performance improvements over a simulation run. During a simulation run, once a statement is covered, it does not need to be measured later in the simulation. For example, the PLI calls responsible for the statement can be disabled. Therefore, as coverage gradually reaches a plateau, simulation performance should gradually improve. At the coverage plateau, simulation speed should approach that without coverage measurement.

Finally, performance can be improved by simulating several tests in parallel and by combining the results. This technique is particularly useful in profiling tests on their coverage scope and using the profiling data to trim redundant tests and order tests.

## 5.6.2  Parameter Coverage

A major shortcoming of code coverage is that it only shows whether code is exercised and provides only an indirect indication of correctness and completeness of the operation of the implementation. As discussed earlier, it is possible to have 100% code coverage on a completely wrong implementation. To gain further insight into operational correctness, parameter coverage is introduced to measure functional units. To motivate the need for parameter coverage, consider verifying a stack. Testing the pushing and popping of an item to a stack is only a point in the operational space of the stack, because the stack has a depth greater than one. A more robust verification will also test the stack at various depths, and when it is empty and full. Generalizing this rationale, we first enumerate all parameters in a functional unit and then we measure the extensiveness that the parameters are stressed. This measure is parameter coverage. In a sense, a high parameter coverage indicates that a particular test of an operation or function is expanded to various possible values of its parameters. In the example of a stack, a parameter is stack depth, and the parameter coverage records the depths encountered throughout a simulation. Hence, a good simulation effort should cover all depths, including empty and full, of the stack.

As another example, consider a 32-bit adder. A parameter is the width of the operands. A test with high parameter coverage operates the adder with operands of different widths. The parameter of operand width does not necessarily increment by one. If the 32-bit adder is constructed from 4-bit adders, then the parameter increments by four, meaning that the 32-bit adder can be verified by verifying the building block, 4-bit adder, and treating each unit of 4-bit adder in the 32-bit adder as a black box, and verifying the interconnection of the black boxes. This brings out the notion of equivalent tests. In this example, the tests verifying only the building blocks of the 4-bit adder are equivalent. If equivalent tests can be derived, the verification effort is drastically reduced. However, there are no known algorithms to deduce equivalent test classes. Obtaining equivalent test classes requires ingenious engineering effort.

In summary, to get parameter coverage, the following steps are suggested:

1. For each functional unit, enumerate all parameters.
2. Determine the ranges of each parameter.
3. Set up monitors to record the value of each parameter.
4. Compute the ratio of the exercised parameter values to the range of the parameter. The collection of these ratios is the parameter coverage.

---

**Example 5.4**

Parameter coverage for a three-queue, two-server system.

To illustrate the previous procedure, consider a client–server system with three queues and two servers, as shown in Figure 5.25. Clients can arrive from any of the three sources and they wait in any of the queues. The two servers retrieve requests from the three queues. If a queue is empty, the server moves to the next queue. Let's assume that a request can have a processing time varying from three to six units, and the queue depth is seven. First we need to identify all parameters in this system. The parameters are filled depths of the three queues, server condition (disabled or enabled), and request processing times. These parameters form a vector, as shown in Table 5.8. The first parameter has a total of four combinations; the second, four; and the third, 512.

*continues*

**Example 5.4    (Continued)**

Monitors are then set up to record the values of the three parameters. For server status, 2 bits record the four possible states. For each task taken from a queue, its processing time is compared with previous recorded processing times. If it has not been recorded, record it; otherwise, discard it. This can be done by recording request processing times into CAM. Data sent to CAM are compared with the contents in memory. If they are already there, the data are discarded; otherwise, they are saved. Alternatively, task processing times can simply be saved to memory to be dumped out for later coverage analysis, which eliminates duplicates. For queue filled depth, ideally all queue configurations encountered should be recorded. For simplicity, we assume that the queues are independent and we only record the maximum depth encountered for each queue.

At the end of a simulation, the recorded parameters are collected and compared with the ranges of parameters. Their respective exercised percentages are the parameter coverage. For example, if a particular simulation yields two server configurations (S1 on, S2 off) and (S1 on, S2 on); processing times of 3, 4 and 5; and the maximum queue lengths are (6,4,8), then the parameter coverage is server, 50%; processing time, 75%; and queue length, 75%, 50%,100%.

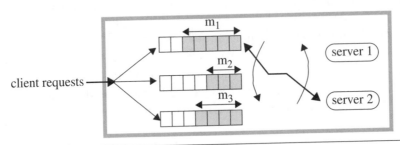

**Figure 5.25** Determine parameter coverage for a 3/2 client–server system

**Table 5.8**    Parameter Coverage Matrix

| Parameter | Range |
|---|---|
| **Server** | $\{(S_1,S_2): S_1, S_2$ is either on or off$\}$ |
| **Task processing time** | $\{3,...,6\}$ |
| **Queue length** | $\{(m_1,m_2,m_3): 0 \leq m_i \leq 7, 1 \leq i \leq 3\}$ |

### 5.6.3   Functional Coverage

Both code and sometimes parameter coverage operate on a particular implementation of the design. Hence, they are only as good as their implementation. Ideal coverage should depend only on specifications, so that it provides information on how well the specifications are implemented correctly. A specification consists of a set of functions to be implemented; functional coverage measures implementational completeness and correctness of this set of functions. A key (and the most difficult) step is to obtain as complete as possible a set of functions from the specifications. Adding to this difficulty is that many difficult bugs come from ambiguous specifications or nonspecifications. As a result, functions derived from specifications become better and better defined as the verification progresses. Once a functional list is in place, monitors are created to detect whether the functions are verified. Then, functional coverage is the percentage of verified functionalities.

To illustrate functional coverage and its calculation, consider verifying a cache system for a two-processor network. As shown in Figure 5.26, each processor has its own level 1 look-through cache, which in turn connects to the main memory through the system bus. A cache reduces data retrieval and storage latency by allowing a processor to interact with the cache memory, rather than with the slower main memory. A requirement for caches in multiple-processor systems is that data in all caches must be coherent. This requirement can be violated if a processor reads from and writes to its cache without interacting with other caches. An instance of cache incoherence occurs when processor A writes to its cache without updating the cache of processor B. The cache coherence protocol ensures consistency among caches.

Let's look at some functions that an MP cache system must possess. First we need to set up monitors to record the invocation of these functions. To understand this topic better, we need to discuss states of cache, the snooping process, and the memory updating process. An entry, called a *data block*, in a cache can have three states: exclusive, shared, and

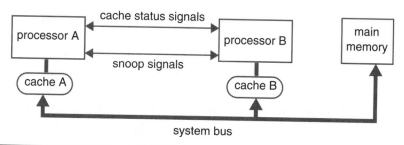

**Figure 5.26**  Cache coherence for a multiple-processor system

modified. A data block is in-state exclusive if no other caches have it. It is shared if another cache also has the data. It is modified if the data were modified, and thus is considered different from the copy in main memory.

When processor A initiates a read and registers a miss in its cache, cache A, it requests processor B, through snoop signals, to check cache B. If cache B also does not have the data, processor B notifies processor A and main memory. The data are then retrieved from main memory and are shipped to processor A. At the end of read, these data are stored in cache A and are marked as exclusive—meaning, no other cache has these data.

Suppose that processor B now initiates a read on the same block of data. It will register a miss on its cache, cache B. So, it activates the snoop signals to processor A and a data fetch command to main memory. Now processor A responds with a hit. Because the data block has not been modified, the data are transferred to processor B, and the data in cache A change state to shared. At the end of the read cycle, processor B puts a copy of the data in cache B and marks it shared.

If processor A now modifies the data block and finds it in its cache, and because the data are in the shared state, it will first notify processor B to invalidate the copy in cache B. If processor B is not notified, cache A and cache B contain different data at the same address—a cache incoherence. Processor B must be notified before processor A modifies, because there is a window of possibility that processor B will read that data block while processor A is modifying it. After processor B responds with a success of nullifying the data block in cache B, processor A writes the data block in cache A and marks it modified.

Now if processor B requests a read on the data block, it will register a miss because its cached copy was nullified. So it asserts the snoop signals to processor A and at the same time sends a read request from main memory. This time, processor A sees a modified copy. Next, processor A responds with an abort signal to main memory to stop it from shipping the data to processor B, then it signals processor B to read after main memory finishes updating. Then processor A transfers the data block from its cache to main memory and changes the data block state to shared. After this is done, processor B reinitiates a read and gets a hit in cache A. The data are then shipped to processor B and are marked shared.

Suppose processor B requests a write instead. Before processor B starts the write cycle, processor A writes back its data to main memory and invalidates its cache copy. Next, processor B writes back to main memory. If the entire block is to be written by processor B, processor A simply invalidates its cache copy without flushing it to memory, because the block will be totally overwritten anyway.

These functions, summarized in Table 5.9, are required of MP caches. In Table 5.9, the data block is represented by D and its state is in parentheses. The first column lists the functions; the second column, the cache contents; the third column, the actions; and the last column, the resulting cache contents. For example, function 1 is that processor A initiates a read on D, but D does not exist in both caches. So the action is that processor A reads from main memory and marks it exclusive. The last column shows that cache A now has D in the exclusive state.

The processor and cache labels can be systematically replaced to create other symmetric scenarios. For example, function 1 has the symmetric counterpart of processor B read D, and both cache A and cache B have no D. For each of the functions in Table 5.9, a monitor is created to detect whether the operation of the function is stimulated. For function 1, the monitor reads in commands from processor A. If the command is READ D, the monitor

**Table 5.9**   Functions to Cover in a Multicache System

| Processor | Cache | Action | Result |
|---|---|---|---|
| 1. Processor A reads D | Cache A:<br>Cache B: | Read from main memory and mark it exclusive | Cache A: D (exclusive)<br>Cache B: |
| 2. Processor B reads D | Cache A: D (exclusive)<br>Cache B: | Copy D to cache B and make both copies shared | Cache A: D (shared)<br>Cache B: D (shared) |
| 3. Processor A writes D | Cache A: D (shared)<br>Cache B: D (shared) | Invalidate all copies of D, update cache A, and mark it modified | Cache A: D (modified)<br>Cache B: no D |
| 4. Processor B reads D | Cache A: D (modified)<br>Cache B: | Deposit D of cache A to memory, copy D to cache B, mark both shared | Cache A: D (shared)<br>Cache B: D (shared) |
| 5. Processor B writes D | Cache A: D (modified)<br>Cache B: | D of cache A is invalidated and deposited to memory if partial block is overwritten, then processor B writes to memory | Cache A:<br>Cache B: |

checks whether D is absent in both caches. If it is indeed absent from both caches, the monitor records that this function has been stimulated. A fragment of a sample monitor for detecting function 1 is shown here. Note that this monitor does not check whether cache A has D in the exclusive state at the end of the transaction, because this monitor is only for functional coverage. However, an assertion monitor would then check the result. Task monitor_function1 takes a 32-bit command and outputs whether function 1 is stimulated. First it checks whether it is a READ opcode. If it is, it takes out the address portion and compares it with the addresses in the two caches by looping through all entries in the caches. For each entry, it determines whether it is valid. If it is, it compares the address with the tag. If there is a match, the data are in the cache. After the comparisons, if the data are absent in both caches, function 1 is invoked:

```
task monitor_function1;
input [31:0] command;
output invoked;

reg ['ADDR:0] address;
reg hit_A, hit_B;
integer i;
begin
    hit_A = 1'b0;
    hit_B = 1'b0;
    if (command['OPCODE] == 'READ) begin
        address = command['OPERAND];
        for (i=0; i<='CACHE_DEPTH; i=i+1) begin
            if (cache_A[i][valid] == 1'b1) // valid data
                if (cache_A['TAG] == address['TAG])
                    hit_A = 1'b1;
        end // end of for loop

        for (i=0; i<='CACHE_DEPTH; i=i+1) begin
            if (cache_B[i][valid] == 1'b1) // valid data
                if (cache_B['TAG] == address['TAG])
                    hit_B = 1'b1;
        end // end of for loop

        if ( hit_A == 1'b0 && hit_B == 1'b0)
            invoked = 1'b1;
        else
            invoked = 1'b0;
    end // end of if (command =...
end
endtask
```

Monitors for the other functions can be similarly constructed. For example, to detect function 2, the task has to make sure that cache A has a hit, that the state of the block is exclusive, and that cache B has no hit.

With these monitors implemented, simulations are run and data from the monitors are collected at the end. The ratio of stimulated functions over all functions is the functional coverage for the simulation.

**Functional verification object.** This method of identifying functionality and creating monitors to detect their stimulation can be generalized and cast in the light of object-oriented programming (OOP). A functional verification object is associated with a functionality to be verified. Within such an object is a generalized state variable, which is a set of signals, variables, states, or a combination thereof. Associated with the object is a coverage method, which can be a function or task, that detects execution of the functionality by constantly monitoring the values or sequence of values the state variable takes. A functionality is defined in terms of a particular value or sequence of values of the state variable. Thus, when the state variable attains the value or sequence of values, the functionality is invoked. Sometimes an assertion for the functionality is also associated with the object to determine whether it passed or failed.

Using the previous MP cache protocol as an example, the generalized state variable consists of command, cache_A, and cache_B. State function1 is reached when the opcode portion of command is READ, and neither cache_A nor cache_B has a tag equal to the address portion of command. When state function1 is reached, function1 is invoked. The task monitor_function1 is the associated coverage method. The following module is code for a functional verification object for the MP cache protocol with a coverage method that detects the invocation of function1. Note that because memory cannot be passed to a module, hierarchical references to memory, cache_A, and cache_B are used:

```
module verification_object (command);
input [31:0] command;
reg invoked;

// generalized object consists of
// reg [31:0] command and memory cache_A and cache_B
// Because memory cannot be passed to module,
// hierarchical references are used.

monitor_function1(invoked); // monitor function1 invocation
if (invoked == 1'b1) $display ("function1 has been called");
```

```
// coverage method
task monitor_function1;
output invoked;
...
reg ['ADDR:0] address;
begin
   ...
   if (command['OPCODE] == 'READ) begin
      address = command['OPERAND];
      for (i=0; i<='CACHE_DEPTH; i=i+1) begin
         if (top.cpu.cache_A[i][valid] == 1'b1) // valid data
            if (top.cpu.cache_A['TAG] == address['TAG])
               hit_A = 1'b1;
      end // end of for loop


      ...

      if ( hit_A == 1'b0 && hit_B == 1'b0)
         invoked = 1'b1;
      else
         invoked = 1'b0;
   end // end of if (command =...
end
endtask

endmodule
```

To use the functional verification object, just instantiate the module:

```
verification_object m1 (command);
```

Unfortunately, Verilog does not allow creation of objects and calling of methods in the standard OOP way. Therefore, Verilog can only simulate the OOP effect with a module. An alternative is to use a language other than Verilog, such as Vera, to create the objects.

Organizing functional coverage using verification objects materializes functionality enumeration and encapsulates coverage and assertion methods with objects. It therefore makes functional verification more manageable and reusable.

### 5.6.4   Item Coverage and Cross-Coverage

Item coverage and cross-coverage are commonly used terms in verification. Item coverage is a scalar quantity that measures how a single entity (such as a variable or parameter) of a circuit has been exercised. An example of item coverage is the utility percentage of the input queue of a network interface. When multiple item coverages of a design are in use, they together form cross-coverage. Cross-coverage is vector coverage with components that are the items. For example, two items—utility of input queue and output queue—form a two-dimensional entity (`input queue, output queue`). The coverage for this new entity is cross-coverage. Intuitively, item coverage measures the exercised portion of a one-dimensional axis, whereas cross-coverage, made of $n$ items, measures the exercised portion of the $n$-dimensional space with points that are the vectors, with components being the items. In short, cross-coverage is a Cartesian product of item coverage. Obviously, cross-coverage fills in more "holes" than the sum of its item coverage.

The components of cross-coverage need not be distinct entities. They can be the same entity sampled at different times. For example, if an item is the instruction of a processor, then cross-coverage can be formed by having its $i$th component be the $i$th instruction. This cross-coverage then measures the thoroughness of instruction sequences exercised.

## 5.7 Summary

In this chapter we first illustrated the interplay among the three verification entities: test plan, assertion, and coverage, and how they contribute to design controllability and observability, and enumerating test corner cases. Next, three levels of hierarchical verification were presented as a way to handle large designs. We then studied test plan generation, which consists of extracting functionality from architectural specifications, prioritizing functionality, creating test cases, and tracking progress. In particular, we used a state space method as a guide to enumerate corner cases systematically. As a complement to directed test generation, we studied the structure of pseudorandom test generators by focusing on a generic random code generator. The next major section emphasized assertion. First, the four components of an assertion were illustrated. Then, common assertion classes were discussed in two categories: combinational and sequential assertions. In combinational assertions we discussed checking for signal range, unknown value, parity, signal membership, and one-hot/cold-ness. In sequential assertions we covered assertions on time window, interval constraint, and unclocked timing requirements, and concluded with container assertion. We then looked at SystemVerilog assertions and discussed sequence construction and operators, along with built-in system functions. In the final major section, we studied verification coverage in three stages: code coverage, parameter coverage, and functional coverage. We illustrated each type of coverage and compared their advantages and disadvantages.

## 5.8   Problems

1. A one-ho multiplexor requires that the signals at its select lines be one-hot. If a selection is not one-hot, the output depends on the specific implementation of the multiplexor. In this exercise, let's assume that if a selection is not one-hot, the input corresponding to the most significant nonzero select bit becomes the output. For instance, in a 4:1 multiplexor, inputs are labeled as 1, 2, 3, and 4. If selection is 0100, then input 2 is connected to the output. If selection is 0101, input 2 is also the output, as a result of this implementation. Consider the circuit in Figure 5.27. Suppose the select signal S of the shaded multiplexor is not one-hot as a result of a design error. For example, to select input 3, S has value 0011, a binary number, instead of 0100.

   a. Define error detection latency (EDL) to be the number of cycles that must have passed before an error is observed during verification. When is the minimum EDL for this error seen at the output? What is the maximum EDL?
   b. How can you improve the EDL of this error by adding an assertion so that both the minimum and the maximum EDLs are equal to zero?

2. In this problem, let's consider a grayscale JPEG image processor. First, an image is partitioned into blocks of 8 × 8 pixels. Then each block is transformed and quantized separately. Finally, the code of the blocks is encoded and compressed to produce JPEG code. A JPEG processing system consists of three major components: a discrete cosine transform (DCT), a quantizer, and an encoder (Figure 5.28A). The DCT components take an 8 × 8 patch of image and compute its two-dimensional DCT. The result, a vector of 64 real coefficients, is then quantized to the bit capacity of the processor. The code from the blocks is then encoded and compressed in the zigzag diagonal order shown in Figure 5.28B. For example, a 1280 × 1024 image has 160 × 128 (or 20,480) blocks, each of which is a vector of 64 numbers, for a total of 1,310,720 numbers. During the encoding and compression stage, consecutive zeros are encoded with run length encoding (for example, seven consecutive zeros is written as 07). The rest is encoded with Hoffman coding, which assigns the shortest codes to the most occurring patterns subject to the so-called prefix constraint. Of the three components, only the quantizer is lossy, meaning output information content is reduced. Assume that we want to verify this JPEG processor hierarchically. Determine which of the specifications belong to the JPEG system or its components.

   a. The total picture quality must achieve at least 99.5% fidelity.
   b. The average file size of a 1280 × 1024 image is 1MB.
   c. Processing time for a 1280 × 1024 image is less than 10 msec.

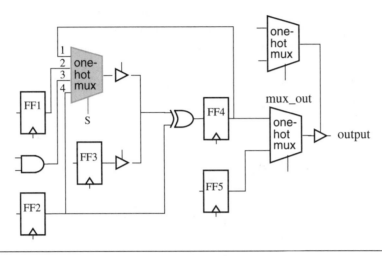

**Figure 5.27** Bug detection latency in a one-hot multiplexor circuit

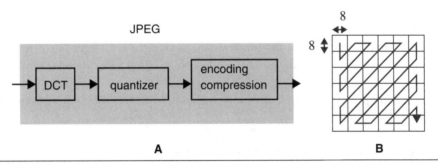

**Figure 5.28** (A) JPEG processing components (B) JPEG encoding order of blocks

3. Using the state space approach, derive a test case in category 3, $S \rightarrow S|_{\bar{i}}$, for the previous JPEG system. Then derive a test case that verifies an instance in each of the following categories: error handling (category 2), power-on test (category 5), and normal operation (category 4).

4. Write a test plan for the previous JPEG system. Use the test cases from problem 3.

5. To verify the JPEG processor, one can feed it random pixel values, compute the expected JPEG file contents, and compare the computed contents with those produced by the processor. For simplicity, let's apply this methodology to verifying the DCT component. We first generate random pixel values, perform DCT on the values, and compare the

resulting coefficients with those from the DCT design. This is called *self-checking*, and all these computations are done in a C/C++ routine that is run concurrently with the simulation of the DCT design and communicate via a PLI interface.

a. Write a C/C++ program that automatically generate blocks of $8 \times 8$ random values between 0 and 1024. For each block of the pixels, compute the DCT coefficients as shown here:

$$DCT(i, j) = \frac{1}{\sqrt{2N}} K_i K_j \sum_{x = 0}^{N-1} \sum_{y = 0}^{N-1} P(x, y) \cos\left(\frac{(2x + 1)i\pi}{2N}\right) \cos\left(\frac{(2y + 1)j\pi}{2N}\right)$$

where $K_i$ is equal to $\frac{1}{\sqrt{2}}$ if x equals 0, and $K_i$ equals 1 if x > 0. P(x,y) is the value of the pixel at (x,y).

b. Design an RTL module that computes DCT.

c. Using PLIs, compare the DCT coefficients from (a) with those from (b) at the end of every block. Whenever a block is done, the next block of pixel values are retrieved from the program in (a).

6. Write a Verilog assertion that the input signal to an encoder is one-hot whenever the encoder is enabled. The input signal can take on any value if the encoder is not active.

7. Write a Verilog assertion that ensures the data written to a memory contain no unknown values.

8. Write an assertion that at most one driver is driving the bus at any time.

9. The Hamming distance between two vectors is the number of bits that differ. Write a Verilog assertion that enforces a Hamming distance of 2 between all states in a finite-state machine.

10. Write a Verilog assertion that issues a warning when the time a device possesses a bus exceeds eight cycles.

11. Write a Verilog assertion that a signal must change within a time window specified by two events, assuming the delimiting events do not align with the signal transitions.

12. Write an assertion on a FIFO that issues a warning when it is full, and a push command is issued.

**13.** Write an assertion that ensures the data written to a memory are not corrupted. That is, data read are identical to data written.

**14.** Based on Figure 5.29, find a SystemVerilog sequence such that

    a. Waveform A satisfies
    b. Waveform B satisfies
    c. Both waveforms satisfy

**15.** In a PCI bus, a deasserted IRDY# means the bus is idle and an asserted GNT# means it has bus acquisition. If GNT# is asserted, the master asserts FRAME#, which is active low, to start the transaction immediately. Coincident with the FRAME# assertion, the initiator drives the starting address onto the address bus. One cycle later, IRDY#, which is active low, is asserted. Write a concurrent assertion to ensure that the temporal relationship among IRDY#, FRAME#, and GNT# is correct.

**16.** For the waveforms shown in Figure 5.30, determine which of the following sequences are satisfied. If satisfied, specify the times when they are satisfied. Assume the current time is time 1.

    a. $S_1 = \#\#1\ x_2\ \#\#[1:3]\ (x_1 + \overline{x_2})\ \#\#2\ \overline{x_4}$
    b. $S_2 = \#\#1\ (x_1 x_2)[*1:2]\ \#\#1\ (\overline{x_3})\ \#\#1\ \overline{x_2}$
    c. $S_3 = x_4\ \#\#[0,3]\ x_3\ \#\#1\ \overline{x_2}$
    d. Can we have $(S_1\ \textbf{intersect}\ S_2)$? Why?
    e. When is the finish time of $\textbf{first\_match}(S_2)$?
    f. $S_3 \mathrel{|\text{-}>} S_1$, excluding vacuously true case when $S_3$ is false
    g. $(x_1\ \#\#1\ \overline{x_3})\ \textbf{within}\ S_1$
    h. $S_2$ ended $\#\#2\ (x_2 + x_4)$

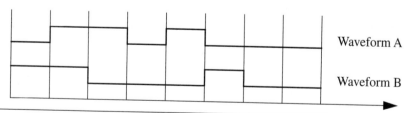

**Figure 5.29** Waveforms for SystemVerilog sequences

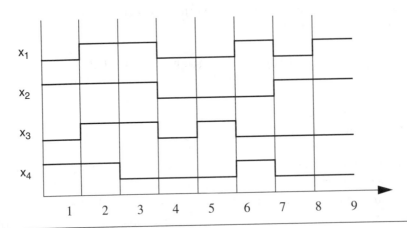

**Figure 5.30** Waveforms for evaluating sequences

**17.** Write a SystemVerilog assertion for the property that vector S is always even parity and the Hamming distance among its feasible values is 2.

**18.** Determine the statement, block, and path coverage of the following code after it is simulated for three clock ticks:

```
initial begin
    x = 0;
    y = 8'b101111;
    p = 8'b01111101;
    count = 1;
    shift = 1;
    mask = 8'b1111001;
end

always @(clock) begin
    x = y & p;
    repeat (5) begin
        if (x > 8) begin
            count = count + 1;
            x = shift >> 1;
        end
        else begin
            count = count - 2;
            x = x + 2;
        end
```

```
        end
    if ( x < 7 )
        step = count & mask;
    else
        step = count ^ mask ;
    @(clock)
    x = x + 2;
    @(clock)
        if (x < 8) begin
        y = x + 2;
        p = mask & y;
        step = count + 3;
    end
end
```

19. Compute the expression coverage for the following expression in a run that has encountered these values for a, b, and c: $(1,1,0)$, $(0,0,1)$, $(1,0,1)$, and $(0,1,0)$.

$$((a \ ? \ \overline{b} \ :c) \ \verb|^| \ (a \ b + \overline{a} \ \overline{c}))$$

20. In this exercise, let's consider a Huffman encoder. The Huffman algorithm uses variable-length code to encode and compress. It assigns code with lengths that are inversely proportional to the frequencies of the symbols (for example, the shortest code to the symbol with the highest occurring frequency). To code the sentence "finding all bugs is ideal," the frequencies of the letters are first determined, and they are space(4), i(4), l(3), n(2), d(2), g(2), a(2), s(2), f(1), b(1), and e(1). Then the two symbols with lowest frequencies are merged to form a new symbol with a frequency that is the sum of the two. Once merged, the two symbols are removed from the set and the new symbol takes their place. This process is repeated until only one symbol is left. Figure 5.31 depicts the process on the example sentence. The numbers on the edges in the first two graphs denote the order in which the nodes are combined. In the last graph, the left edges are assigned 1s and the right edges are assigned 0s. Now the code for the letters are read off the paths from the root to the letters. For example, letter i is encoded as 110, f as 01110.

   a. Define the parameters for parameter coverage of a Huffman encoder. A good parameter affects the algorithm and hence affects its output. Give an example of a good and a bad parameter.
   b. Define the operational features for functional coverage of a Huffman encoder. An operational feature or functional feature is a property that must or must not hold if the algorithm operates correctly. (Hint: Consider relationships between sentence characteristics and coding properties, such as letter frequency and code length.)

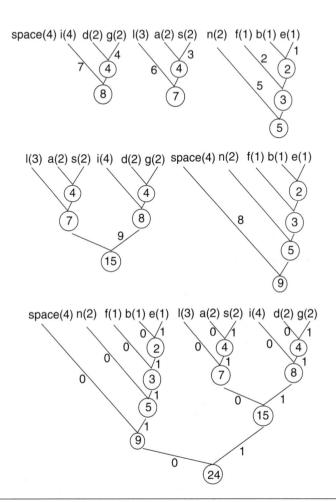

**Figure 5.31** Huffman encoding process

## CHAPTER 6

# Debugging Process and Verification Cycle

## Chapter Highlights

- Failure capture, scope reduction, and bug tracking

- Simulation data dumping

- Isolation of underlying causes

- Design update and maintenance: revision control

- Regression, release mechanism, and tape-out criteria

The ultimate goal in setting up a test bench, adding assertions and monitors, and running simulations is to find bugs. A typical cycle of bug hunting consists of detecting bug symptoms, investigating the root causes, fixing the bugs, updating the design, and preventing the bug from appearing in future revisions. In this chapter we will study methodologies and procedures practiced at various junctions of the bug-hunting cycle.

The first hint of a bug comes when simulation outputs differ from expected results. When this happens, the symptoms should be preserved and made reproducible for debugging. To accomplish this one must capture the environment producing the bug and register the bug with a tracking system for bookkeeping purposes. To make debugging easier, an effort should be made to reduce the circuit and test case size while preserving the error. Then, simulation data are dumped out to trace the root cause of the problem. In a large design, it is common to have a bug appear after days of simulation, so it is not practical to record every node value for the entire simulation interval. The crux is deriving an optimal strategy for recording just enough data. Ideally, data should start to be recorded at a time as close as possible to and before the time the problem is triggered, not observed. In this chapter we will study the check pointing strategy. When a bug is detected, a mechanism is needed to keep track of its status. Next, dumped data are analyzed to determine the root cause of the error. We will look at the key concepts, techniques, and common practices in debugging. After a bug is fixed, a policy and methodology should be imposed to ensure that the updated design is visible to the whole team and to prevent the same bug from recurring in future revisions of the design. We will conclude our study with a discussion of regression, computing grid, release, and tape-out criteria.

## 6.1   Failure Capture, Scope Reduction, and Bug Tracking

All the facilities discussed in previous chapters, such as assertions, monitors, and test cases, are used for the same goal of revealing design errors, by enhancing observability and expanding input space. When an error surfaces, three actions are in order before debugging can begin. First, the error must be made reproducible; second, the scope of problem should be reduced as much as possible. The scope of the problem can be either the portion of the circuit and test case that may cause the error, or the time interval in which the error is first activated. Third, the error should be recorded to track its progress. The first and the third actions are essential; the second action is necessary only for large simulation tasks.

### 6.1.1   Failure Capture

To capture an error, the minimum environment necessary to reproduce the error should be preserved. The environment includes the version of the design, test bench, test case, simulator, and host parameters. Host parameters consist of OS version, physical memory, environment variable setting, and machine (such as workstation, personal computer, or a computing grid) characteristics such as speed and type. As an example, in the UNIX environment, environment variables are a component of the simulation environment and can be captured by the UNIX command `printenv`. Software programs, such as simulators, also have settings to be saved, which often are in an `rc` file (such as `simulator.rc`), which contains various parameter values of the simulator. These `rc` files should be saved as a part of the environment. When there are doubts regarding whether an environment parameter affects the error, save it. Saving the environment is only half the process; the other half is to provide a facility to restore the environment from the given environment parameters. A good practice is to design a pair of scripts—saver and restorer—one that saves the environment parameters and the other that restores them from a saved record. For example, if a design is under revision control (software that dictates what version of the design be visible), the saver script will inquire about the revision control of the version in use and will store the version number, whereas the restorer, when given the revision number, will pull out the required version of the design.

## 6.1.2   Scope Reduction

**Circuit reduction.** Scope reduction reaps the most benefit when it comes to debugging large systems. It is often the case that the simulation speed in a debug environment runs much slower than in a normal simulation environment, because of additional data visibility in the debug environment. Therefore, engaging directly in debugging without first reducing the scope of the circuit can prove to be unproductive. Scope reduction falls into two categories: reducing circuit and test case size, and limiting the simulation time. Reducing circuit and test case size removes as much as possible the parts of the circuit and test case while retaining the manifestation of the bug. To be effective, it requires insight into the design of the circuit and the content of the test case. However, sometimes it can be accomplished by trial and error. For example, an error occurring in an instruction fetch unit may mean that the I/O unit and the instruction decode unit can probably be eliminated from the circuit without affecting the error. However, there is no guarantee that these units can be safely discarded, because it could be that the error symptom surfaces in the instruction fetch unit but the root cause lies in the I/O and instruction decode unit (for example, erroneous instruction decoding causes an incorrect speculative instruction `fetch`). Therefore, every time a part of the design is eliminated, the simulation should be run to confirm that the nature of the error is preserved.

When a part of the design is cut, some severed lines (inputs) require that they be supplied with signal values that existed before the surgery. This is required because the rest of the design still gets input from the severed unit. To emulate the severed unit, all input values from the unit are recorded into memory during a simulation and are replayed from memory after the unit is removed. That is, during a simulation without the severed unit, the values at the interface are read from memory. As an example, before cutting out the I/O unit, the values across the interface between the I/O unit and the rest of the circuit are recorded for the simulation, as shown in Figure 6.1A. Then the I/O is removed, and memory storing the recorded values is instantiated as a part of the test bench and it supplies the values. In this example, during a simulation without the I/O unit, at time 20, value `1010` is read from

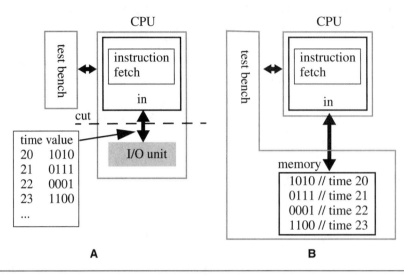

**Figure 6.1** Preserving I/O behavior with memory. (A) Record the I/O values around the unit to be cut. (B) Replace the eliminated unit with memory supplying the recorded I/O values.

memory to emulate the I/O unit. Similarly, at times 21, 22, 23, and so on, values `0111`, `0001`, `1100`, and so on, are output from the memory to the rest of the CPU. The following code demonstrates the part of the test bench emulating the I/O unit:

```
module testbench;

cpu cpu(.ioport1(stub1),.ioport2(stub2),...);
// the io ports, external ports, of I/O unit are stubbed.

// the following emulates the I/O unit
initial begin
    #20 testbench.cpu.in1 = 4'b1010;
    #21 testbench.cpu.in1 = 4'b0111;
    #22 testbench.cpu.in1 = 4'b0001;
    #23 testbench.cpu.in1 = 4'b1100;
    ...
end

// other parts of test bench
...
endmodule // end of test bench
```

**Test case reduction.** Reducing the design size can be regarded as spatial reduction. Its counterpart, temporal reduction, minimizes the test case while preserving the error. An error observed after the simulation has run ten million cycles will probably still show up by rerunning the simulation only during the last 10,000 cycles. If the same error can be reproduced with a shorter test case, debug time is reduced. Cutting down test cases requires an understanding of the functions of the test cases. If a long test case is made of independent, short diagnostic tests, then only the last diagnostic test is needed to reproduce the error. Furthermore, if the state of the design can be determined to be correct at a time, say $T$, before the time the error happens, then the error must be triggered after time $T$. Then the error can be reproduced by running the simulation from time $T$ only.

A brute-force method is to perform a binary search on the test case. If the error happens at time S, then run simulation from time S/2 to S, with the circuit initialized to the state at time S/2. If the error occurs, the length [S/2,S] is further cut in half and one repeats the procedure. If the error does not occur, the starting time is selected halfway through the interval [0, S/2]. This process continues until a reasonably short test case is obtained. This binary reduction is summarized here:

---

**Binary Test Case Reduction Algorithm**

1. Let $A$ be the maximum starting time that produces the error. Let $B$ be the minimum starting time that does not produce the error.
2. Initially, $A$ is zero and $B$ is the time the error was observed.
3. Run the simulation starting at time $N = (A + B)/2$ and initialize the circuit to start at time $N$.
4. If the error occurs, $A = N$; otherwise, $B = N$.
5. Repeat step 3 until $B - A$ is insignificant for further reduction.

---

When a test case is cut at time $T$, the state of the design must be initialized to the state at time $T$, which can be obtained by simulating the test case using a higher level and faster simulator, such as an architectural simulator from time 0 to $T$. Then the state at time $T$ is transferred to the design. For example, if a processor runs a C program, as shown here, and assumes that each statement takes the same amount of time to execute, then a search in time can be done equivalently on the statements:

```
if (x == y) record(a, b);
a[0] = v[0];
b[N] = w[N];
```

```
for (i=0; i< N; i++){
    prod [i] = a[i] * b[N-i]; // midpoint statement M
    ind[i] = v[i] + w[N-i];
}
x = prod(N/2);
y = ind(N/3);
function(a,b);
```

If statement M is the midpoint of the program when i = N/2, assuming N is even, to run from statement M for i ≥ N/2, the values of variables a, b, v, and w referenced when i ≥ N/2 must be computed a priori (for example, by executing the program up to statement M for i < N/2). The reduced test case is

```
//These values were computed a priori:
//a[0], ..., a[N], v[N/2], ..., v[N],
//b[0], ..., b[N], w[N/2], ..., w[0]
//ind[N/3]
for (i=N/2; i< N; i++){
    prod [i] = a[i] * b[N-i]; // midpoint statement M
    ind[i] = v[i] + w[N-i];
}
x = prod(N/2);
y = ind(N/3);
function(a,b);
```

**Check pointing.** Hard bugs often surface only after millions or even billions of cycles of simulation. To obtain a short test vector for debugging, states of the simulation are saved at regular intervals so that when an error occurs, the state last saved can be used to initialize a simulator, and debugging can start from that point in time instead of the very beginning of the entire simulation. Saving the state of a simulator or, in general, of a machine, is called *check pointing*. If the observed error had already occurred at the time of last check pointing, the next-to-the-last check point has to be used. The idea is to select the most recent check point before the error is activated.

Check pointing can be implemented using custom routines or check pointing commands built into the simulator. If a simulator has a check pointing command, it also comes with a command restoring a check point. Check pointing takes on two forms. One form is to save the state of the design, which includes states of all sequential components and memory contents. When saved in this form, simulation of the design can be resumed on any other simulator by simply initializing the state of the design to the saved values. To initialize, the user needs to write a task or function—in other words, the restore routine is user defined. The other form of check pointing saves the simulator's internal image of the design. For this kind of check point, the simulation can only be restored to a similar type of simulator. An advantage is that the restore routine comes with the simulator.

The interval to check point is a function of design complexity and tolerable simulation performance. The more frequent one "check points," the slower the simulation runs, but the closer a check point is to the error occurring time, which gives a shorter debugging time. A typical interval for a CPU design is around several thousand clock cycles. Saved check points use disk space and can become a problem for long simulation runs. In practice, only the few most recent check points are saved (for instance, the last three check points).

Another typical application of check pointing is seen during simulation on a hardware accelerator. A hardware simulator runs orders of magnitude faster than a software simulator and thus is used for finding bugs that hide deep inside a system's state space. Once a bug is detected, the state of the hardware simulator or the state of the design is check pointed and the image is restored on a software simulator for debugging. There are several reasons for check pointing from hardware to software. First, a hardware simulator is usually shared by several projects; hence, debugging on a hardware simulator occupies much time and should be strictly prohibited. Second, long simulation runs on a hardware simulator prevent one from dumping the entire simulation run for debugging or rerunning it on a software simulator. Finally, software simulators provide much more visibility to the internal circuit nodes than hardware simulators, making it easier to debug.

IEEE standard 1364 provides a pair of system tasks for check pointing: $save and $restart. $save("filename") system task saves the complete state of the simulator into a file named filename. On the other hand, $restart("filename") restores a previously saved state from file filename. $incsave saves only what has changed since the last invocation of $save, and hence has less of a performance impact. An example of check pointing using the IEEE commands is as follows:

```
initial
#1000 $save("full_state");

always begin
   // save incrementally every 10,000 ticks
   //but only the last two check points are preserved.
   #10000 $incsave ("inc1_state");
   #10000 $incsave ("inc2_state");
end
```

```
`ifdef START_FROM_SAVED
    // restore to the beginning of simulation
    initial $restart("full_state");
`endif

`ifdef START_FROM_LAST_SAVED
    // restore to the last check point of simulation
    // assume the last saved file is inc2_state
    initial $restart("inc2_state");
`endif
```

This code saves a full state at time 1000. Then it saves an incremental image every 10,000 cycles. Furthermore, only the most recent two check points are kept. Figure 6.2 shows the actions of the previous save and restart commands. When `$restart("full_state")` is executed, the state image saved at time 1000 is loaded into the simulator and the simulation runs from that time on. When `$restart("inc2_state")` is executed, the state image saved at time 41000 is loaded into the simulator and the simulation starts from there. To start the simulation at time 31000, `$restart("inc1_state")` is used. However, a simulation cannot start at time 21000 or 11000, because the saved states were already overwritten.

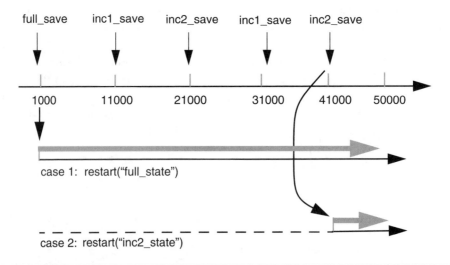

**Figure 6.2**   Save and restart from a state of simulation

### 6.1.3    Error Tracking System

Before an error is sent out for debugging, it should be entered into a tracking system. The purpose of a bug tracking system is to prevent bugs from being lost or not getting fixed in a timely fashion. Bug tracking software, which can be invoked by anyone from the project team, has the following five states: initial, assignment, evaluation, confirmation, and resolution. The states and their transitions are best described using a finite-state machine paradigm. Figure 6.3 shows a state diagram for a bug tracking system, or issue tracking system. When an error is first observed, the test or design engineer enters it into the system and the bug starts in the initial state. The data entered consist of the date the bug was discovered, the environment setup to reproduce the error, and a log file showing the error. Once the error is registered, the manager or project lead is notified. The manager takes the error to the second stage, assignment, where she prioritizes the error and assigns it to the responsible engineer. After this stage, the error has been tagged with a priority and the name of the responsible engineer. Then, the bug tracking system notifies the responsible engineer. When the engineer gets the notification, he moves the bug status to the evaluation state and investigates it. The engineer uses the setup environment and the log file to determine the nature of the error. If it is indeed an error, he uses the setup environment to reproduce the error, corrects it, documents the root cause, and transits the bug tracking system into the next state: confirmation. If it is not a real error (such as a user error), the engineer records an explanation and moves the bug to the confirmation state. At the confirmation state, the initiator reruns the test using the new fix from the design engineer to

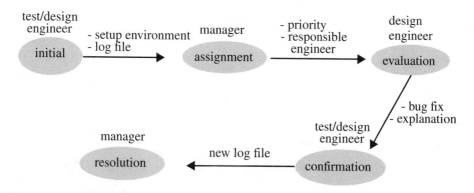

**Figure 6.3**    States and transitions of an issue tracking system

confirm the bug is indeed fixed. If it is, he affirms the fix, attaches the new log file, and moves the system to the resolution state. If it is not a real error, and the initiator agrees, the initiator advances the system to the resolution state. If, for any reason, the initiator disagrees with the design engineer, the design engineer is notified. In this case, the bug stays in the confirmation state. At the resolution state, the manager assesses the root cause of the bug and the fix. If it is satisfactory, the bug is marked as closed and the test case may be used for regression. Once a bug is entered into a bug tracking system, it cannot be removed by anyone. The only satisfactory outcome of a bug entered is at the resolution state.

Furthermore, bugs are prioritized according to their impact on the project's functionality, schedule, and performance. Bug priority tracks the functionalities in the project. A high-priority bug disables a high-priority functionality. There are bugs that are not design errors per se, but are features that customers prefer in the future. In this case, these "bugs" are also prioritized according to effort to implement and market demand, and they may be swept into a future enhancement category. An issue tracking system can sort entries by priority, assignee, status, date, or combinations thereof. A typical bug entry in an issue tracking system looks like that shown in Table 6.1.

**Table 6.1**    Sample Bug Entries in an Issue Tracking System

| Problem | Priority | Assignee | Date/Filer | Status | Comment |
|---------|----------|----------|------------|--------|---------|
| System hangs whenever INT and REQ are both active. test-case: /proj/test/ hang_int | 1 | J. Smith | 11.2.03/ M. Hunt | Evaluation | Error reproduced; it appears to be bus contention |
| Receiver counter is off by 1. test-case: /proj/test/ CNT | 2 | A. W. | 2.3.04/ N. Lim | Confirmed | Fixed |
| Cannot take 8 clients at the same time | 4 | M. Manager | 9.4.03 | Resolved | Future enhancement |

## 6.2   Simulation Data Dumping

Debugging can be done in interactive or postprocessing mode. In interactive mode, users debug on a simulator. They run the simulator for a number of cycles, pause it, examine the node or variable values in the circuit, continue simulating for a number of cycles, and repeat these steps until the problem is resolved. The following is a sample interactive debug session. The simulator was paused after running a specific number of cycles. When it paused, it entered into an interactive mode with prompt >. The commands typed by the user are in bold. The user first prints the value of the hierarchical node top.xmit.fifo.out. Then she sets the current scope to be top.xmit.fifo. Next she finds all loads driven by node out, for which the simulator returns three loads: in_pipe_a, x1_wait, and loop_back. Then the value of loop_back is printed. When data examination is completed at cycle 32510, the simulation resumes for 10,000 cycles:

```
...
simulation paused at cycle 32510.
> print top.xmit.fifo.out
> 32'h1a81c82f
> scope top.xmit.fifo
> find_load out
> in_pipe_a, x1_wait, loop_back
> print loop_back
...
> continue 10000
resume simulation from cycle 32510
...
```

In interactive mode, because the simulator does not save values from past times, the user can only look at values in the current and future cycles. In postprocessing mode, a simulation is first run to record the values of the nodes and variables into a file. Then debug software is invoked to read in data from the file. It displays the values of nodes and variables at the time specified. Postprocessing debugging relinquishes the simulator and hence more users can share the simulator. On the other hand, large amounts of data are dumped, costing disk space and time. In practice, almost all hardware debugging is done in postprocessing mode. In the following discussion we study data dumping in more detail.

### 6.2.1   Spatial Neighborhood

As discussed earlier, an effort should be made to reduce circuit size and shorten the test case to speed up debugging. After a circuit and a test case of manageable size are in hand, simulation values are traced or dumped to determine the root cause of the bug. Instead of dumping out values for all nodes at all times during the simulation, which has a drastic impact on performance, it is wise first to gauge the proximity of the bug in location and time, and dump out node values within the neighborhood and time interval. This initial judgment of the bug is an art that requires knowledge of the design, and intuition. The nature of the bug provides a hint regarding physical proximity, and the method of locating it resembles that of circuit scope reduction, but differs in that the neighborhood in this case can be arbitrarily small, whereas the scope in circuit reduction must satisfy the goal of preserving the bug. As debugging progresses, the dumping neighborhood can widen or shift. When a neighborhood moves, a simulation needs to be run to obtain data. If data dumped in several sessions can be merged, which depends on the format of the dumped data, then only the incremented scope needs to be dumped. The complete scope of data can be obtained by merging the previous incrementally dumped data. An optimal dumping neighborhood becomes clearer as experience grows.

### 6.2.2   Temporal Window

Determining the dumping interval, or window, can be done via trial and error or it can be derived from a statistical mean. For example, based on past bugs' data, one can compute according to the functional unit to which a bug belongs, the time interval between the moment a bug is triggered to the time it is detected, compute the average time interval for each functional unit, and use the average interval of a functional unit as the initial dumping interval for the bug currently under investigation. The functional unit in which the bug is presumed to reside may change as debugging progresses. If a bug leads to signal values outside the dumping interval, another signal dump simulation run is required. The new interval is further back in time and ends where the last time starts.

Figure 6.4 illustrates a typical dumping sequence dictated by the backtracking movement of a bug and the interaction of spatial neighborhoods and temporal windows. The two regions represent two neighborhoods. $N_1$ is the first dumping neighborhood. There are four signal dumps labeled on the arrows. The bug is first observed at node a at time interval $I_1$. Scope $N_1$ is dumped for interval $I_1$. As the bug is traced backward across multiple FFs, the time goes outside interval $I_1$ but remains within the same neighborhood. Thus, a second dump is executed for interval $I_2$. As backtracking continues to node b in the circuit, the signal goes outside neighborhood $N_1$. Thus neighborhood $N_2$ is dumped for interval $I_2$. Again, backtracking across multiple FFs moves the time past interval $I_2$, causing another dump for interval $I_3$. In interval $I_3$, the root cause of the bug is found.

The mechanism for dumping node values should be methodologically designed into the RTL. A basic feature of a dumping mechanism is the ability to select the dumping scope either at compile time or runtime. The scope can be a module, a task, or a function.

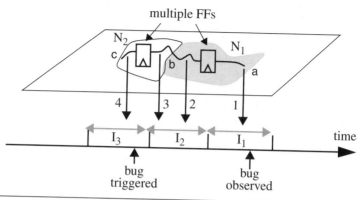

**Figure 6.4**   Signal dumping sequence in a debugging process

Compile-time dumping selection can be implemented using `ifdef` directives to select the desired dump commands. Runtime selection, on the other hand, can be implemented by scanning in runtime arguments. Many simulators have built-in system tasks for checking command arguments on the command line (for example, `$test$plusargs()`). Compile-time selection implementation runs faster, but the user must know what to dump at compile time. In the case when dumping is activated only when an error is detected, runtime selection is more convenient. Based on the nature of the error, an appropriate scope is selected to dump out its node values.

To dump out node values inside a scope, most simulators have built-in system tasks so that the user does not need to write a long `$display` statement for all nodes to be dumped. An example of a dump command is IEEE standard command `$dumpvar(1, module_name)`, which saves the values of all nodes at level 1 inside scope called `scope_variable` into a file which can be specified by `$dumpfile("filename")`. The following is an example showing a compile-time selection implementation and a runtime selection implementation. To invoke the compile-time selection, just define the corresponding variable, such as `define DUMP_IDU`. To invoke the runtime selection, use `plusarg` at the simulator command line, such as `simulator +dump_IDU+dump_L2`:

```
// compile-time selection
`ifdef DUMP_IDU
   $dumpvar(1, instruction_decode_unit);
`endif
`ifdef DUMP_L2
   $dumpvar(1, L2cache);
`endif
```

```
// runtime selection
if ($test$plusargs("dump_IDU") )
    $dumpvar(1, instruction_decode_unit);
if($test$plusargs("dump_L2");
    $dumpvar(1, L2cache);
```

Lightweight dumping records only select strategic variables and hence has a minimal performance impact while sacrificing full visibility. Lightweight dumping is beneficial for error scouting simulation runs when errors are not known to happen a priori. However, when errors appear, the information from the lightweight dump narrows the scope for a full-fledged dumping.

## 6.3 Isolation of Underlying Causes

In this section we will look at the basic principles and practices in debugging hardware designs. First we will discuss using expected results as a guide to trace errors, and then we will study how erroneous signals are traced forward and backward to locate their root cause. There are many forking points during tracing, and we will introduce a branching diagram to keep track of the paths. As sequential elements are traced, the time frame changes. We will consider the movement of the time frame for latch, FF, and memory. Next we will look at the four basic views of a design and their interaction in debugging, along with some common features of a typical debugger.

### 6.3.1 Reference Value, Propagation, and Bifurcation

Debugging starts with an observed error, which can be from a $display output, failed assertion, or a message from a monitor, and ends when the root cause of the error is determined. Before the user can debug, he must know what is correct (called the *referenced value*) and what is not. The definition of correctness is with respect to the specifications, not with respect to the implementation. For example, if a 2-input AND gate is mistakenly replaced by an OR gate and the inputs to the OR gate are 1 and 0, then the person debugging must know that the expected output should be 0 instead of 1. That is, the output value of 1 is incorrect even though the OR gate correctly produces 1 from the inputs. It should be stressed that this distinction between the correctness of the implementation and the correctness of the implementation's response be fresh in the mind of the person who is debugging. When one gets tired after hours of tracing, it is very easy to confuse these two concepts and get lost. The reference behavior is the guide for a designer to trace the problem. Without knowledge of the reference behavior, one will not be able to trace at all. The questions to ask in debugging are the following: What is the reference value for this node at this time? Does the node value deviate from the reference value? If the values are the same, tracing should stop at the node. If they are different, one would follow the drivers or the loads of the node to further the investigation. The value of a node that is the same as the reference value for the node is called *expected*.

The root cause is the function that maps expected inputs to unexpected outputs. Therefore, a module that accepts expected inputs but produces unexpected outputs must have an error in it. Using the previous example of a misplaced OR gate, the OR gate takes in the expected inputs, 0 and 1, and produces 1, which is unexpected. Therefore, the OR gate is the root cause. On the other hand, a module accepting unexpected inputs and producing unexpected outputs may or may not have an error. Furthermore, an error can have multiple root causes.

As debugging progresses, the reference behavior can take on different but equivalent forms. For example, if a reference value on a bus is 1010, then as we trace backward to the drivers of the bus, the reference behavior becomes the following: Exactly one bus driver is enabled and the input to the driver is 1010. Similarly, if we trace forward and see that this reference value is propagated to a decoder, then the reference value for the output bits of the decoder becomes the following: Only the tenth bit is active. Furthermore, a reference value can bifurcate and become uncertain, creating more possible routes to trace. A case in point is that the reference value of the output of a 2-input AND gate is 0, but the actual value is 1. Moving toward the inputs, the reference behavior bifurcates into three cases: either or both inputs are 0. To investigate further, you must assume one case of reference behavior to proceed. If you end up at gate or module with outputs that are all expected, the assumption is wrong. Then the next case of reference behavior is pursued. This phenomenon of uncertainty and bifurcation is the major cause of debugging complexity. Therefore, a key to effective debugging is to compute correctly the reference values during tracing.

## 6.3.2 Forward and Backward Debugging

There are two methods of debugging: forward tracing and backward tracing. Forward tracing starts at a time before the error is activated. Note that the effect of an activated error may not be seen immediately, but only after cycles of operation. Therefore, a critical step in forward tracing is finding the latest starting time before the error is activated, and there is no general algorithm to determine such times. Assuming that such a time is given, we must assume all node values are correct, or expected, and move along the flow of signals, forward in time, to get to the point when the error is activated. During this search, the first statement, gate, or block producing unexpected outputs contains a root cause. Besides finding a good starting time, another difficulty in forward tracing is knowing where in the circuit to start that will eventually lead to the error site. Figure 6.5 shows forward tracing paths for node B. The shaded OR gate is the root of the problem. Of the two forward tracing

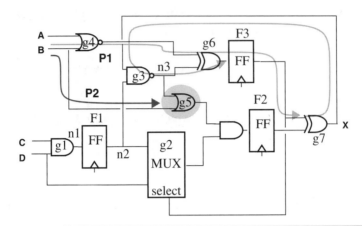

**Figure 6.5**    Forward tracing paths for node B

paths, one leads to the error site and the other does not. When we come to a node with multiple fanouts, we must decide which paths to pursue, and there are exponentially many such paths. The ability to locate the starting point and making wise decisions at multiple-fanout forks can only be acquired through understanding the design and the nature of the bug.

Backward tracing proceeds against the flow of signals and backward in time to find the first statement, gate, or block that yields unexpected outputs from expected inputs. Unlike the uncertainties faced in forward tracing, the starting site and time are the location and time the error was observed, and one moves backward toward the fanins. With this method, the error is an unexpected behavior. The person debugging must know the reference behavior and be able to translate the reference behavior as he proceeds backward. The major difficulty in backward tracing, shared with forward tracing, is that when a gate, statement, or a block has multiple fanins, a decision must be made regarding which fanin to follow next, and there are exponentially many possible paths. When a multiple-fanin gate is encountered, the path or paths to pursue are the ones that show unexpected values. However, it is often possible that several paths show unexpected values.  Figure 6.6 shows three backward tracing paths from node X.

### 6.3.3   Tracing Diagram
With either tracing method, the fanin and fanout points are branching points that require making decisions. If a selection does not turn up root causes, we need to backtrack to the decision points to select other paths. To keep track of what has been visited and what has not, a tracing diagram comes in handy. The branching points in a tracing diagram

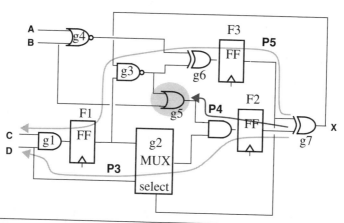

**Figure 6.6** Backward tracing paths from node X

systematically enumerate all possible selections and guide the selection decision in back-tracking. Tracing diagrams are usually generated by a software program instead of being created by hand.

A node in a tracing diagram is either a primary input, port of a gate, module, or user-defined node. A user-defined node is a net that terminates tracing, e.g. a known good net. An arrow from node A to node B means that there is a path from A to B. The path is a forward path in forward tracing, and is a backward path in backward tracing. A reduced tracing diagram contains only nodes with more than one outgoing arrow, except for primary inputs and user-defined nodes.

Figure 6.7 shows two reduced tracing diagrams: one for forward tracing from primary input B and the other for backward tracing from net X. The convention used here is that the input pins of a gate are numbered from top to bottom starting from 1. Outputs are similarly numbered. A node labeled as G.i represents the ith input of gate G in forward tracing, and the ith output in backward tracing. The rectangular nodes are user-defined nodes that, in this case, are the fault site. Fault sites are not known in advance in practice; they are shown here for illustration. The shaded nodes are primary inputs.

When obtaining a reduced forward tracing diagram, gates with only one fanout are not represented in the tracing diagram because these gates have only one outgoing arrow. Similarly, nodes having only one fanin in a reduced backward tracing diagram are not shown. Forward tracing starts from primary input B. At the outset there are two fanouts: g4.2 and g5.2. Thus, node B in Figure 6.7A has two branches: one leading to node g4.2 and the

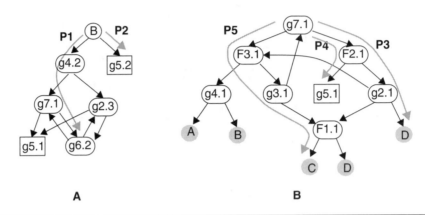

**Figure 6.7**   Tracing diagrams for forward tracing of primary input B (A) and for backward tracing of wire X (B)

other leading to node g5.2. The node inside the box, g5.2, is the root cause of the problem, and we assume that the debugging process ends when that node is reached.

If there are loops, the loop may be traversed several times. Each time a sequential element is crossed, the time frame may change. For instance, the loop in Figure 6.7A, consisting of g7.1 and g6.2 can be traversed multiple times, and each traversal advances the time by one cycle because the loop contains FF F3. Similarly, the loop in Figure 6.7B, consisting of g7.1, F2.1, g2.1, F3.1, and g3.1, contains two FFs, and therefore time retracts by two cycles whenever the loop is traversed once.

### 6.3.4   Time Framing

In tracing, when a combinational gate is traversed, either forward or backward, the current time of the simulation does not change. When a sequential element is traversed, the time of the simulation changes depending on whether it is forward or backward tracing. For example, when forward traversing an FF (such as from data to output), the time of the simulation advances by one clock cycle because the value at the output happens one cycle after the data input. On the contrary, in backward traversing (from output to data), the time of the simulation retracts by one cycle. Consider forward tracing from node n1 of Figure 6.5, and suppose the current time of the simulation is *N*. When we arrive at node n2, time advances to *N* + 1, because FF1 has been traversed. When we continue to node n3, the simulation time stays at *N* + 1 because the NOR gate is a combinational gate. In general, to compute the amount of time movement when traversing from node *A* to node *B* across a sequential circuit, we determine the time for data to propagate from *A* to *B*. Time moves forward in forward tracing and backward in backward tracing.

In a circuit with multiple clocks, time advance is with respect to the clock of the sequential element that has just been traversed. Consider the multiple-clock domain circuit in Figure 6.8. Suppose we are looking at node D and we want to determine the time at node A, which affected the current value at node D. Assume the current time at node D is the last rising edge of clock clk2 at time 19. Moving to the input, the time at which the value at node C might have changed can be anywhere between 9 and 14, during which the latch was transparent. To determine exactly when, we need to examine the drivers to the latch. Going over the AND gate does not change time. Node B could change at a falling transition of clock clk2. Therefore, node C might change at time 6. Moving backward further, node A might change only at a rising transition of clock clk1. Therefore, the value of node A at time 1 affects the current value at node D. If the current value of D is erroneous, the value of A at time 1 is a candidate to be examined.

The same principle can be applied to circuits in RTL. Consider the following sequential element:

```
DFF g1 (.clk(clk1), .Q(A), .D(D), ...);

always @(clk2) begin
    data = A;
    @ (posedge clk2) begin
        state <= data << guard;
        out <= state ^ mask;
    end
end
```

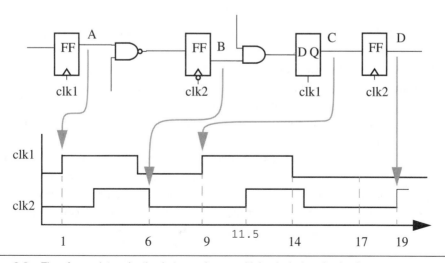

**Figure 6.8**   Time frame determination in traversing a multiple-clock domain circuit

We want to determine the time of variable D that affected the current value of out. The current time is 17, using the waveforms in Figure 6.8. To trace backward, we need to determine the last time clock clk2 had a positive transition, which, based on Figure 6.8, last changed at time 11.5. The assignment to data was executed when clk2 changed at time 6. Hence, the value assigned to data from A is the value of A at time 6. Because A is the output of the DFF g1, which is clocked by clk1, the time of D that affected A at time 6 is 1. Therefore, the time of D that affected out is 1. Any error in variable D at time 1 will be observed in variable out at time 17.

### 6.3.5   Load, Driver, and Cone Tracing

To understand the cause of a symptom at a node, the logic or circuitry potentially contributing to the node needs to be traced. Three common items are traced in practice: load, driver, and cone. Load tracing finds all fanouts to the node and is often used in forward debugging. Finding all fanouts of a node, which can be difficult in a large design in which the fanouts are spread over several files and different directories, is done with a tool that constructs connectivity of the design. Similarly, such a tool is used to find all drivers, or fanins, of a node. Tracing fanins or fanouts transitively (finding fanins of fanins) is called *fanin* or *fanout cone tracing*. A fanin cone to a node is the combinational subcircuit that ends at the node and starts at outputs of sequential elements or PIs. Similarly, a fanout cone is the combinational subcircuit that starts at the node and ends at inputs of sequential elements or POs.

Let's consider an example of debugging that requires driver and cone tracing. Consider the circuit in Figure 6.9 in which a data bit at node a has an unexpectedly unknown or indeterminate value *x* at the current time 5. Assume that all FFs and latches are clocked by the same clock clk, with the waveform shown. For simplicity, let's assume that all clock pins are operational, free of bugs. To debug, we trace all drivers to node a. Because node a is the output of a transparent low latch, the time of the latch's input that affected node a at time 5 is between 4 and 5. Therefore, as we backtrack across the latch, the current time frame changes from 5 to 4. The driver to the latch is an XOR gate with an output value that is unknown. The XOR gate has two fanins, both of which have unknown values. Selecting the lower fanin, we arrive at an OR gate. One of its fanins, node f, has an unknown value. Node f is driven by an FF that had an unknown input value. Because this FF is positive-edge triggered, crossing it backward moves the current time frame from 4 to 3. The driver to the FF is a tristate buffer that is enabled at time 3; thus, the unknown value comes from the bus. The value on the bus indeed had an unknown value. Now we find all drivers to the bus and determine which ones are active at time 3. There are two active drivers to the bus and they are driving opposite values because their inputs are opposite. Further investigation

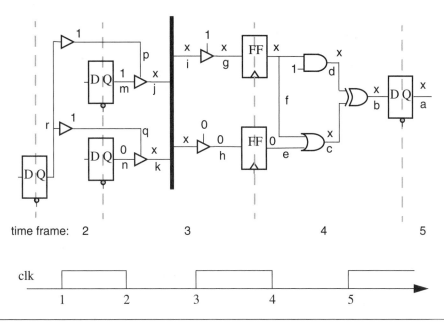

**Figure 6.9**   Debugging an unknown data bit via driver and cone tracing

into why both drivers are turned on at the same reveals the root cause: one of the buffers to the bus drivers should be an invertor.

In a large circuit, instead of tracing drivers one gate level at a time, the entire cone of logic can be carved out for examination. Cone tracing is not limited just to combinational cones; it can be a cone of logic spanning a couple cycles. Three fanin cones for nodes x and y are shown in Figure 6.10. One-cycle cones are just combinational cones. Multiple-cycle cones are derived by unrolling the combinational logic multiple cycles and removing the sequential elements between cycles. For example, the two-cycle cone consists of all gates that can be reached from node x or y without crossing more than one FF. Similarly, the three-cycle cones include all gates reached without crossing more than two FFs. The primary inputs to a cone are the original primary inputs and outputs of the FFs. The cone's primary inputs are marked with the cycle numbers. For example, P3 means the value of P three cycles backward from the current cycle. The current cycle is 1. Note the fast growth of cone size as the number of cycles increases; thus, in practice, only a small number of cycles are expanded for logic cones.

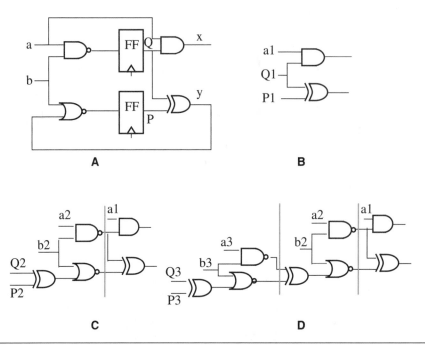

**Figure 6.10** Unrolling to obtain a multiple-cycle logic cone. (A) Original circuit (B) Combinational cone (C) Two-cycle cone (D) Three-cycle cone

### 6.3.6  Memory and Array Tracing

Whenever an FF or a latch is crossed, time progresses or regresses. When memory or an array is crossed, the number of cycles that the time changes is a function of what data are being traced and can be deduced as follows. Suppose that we find out that the output of memory is wrong and we want to back trace to the root cause. Assuming the current time is $T$, we first determine whether the address and control signals to the memory (such as read, write, and CS) at time $T$ are correct. If any of these signals is not correct, tracing continues from that line and time does not change. However, if the address and control signals are correct, then the wrong data were caused by either a bug in the memory model itself or by writing wrong data to that address. We search for the most recent time at which that address was written. Let this be time $W$. If the data at time $W$ are not identical to the output at time $T$, the memory model has a problem. If the data are the same, the input data are wrong, tracing follows the input data, and the time frame becomes $W$. That is, the amount of time of the backward time lapse is $T - W$. To illustrate this algorithm with the memory and waveforms in Figure 6.11, let's assume that the output of memory at address 8'h2c is expected to be 32'hc7f3 at time 1031, but is 32'ha71 instead. We back trace across

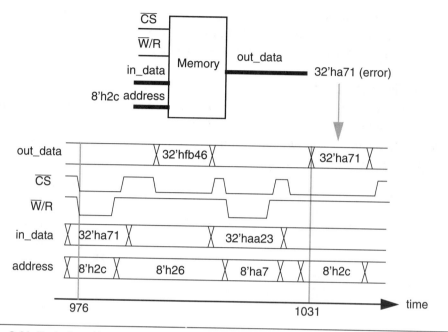

**Figure 6.11** Illustration of back tracing across memory

memory. From the waveforms, the control signals, CS, W/R, and address are correct. That is, CS is active, W/R is READ, and the address is 8'h2c. So we search for the last time the memory was written at address 8'h2c, and the time was 976. Because the input data, value of in_data, are identical to the output data at time 1031, the memory model is fine and the error tracing continues from time 976 to determine why the input data had the wrong value of 32'h0a71. The algorithm for back tracing memory is shown here:

---

**Memory Back Tracing Algorithm**

1. Assume the current time is $T$. Examine all inputs except data for correctness. If any of these is incorrect, back trace from that input and time remains $T$.
2. If the inputs are correct at time $T$, search for the last time the address was written and mark this time $W$.
3. If input data at time $W$ are not the same as the output at time $T$, the memory model has a fault; otherwise, back trace from the data input and the time changes to $W$.

In summary, forward tracing is just simulation. Backward tracing is searching the current or last input combination or condition that produced the current output. If found, time moves to that time and back tracing continues from there.

### 6.3.7  Zero Time Loop Constructs

Loop constructs occur most often in test benches, as used when iterating array elements, and usually do not contain delays. Hence they are executed in zero simulation time. That is, variables of the loop are computed at multiple times at the same, current simulation time as the loop is iterated. An example of such a loop is as follows:

```
always @(posedge clock) begin
    if (check_array = 1'b1)
        for (i=0; i<= `ARRAY_SIZE; i=i+1) begin
          var = array[i];
          if( var == pattern ) found = 1;
          . . .
        end // end of for loop

    if (found) ...
end // end of always block
```

The loop is computed with no simulation time advancement. Variable `var` is assigned to `array[i]` the number of times equal to `ARRAY_SIZE` at the current simulation time.

Multiple writes and reads to the same variable at the same simulation time cause difficulties in debugging, because when the simulation is paused, the variable value displayed is that of the last write. For example, the value of `var` displayed when the simulation is paused is `array[`ARRAY_SIZE]`. If a bug is caused during the loop computation, seeing only the last value of the variable is not enough. To circumvent this problem, variables inside a zero time loop need to be saved for each loop iteration so that their entire history can be displayed at the end of the simulation time. For the previous example, the intraloop values of `var` can be pushed to a circular queue every time `var` is written:

```
always @(posedge clock) begin
    if (check_array = 1'b1)
        for (i=0; i<= `ARRAY_SIZE; i=i+1) begin
          var = array[i];
          queue_push(var);
          if( var == pattern ) found = 1;
          . . .
        end // end of for loop

    if (found) ...
end // end of always block
```

Some debuggers show all intraloop values of loop variables (such as `var[1]`, ..., `var['ARRAY_SIZE])` when the variables are displayed.

## 6.3.8 The Four Basic Views of Design

In RTL debugging, four views of a circuit are essential—RTL, schematic, finite-state machine, and waveform—although other views exist, such as layout and DFT. A circuit and waveform viewer displays these four views and allows the user to switch among views. An RTL view shows the design code. A schematic view is a circuit diagram representation of the design code. The viewer creates the schematic by mapping simple code constructs in the design to a library of common gates. For example, a quest operation `x ? y : z` is mapped to a multiplexor. Other simple constructs are `AND`, `OR`, multiplexor, bus, and tristate buffers. Finite-state machines and memory, if they conform to a set of coding guidelines, will also be recognized. The mapper attempts to recognize as many common constructs as possible. If recognized, the constructs are represented with graphical symbols in the schematic view. The constructs not recognized are "black boxed." A schematic view preserves the module boundaries of the design so that a module instantiation is represented as a box labeled with the module instance name. To go inside the module, simply click on the box. The finite-state machine view shows state diagrams of finite-state machines. To recognize finite-state machines, many viewers assume certain finite-state machine coding styles. Finally, a waveform viewer displays waveforms of nodes. The waveforms are created from dumped data files in either standard format, such as VCD, or vendor-specific format, such as fsdb. Figure 6.12 shows an example of the four views. In the schematic view, the reduction `XOR ^` is not recognized as a common construct and hence is black boxed (the shaded box labeled `^ OP`.) All other constructs are recognized and are represented by standard circuit symbols. The coding style of this example conforms to the finite-state machine's coding style: hence, it is recognized as a finite-state machine and its state diagram is shown in the state machine view. The waveform view displays signals or variables specified by the user.

For most viewers, the different views of a circuit are coordinated by drags and drops. For example, to switch from RTL to schematic view, click on a variable or signal and drag it to the schematic view. The schematic view will display the scope (such as a module) in which the variable or signal resides. To see the waveform of a signal, simply drag and drop the signal to the waveform viewer. The different views offer their unique benefits. The RTL view shows the exact functionality of the design unit, the schematic view best displays connectivity, the state diagram view offers a functional and graphical description of the RTL code, and the waveform view reveals the temporal behavior of signals.

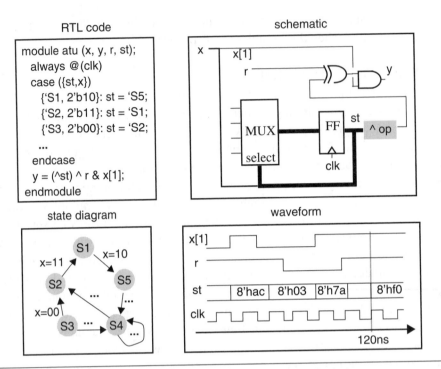

**Figure 6.12** RTL, schematic, finite-state machine, and waveform views of a design

## 6.3.9 Typical Debugger Functionality

Let's discuss some typical functionality in a debugger. The most basic functionality is tracing of drivers and loads of a node in the RTL and schematic views. With the schematic view, a command to trace a signal highlights all drivers or loads, depending on whether the driver or load option is set. Such a command can be a simple click on the signal. A continual command on a highlighted driver or load effects transitive tracing. With RTL view, a list of drivers or loads is shown when a net is traced. It is also possible to select a cone tracing option and have the debugger show a fanin or fanout cone of a variable or a net.

Tracing must be coupled with simulation values to be useful. When the user comes to a decision point—a multiple-fanin point for backward tracing or a multiple-fanout point for forward tracing—she needs to know the fanin or fanout that has the wrong value to continue. A convenient feature is annotation of simulation values in the RTL and schematic views (that is, signal values at the current time are appended to signals or variables). An example is shown in Figure 6.13, in which the annotated values are in bold. At the time shown, clock `clk` is in a falling transition. If the current time is changed, the values will change to reflect the simulation results. Based on the annotated values, the branches with

RTL code                                          schematic

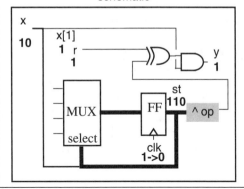

```
always @(clk)
        1->0
 case ({st,x})
        {110,10}
    {`S1, 2'b10}: st = `S5;
        110
    {`S2, 2'b11}: st = `S1;
        110
    ...
 endcase
 y = (^st) ^ r & x;
 1   110  1   1
 endmodule
```

**Figure 6.13** Annotation of simulation values to RTL and schematic views

expected values are pruned. Tracing follows the paths with unexpected values. To assist in keeping track of a tracing, branching points can be bookmarked and later revisited. An application of bookmarking is that after the current selection at a branching point turns out to be a deadend, the saved branching point is reverted so that another path is pursued.

With waveform view, waveforms can be searched for values or transitions. For instance, a waveform on a bus can be searched to find the time the bus takes on a specific value. Furthermore, two sets of waveforms can be compared, and the differences are displayed at the times they differ.

Finally, a debug session can be saved to a file and can be restored later. This is useful when a debug session needs to be shown to another person at a remote location. Then the saved session file is sent to that person.

---

**Example 6.1**

The following example illustrates a typical debugging process using the backward tracing method. The first sign of a bug is an error message from a simulation run. Suppose error message "Missing packet" appears at time 3000. The following sequence of actions will then take place:

1. Determine from the nature of the error whether the error can be reproduced with only a port of the circuit and with a smaller test case.
2. If the test case simulation is long, rerun the simulation from the last check point to determine whether the same error message occurs. If it does, debug from this check point; otherwise, try the previous check point until one is found that produces the error.

*continues*

**Example 6.1    (Continued)**

3. Determine the neighborhood around the site of the error for the purpose of signal
dumping. The site of the error is the statement that causes the printing of the error
message, as shown in the following code:

```
reg [3:1] index;
...
always @(clock) begin
        0->1
  for (i=0; i<=128; i=i+1) begin
      item = in_queue[i];
    32'b0ffa3            128
        if (item == packet) packet_received = 1'b1;
        32'b0ffa3   32'h9ba0
  end
  if (packet_received!=1'b1) $display ("Missing packet"); // error site
        1'b0
end // end of always
...
always @(clk) begin
        0->1
  if (ready_xmit) begin
        1'b0
    in_queue [index] = data_out;
                5           32'hffff
  end
  ...
  index = base + inc ;
      5       2       3
end
```

The smaller the scope, the faster the simulation will be. However, too small a scope
runs the risk of having to rerun the simulation for a larger scope if a traced signal goes
out of bounds.

4. Simulate the reduced circuit and test case and dump out the data from the selected
scope.

5. Load the dumped data into a debugger and annotate the signal values into the RTL or
schematic views. The annotated values are shown in italics in the previous code.

6. Search for the site of the error in the RTL or schematic views and back trace the drivers
that trigger the error.

---

**Example 6.1     (Continued)**

---

**7.** Based on the statement at the error site, the error message was triggered by signal `packet_received` equal to `1'b0`. Tracing drivers to `packet_received`, we find the `for` loop just above the error site to be a driver. Because of the zero simulation time of the loop, only the last values of the loop variables are shown. Signal `packet_received` is driven by the condition `if(item == packet)`, where `item` is an item of `in_queue` and `packet` is the expected packet. If this condition fails, the array `in_queue` does not contain the data equal to `packet`. To find out why `in_queue` does not have the data, we trace the drivers of `in_queue` and locate the sole driver to be `in_queue [j] = data_out`. Then we search in the waveform of `data_out` for `32'h9ba0` and determine whether the expected packet was ever sent (in other words, assigned to `in_queue`). A search for `data_out` for `32'h9ba0` turns up the data at time 1200, as shown in Figure 6.14. The index value at 1200 was 6, meaning `in_queue[6]` had `32'h9ba0`, but somehow it was lost. The error must have occurred between time 1200 and time 3000. So we need to search for any write to `in_queue` at index 6. We find that at time 2200, location 6 of `in_queue` was overwritten by another value. Suppose it is correct to have a write to `in_queue` at time 2200, but the location should not be 6. We then trace the drivers of `index`, which is `index = base + inc`, as shown in the previous code. Moving time to 2200, annotated values reveal that both variables `base` and `inc` take on a value of 7, but `index` takes on 6. Checking the declaration of `index`, we discover that `index` was declared to be `reg[3:1]`, a 3-bit variable that should have been `reg [3:0]`. Thus the MSB was truncated, producing 6 from 14. This is the root cause of the error.

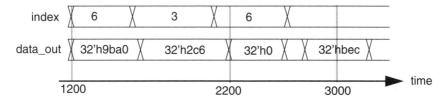

**Figure 6.14** Array waveforms used in back tracing an error

## 6.4   Design Update and Maintenance: Revision Control

When a group of engineers work on a project consisting of many files, files change constantly while being accessed. It is imperative to have a system to manage multiple variants of evolving files and track revision, as well as manage the project environment so that only the stable and correct version combinations of the files are visible. This is where a revision control system comes in. A revision control system has two objectives: to grant exclusive write access to files and to retrieve versions of files for read consistent with a user configuration specification. The exclusive write access feature ensures that a file can only be modified by one person at any time. This prevents a file from being edited simultaneously by more than one user, where only one user's result is saved and the others' are lost. A file must first be checked out before it can be modified. Once a file has been checked out, it can no longer be checked out again until it is checked back in. A file can have many revisions.

When a file is accessed, the version of the file needs to be specified. A view of a project is a specification of versions, one version for each file. This specification determines a view and is sometimes called a configuration file. To access a particular version of a file, the user simply specifies the version number for the file in a configuration file. When a configuration file is activated to provide a view, only the files that meet the specifications in the configuration file are accessible. An example configuration is shown below. The first item is the name of a file, followed by the version used in this configuration. File all_RTL.h in this configuration has version 12.1. The last line indicates that any file without a specified version is assigned the version labeled RELEASE_ALPHA.

```
all_RTL.h version 12.1 // header file
CPU.v version 15.3 // top level CPU
itu.v version 15.3 // ITU block
* version ALPHA_RELEASE
```

Conceptually, a revision control system has the following key components and architecture, as shown in Figure 6.15. All files are stored in a centralized database. Each file is stored in a layered structure, with the bottom layer being the full content of the file. On top of it is a collection of incremental changes to the file. Each layer is labeled with a version number. A user environment is determined by a view with a configuration specification. To reconstruct files of particular versions, the view manager in a revision control system takes in a configuration file and dispatches each file and its version to a version handler, which reconstructs that version of the file from the layers in the centralized database.

When a file is checked out, a copy of the file is placed in the local file storage area and can be modified by the user. Any modification to that file is only visible in that view, and thus does not affect other users using that file in other views. When changes are finalized, the file is checked in. When a check-in command is issued to the view manager, it removes the file from the local file storage area and passes it to the version handler, which attaches a new version number to the file and stores it incrementally to the centralized file database.

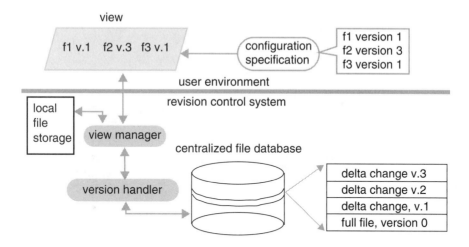

**Figure 6.15** Architecture and key components of a revision control system

Once a file is checked in, its newly updated contents can be visible by specifying in the configuration file the new version number assigned to the file.

In practice, a configuration specification file does not indicate a version for every file explicitly. Instead, files without an explicit version are assumed to be retrieved from a default version, such as the latest version. Furthermore, when a project arrives at a milestone release, all checked-in files at that milestone can be tagged with a label, say ALPHA_RELEASE, so that a view of this milestone release can be invoked by simply using label ALPHA_RELEASE, as opposed to the version numbers for the files.

In a large project, restricting editing of a file only to one user serializes development and hampers progress. When several engineers need to modify a file during the same period of time, copies of the file can be created through a branching process so that the engineers can work on it simultaneously. When a branch is created, the original version becomes the main branch. Files on any other branches are revision controlled in the same way as the main branch. When a branch of the file has reached stability, it can be merged with the main branch. When merging two branches, the version control system displays the two versions of the file, highlights the differences, and prompts the user to make decisions regarding which of the differences should go to the merged version. Figure 6.16 is a diagram showing a file's revision history, in which the nodes represent versions and the file has been branched three times. The main branch forks at a side branch for DFT team at version 1.0. Branch DFT has two versions, D1.0 and D1.1, before it merges with the main

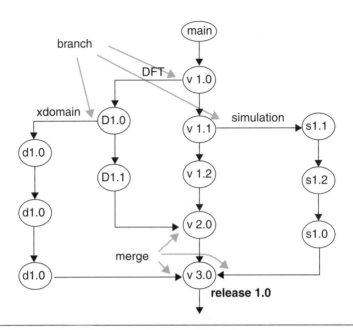

**Figure 6.16** Version tree of a file showing versions, branches, and merges

branch at version 2.0. In other words, version 2.0 has all updates from the DFT branch. The DFT branch has its own subbranch for another group doing cross-clock domain enhancement. This xdomain branch merges with the main branch at version 3.0. Similarly, at version 1.1, a branch is created for simulation work that itself has three versions and later merges with the main branch at version 3.0. Version 3.0 is also labeled as RELEASE 1.0, denoting a milestone version. A branch is not restricted to merging only with the main branch; it can be merged with any other branch.

## 6.5   Regression, Release Mechanism, and Tape-out Criteria

The centralized file database must maintain high-quality code. To prevent bugs from being checked into the centralized file database, a set of tests, called *check-in tests*, must be run to qualify the to-be-checked-in code. If the tests pass, the code can be checked in. A check-in test may not detect all bugs; therefore, all files in the centralized file database should be run on a larger suite of tests, called a *regression test*, at regular intervals (such as weekly). A check-in test can be regarded as a smaller scale regression test. Its runtime is much shorter, so that code can be checked in without much delay.

Large regression tests can be layered further. First, the full regression test suite is run only occasionally (for example, biweekly or before a major release). Second, the regression suite of the next scale can be run for patch releases or for a weekly release. Third, a nightly regression suite is run to catch as early as possible bugs in newly checked-in files. Finally, a check-in test, when considered a regression test, is run whenever there is a file to check in. The sooner a regression is run, the sooner the bugs are detected; however, regression tests place a heavy burden on computing resources. A full regression test suite can take as much as a day on a computer farm, and a nightly test can take as long as 12 hours on a project server. Therefore, regression tests are often run after hours. If a regression suite takes more than 12 hours on the project server, it should be run either on a weekend or on a computer farm to avoid slowing down engineering productivity during the day. The frequency of running a regression test may increase as a milestone release approaches.

A regression suite, whether for a check-in test or the entire code database, collects test cases from diagnostics targeted at specific areas of the code, randomly generated programs, and stimuli that had activated bugs. A well-designed regression suite has good code and functional coverage, and a minimum number of overlapping tests. To verify a design using a regression test, the output from the regression run is compared with a known, good output. Because code and even project specifications can change over time, the known, good output also changes. Therefore, regression suites require maintenance. Mismatches from a regression run should be resolved in a timely fashion to prevent error proliferation. All errors from the current week's regression run must be resolved by the end of the week so that these errors will not create secondary errors. An objective of the regression run is to minimize the number of errors per root cause while maximizing detection of errors.

Large regression suites are run distributively on a computing grid, also called a *computer farm*, which consists of hundreds and thousands of machines. The individual tests in a regression suite are run simultaneously on separate machines. A computing grid has a multiple-client/multiple-server queue as its interface. Jobs submitted are queued up and served by available machines from the farm. A job submission entry usually contains a script for job execution, a list of input files, and a point to store output files. When a job is finished, the submitter is notified of the status and the time of completion.

Release mechanism refers to the method by which code is delivered to customers. In a hardware design team, products can be RTL code, PLI C programs, CAD tools, and test programs. The most primitive release mechanism is to place files in a specific directory so that customers can download them. A key consideration in determining a release mechanism is to understand how the code will be used and to craft a release mechanism to minimize the impact on the customers' application environment. For example, if the release product is an executable program that resides in the customer's environment in the directory `/application/bin/`, one release mechanism is to put the release program in that directory under the customer's revision control system and attach a new version number to it. To use it, the customer simply changes the version number of the product in his view. In this example, copying the released product directly to `/application/bin` may interfere with the customer's operation. For instance, the older version of the program may be in use at the time of copying. A script that invoked the program multiple times might end up using the older version in the first invocation and the new version in the later invocations.

Because there are no direct ways to know that a design is free of bugs, several indirect measures such as tape-out criteria, are used in practice. One is the coverage measure. Usually, nearly 100% code coverage must be achieved. On the other hand, numerical measures for parameter and functional coverage may not be accurate enough to cover the complete functional spectrum and hence may be subject to interpretation. An alternative practical approach is to have project architects to review all functional tests to determine whether sufficient coverage is achieved. Another tape-out criterion is bug occurrence frequency or, simply, bug rate, which is the number of bugs found during the past week or weeks. If bug rate is low, it may indicate that the design is relatively stable and bug free. Of course, it may also mean that the tests are not finding any new bugs; therefore, they should be used in conjunction with a coverage measure. Another criterion is the number of cycles simulated, which offers some insight into the depth in state space in which the design is explored, but this lacks any proved accuracy. Figure 6.17 shows plot of the three tape-out criteria. The

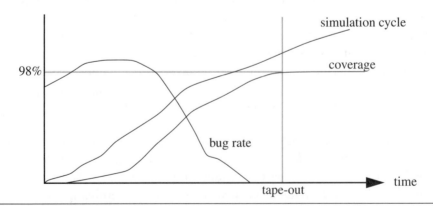

**Figure 6.17** Tape-out criteria based on coverage, bug rate, and simulation cycle

bug rate curve eventually becomes zero and the coverage metric reaches a plateau of 98% regardless of the increases in the simulation cycle. The tape-out time in this example is dictated by the coverage, because it reaches a plateau last. The simulation cycle can never reach a plateau.

## 6.6  Summary

In this chapter we examined the debugging process and the verification cycle. During the debugging process, the environment in which a bug is detected must be preserved to make the bug reproducible. We listed several common environment variables to be saved. To make debugging traceable, the circuit size should be reduced, which amounts to estimating the block in which the error might initially occur and, if a block of the circuit is carved out, replacing the rest of the circuit with a test bench that drives the block using the waveforms from the original simulation. Besides circuit size reduction, the test case can also be trimmed. This can be done with a binary search algorithm. When a test case is cut, the preserved portion of the test case may have to start from the same state as before, which can be obtained by simulating the beginning portion of the test case. Check pointing is a technique to save a system's state so that the system can be run from the saved state instead of the beginning. This technique is often used to reduce test cases and to transfer a circuit from one simulator to another for debugging.

Next we studied the states of an issue tracking system and the process of filing and closing a bug. Then we looked at mechanisms of simulation data dumping and determined the window of data dumping. With regard to the debugging process, we first introduced forward and backward signal tracing and a branching diagram to keep track of decisions made during tracing. As sequential elements are traced, simulation times change accordingly. In particular, we studied tracing across FFs, latches, and memory arrays. To conclude tracing, we examined driver, load, and cone tracing as a basic step in debugging. In passing, we talked about zero simulation time constructs that require special attention to view their value progression. We looked at the basic four views of a design and some typical features in a debugger. When a bug is fixed, it needs to be checked into a centralized database. We then discussed the revision control system and its basic architecture and use, particularly code branching for paralleling design effort. Finally, we studied some aspects of a verification infrastructure: regression, computing grid, release mechanism, and tape-out criteria.

## 6.7   Problems

**1.** If a bug is detected while running a simulation under a revision control system, the environment variables required to reproduce the bug must include the versions of all the files used in the simulation.

    a. List three revision control software programs, whether public domain or commercial.

    b. Select one revision control software program from the previous answer. What are the commands to capture the view of the simulation?

    c. (Optional) Write a script that saves a view to file `view.sv`. Write another script to restore the view from `view.sv`.

**2.** To reduce circuit scope in debugging, it is often necessary to carve out a block of the circuit from a full chip and focus debugging on that block. To do so, a test bench modelling the surrounding circuit around the block has to be created. The test bench instantiates the block and drives the inputs of the block with the waveforms of the inputs captured in a full chip simulation run.

    a. Construct a test bench for the following block. Assume that captured waveforms for the inputs are shown in Figure 6.18 and the block has the following interface:

```
module buggy_block (in1, in2, in3, out1, out2, out3)
input [31:0] in1;
input [3:0] in2;
input in3;
output [31:0] out1;
output out2, out3;
```

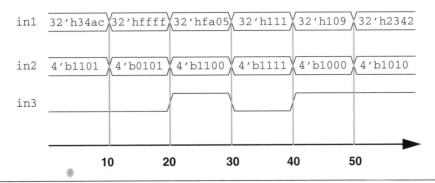

**Figure 6.18** Captured input waveforms for block test bench

b. Suppose input `in3` is an I/O port. What difficulty may arise from your test bench? How could such a difficulty be dealt with?

3. In practice, midpoint statements in a test case are often approximated for various reasons. For example, the exact midpoint is difficult to compute or the state at the exact midpoint statement is complicated. Besides, if an approximation yields only a few more lines of statements while making restarting much easier, it is well worthwhile. In this exercise, assume that a microprocessor runs a C program shown in Figure 6.19 and an error occurs. Furthermore, let's assume the inner loop, L3, is indivisible, meaning if an approximation for a midpoint statement is to be used, it should not happen inside L3. Use a binary test case reduction algorithm to trim down the test case. For simplicity, assume all assignment statements are of equal weight.

a. Formulate an equation to find a midpoint statement of the code. If you are to solve the equation, do you expect the value from the equation to be an integer?
b. To simulate from the midpoint statement found in the previous answer, what variables must be saved to enable restarting from the midpoint statement?
c. (Optional) If loop L3 is not indivisible, formulate the midpoint statement equation.

```
n = 16;
dd = vector(0,n-1);
for (j = 0; j< n; j++) {        ⎫
    d[j] = 0.0;                  ⎬ L1
    dd[j] =0.0;                  ⎭
}
d[0] = c[n-1];
for (j = n; j>0 ; j--) {                      ⎫
    for (k=n-j; k>0; k--) {       ⎫           ⎪
        sv=d[k];                   ⎬ L3       ⎪
        d[k]=2.0*d[k+1]-dd[k];     ⎭          ⎬ L2
        dd[k]=sv;                             ⎪
    }                                         ⎪
    sv=d[0];                                  ⎪
    d[0] = -dd[0]+c[j];                       ⎪
    dd[0] = sv;                               ⎭
}
```

**Figure 6.19** A C program test case for a bisection search

**4.** Let's consider the optimal check pointing interval. In a design, the average number of simulation cycles from the time an error is triggered to the time it is observed is $P$. Assume an error can occur at any cycle with equal probability, and simulations are run with a check pointing interval of $L$ (in other words, the design is check pointed every $L$ cycles).

   a. Define the debug interval of an error to be the number of cycles from the last check point at which the error was triggered to the time it is observed. Show that the mean debug interval can be expressed as shown here, where $E$ is the expectation operator on variable $t$ over interval $[0,L]$, and $t$ is the time of error occurrence modulo $L$. Assume $P$ is greater than or equal to $L$:

$$DI = \underset{t \in [0,L]}{E} \left( \left\lceil \frac{P-t}{L} \right\rceil \cdot L + t \right)$$

   b. What should DI be if $P$ is less than L?
   c. Assume $P$ is at least $L$ and that the cost associated with simulating a cycle in the debug interval is ten and the cost of generating a check point at interval $L$ is $15/L$. The total cost of debugging consists of the cost of generating the check points up to the point an error is observed plus the cost of simulating the debug interval during debugging. Derive a total debug cost function. For $P = 5$, find an optimal $L$ that minimizes the total debug cost.

**5.** Referring to the five states of issue tracking, describe the roles of the manager and engineers, and the transitions of the states in each of the following scenarios.

   a. A bug was filed because of a user error.
   b. A bug cannot be reproduced because by the time it got to an engineer, the design was already improved not to produce the bug.
   c. A bug is about inconsistency between design and its documentation.
   d. A bug cannot be reproduced, possibly an intermittent bug.
   e. A bug is another manifestation of another bug being worked.

**6.** For the circuit in Figure 6.20, construct a forward tracing branching diagram with a depth of 3. The error site is $X$. Repeat for backward tracing.

**7.** Derive a two-cycle fanin cone for node $X$ in the circuit shown in Figure 6.21.

**8.** Consider the circuit and clock waveforms shown in Figure 6.22.

   a. If the current time is just before 100, when is last clock edge before time 100 that may cause a transition at node $X$?

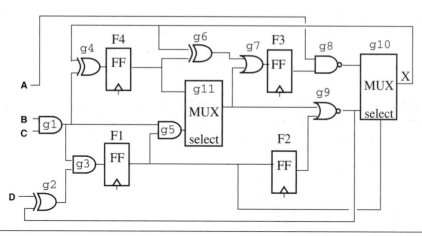

**Figure 6.20** Circuit for creating branching diagrams

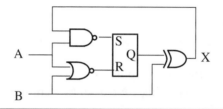

**Figure 6.21** Circuit for fanin cone unrolling

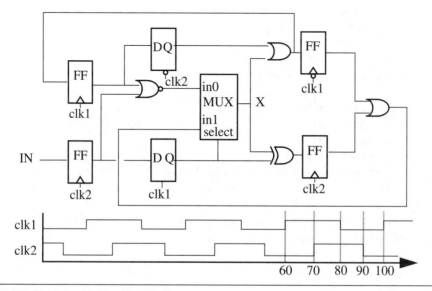

**Figure 6.22** Circuit for driver and load tracing

b. Show by tracing signals that both inputs of the multiplexor being 0 cannot be a steady state. A steady state has settled in the clock cycle.

c. Show that in1=0 and select=1 cannot be a steady state.

d. Show that in0=0 and select=0 can be a steady state.

9. Consider the following RTL code. If message "error found" is displayed, how do you go about finding the root cause using the backward tracing method? What problem may be encountered during your debugging process?

```
always @(posedge clk) begin
    hit = 1'b0;
    for(i=0; i<=10; i=i+1) begin
        vt = target[i];
        if(vt == checked) hit = 1'b1;
    end
end

always @(negedge clk)
if(hit == 1'b1) $display("error found");
```

10. In this problem, you learn to check point and restore from a check point, and dump out signals to debug. The following code is an asynchronous queue. First get familiar with the code and the functionality of the queue:

```
module top;
    parameter WIDTH = 32;
    reg [WIDTH-1:0] in;
    wire [WIDTH-1:0] out;
    wire full,empty;
    reg enq, deq, reset;

    initial begin
        enq=1'b0;
        deq=1'b0;
        reset <= 1'b1;
        // add your input stimuli below
    end

    queue m(.in(in), .out(out), .enq(enq), .deq(deq),
.full(full), .empty(empty), .reset(reset));
```

```
    endmodule

    module queue(
        out,        // queue head
        full,       // 1 if FIFO is full
        empty,      // 1 if FIFO is empty
        reset,      // reset queue
        in,         // input data
        enq,        // enqueue an input
        deq         // dequeue the head of queue
    );

        parameter WIDTH = 32; // queue width parameter
        DEPTH = 3; // length of queue

        output [(WIDTH-1):0] out;
        output full, empty;
        reg full, empty;
        input reset, enq, deq;
        input [(WIDTH-1):0] in;
        integer ic; // item count
        integer i; // index

        reg [(WIDTH-1):0] item [0:(DEPTH-1)]; // queue

    // output is the head of the queue
    assign out = item[0];

    always @(posedge reset)
        begin // reset queue
            ic = 0;
            full <= 1'b0;
            empty <= 1'b1;
        end

    always @(posedge enq or posedge deq)
        begin // enqueue or dequeue
        case ({enq, deq})
          2'b00: ;
          2'b01: // dequeue head of queue
```

```
        begin
           for (i=1; i<ic; i=i+1)
              item[i-1] <= item[i];
           ic = ic-1;
        end
        2'b10: // enqueue input item
        begin
           item[ic] <= in;
           ic = ic+1;
        end
        2'b11: // enqueue and dequeue simultaneously
        begin
           for (i=1; i<ic; i=i+1)
              item[i-1] <= item[i];
           item[ic-1] <= in;
        end
     endcase

     full <= (ic == DEPTH) ? 1'b1 : 1'b0;
     empty <= (ic == 0) ? 1'b1 : 1'b0;
     end // always

   endmodule
```

a. Add stimuli to the initial block so that the three items 32'h12345678,
   32'h90abcdef, and 32'hfeedbeef are added to the queue at times 20, 40, and 60
   respectively. Then dequeue them at times 100, 120, and 140 respectively. Simulate
   the design to demonstrate that your stimuli indeed enqueue and dequeue as
   expected by printing out the queue contents at those times.
b. Check point the circuit at time 80 and exit the simulator. Then, restart the simulation using the check point. Show that the three items are dequeued as expected.
c. Restart a simulation from the second-to-last check point and dump out all nodes in
   VCD format.

11. Use the following stimuli to simulate the asynchronous queue in the previous problem:

```
     enq <= 1'b0;
     deq <= 1'b0;
     reset <= 1'b1;
     #5 reset <= 1'b0;
     #5 in = 32'habcdef;
```

```
        enq <= 1'b1;
        #5 enq <= 1'b0;
        #5 deq <= 1'b1;
        #5 deq <= 1'b0;
        #5 deq <= 1'b1;
        #5 deq <= 1'b0;
        #5 in = 32'h12345;
        enq <= 1'b1;
        #5 enq <= 1'b0;
        #5 in = 32'hbeef;
        enq <= 1'b1;
        #5 enq <= 1'b0;
        #5 deq <= 1'b0;
        #5 deq <= 1'b1;
        #5 deq <= 1'b0;
```

a. What operations do these vectors perform? Does the queue function as expected? If not, dump out all nodes in VCD format and use a debugger to determine the root cause.

b. Fix the bug and rerun the simulation to confirm.

c. What debugger did you use? Were you able to dump out the contents of an array item? Was the debugger able to annotate the contents of the array item to RTL?

12. For each of the following tasks, decide whether a branch in a revision system should be created to accomplish the task for the files in question.

a. Fix a bug in the FPU in a CPU design project.

b. Create a model for hardware acceleration on top of the synthesis model.

c. Based on the RTL model, create an algorithmic model for a formal verification tool.

d. Add postsilicon debugging facilities to an RTL model.

13. Consider the plots of bug rate, coverage, and simulation cycles in Figure 6.23.

a. Plot A displays bug rate versus coverage. Give an explanation for the behavior of the coverage metric in the shaded region.

b. Plot B shows bug rate versus simulation cycles. Is it possible for the bug rate to have a "hump" as shown in the shaded area? Give a possible scenario for the hump.

c. What can you say about the test vectors and the coverage metric in the shaded area of plot C?

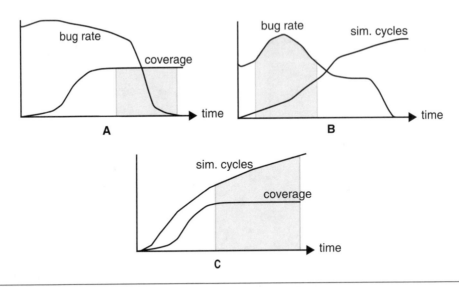

**Figure 6.23** Interpretations of bug rate, coverage, and simulation cycles

14. A computing resource for a project is usually tiered in terms of response time and computing capacity, such as performance, memory, and disk space. Three tiers are possible: computer farm, project server, and engineer's workstation. A computer farm is shared with other projects and jobs are queued, and it usually has the largest capacity among the three. Large jobs that do not require fast turnaround times are sent to a computer farm. Project servers are used for running large jobs from the project that need a response faster than that given by a computer farm. An engineer's workstation is reserved for the quickest response and small computing jobs. For each of following tasks, decide to which computing resource to send the task.

   a. Run a check-in test to verify a bug fix is correct
   b. Debug a unit
   c. Run a nightly regression
   d. Run a major release regression
   e. Check in a file

## CHAPTER 7

# Formal Verification Preliminaries

## Chapter Highlights

- Sets and operations

- Relation, partition, partially ordered set, and lattice

- Boolean functions and representations

- Boolean functional operators

- Finite-state automata and languages

S imulation-based verification relies much on one's ability to generate vectors to explore corner cases. In reality, two major problems exist. First, it is almost impossible, especially in large designs, to consider all possible test scenarios and, as a result, bugs arise from scenarios that were never anticipated. Second, even if one could enumerate all the possible corner cases in a design, it must be done intelligently so that the computation time to verify them all is realistic. Formal verification is a systematic way to traverse exhaustively and implicitly a state space or subspace so that complete verification is achieved for designs or parts of designs. However, formal verification is not a panacea. As seen in later chapters, formal verification achieves complete verification only up to the accuracy of the design representation and the property description. A case in point, formal verification cannot prove that all properties of a design have been enumerated, although for a given property it can prove whether the property is satisfied.

Formal verification, in contrast to traditional simulation-based verification, requires a more rigorous mathematical background to understand and to apply effectively. Key concepts in formal verification can be stated in intuitive plain English, but mathematics makes the concepts more precise and concise. Therefore, the crux to being proficient with mathematical symbols and manipulations is to go beyond their complicated appearance and understand the ideas they represent. Once an intuitive understanding has been developed, the mathematics comes naturally. Complicated mathematical expressions may seem daunting at times. It is comforting to know that the great mathematician Godfrey H. Hardy once said, "There is no place for ugly mathematics." ("A Mathemetician's Apology," G. H. Hardy, Cambridge University Press, Reprint Edition, 1992.) So, a complicated mathematical expression is more often than not a result of an ill-advised formulation, rather than a lack in the reader's mathematical maturity. The message is: Do not get discouraged easily.

## 7.1    Sets and Operations

A set is a collection of distinct objects, and the objects, called the *members* of the set, are enclosed in a pair of curly braces. Set $S$, consisting of alphanumeric symbols $x$, 2, \$, 3, %, and $L2$, can be written as $S = \{x, 2, \$, 3, \%, L2\}$. Inside the braces can be a qualification expression for the set members. For example, $F = \{p : p \ is \ a \ prime \ and \ p = 2^i - 1\}$ represents the set of all prime numbers that can be expressed in the form $2^i - 1$ for some integer $i$. The expression after the colon is the qualification expression. Sometimes a vertical bar is used in place of the colon: $F = \{p | p \ is \ a \ prime \ and \ p = 2^i - 1\}$. Set $U$ is often denoted as the universal set that consists of everything in the context. For example, in the context of integers, $U$ consists of all integers. Set $Z$ usually denotes the set of all nonnegative integers, whereas $R$ represents the set of all real numbers. $Z^+$ and $R^+$ are positive integers and positive real numbers respectively. The number of elements in set $S$, called the *cardinality of S*, is denoted by $|S|$.

*Membership* in a set, such as $x$ is a member of $S$, is denoted by $x \notin S$. Similarly, $y$ is not a member of $S$ is denoted by $y \notin S$. Using the previous definition of set $F$, we have $7 \in F$ and $11 \notin F$. Symbols $\forall s$ and $\exists s$ mean "for all variable $s$" and "for some variable $s$" respectively. Set $V$ is a subset of set $W$, denoted as $V \subseteq W$, if $\forall s \in V \Rightarrow s \in W$, which says, for every element $s$ in set $V$, $s$ is also in $W$. Symbol $x \Rightarrow y$ means "$x$ implies $y$." If there is at least one element in $W$ that is not in $V$, $V$ is said to be a proper subset of $W$ and is denoted as $V \subset W$. The equality sign is removed.

Compositions of sets can be defined in terms of the previous notations. If sets $A$ and $B$ are sets, then

$$A \cap B = \{s | (s \in A) \ and \ (s \in B)\}$$

This is called the *intersection* of $A$ and $B$, which is the set of all elements common to $A$ and $B$. Using sets $S$ and $F$, we see that $S \cap F = \{3\}$. If $A$ and $B$ do not have common element, the intersection is empty: $A \cap B = \varnothing$. Two sets with an empty intersection are said to be *disjoint*.

The *union* of $A$ and $B$ is defined as

$$A \cup B = \{s | (s \in A) \ or \ (s \in B)\}$$

which is the set of all elements in $A$ or $B$. Obviously, $A \subseteq (A \cup B)$, $(A \cap B) \subseteq A$, and $(A \cap B) \subseteq (A \cup B)$.

The *Cartesian product* $A \times B$ is defined as

$$A \times B = \{(a, b) | (a \in A) and (b \in B)\}$$

The Cartesian product is a common method used to construct higher dimensional space from lower dimensional spaces. For example, the set of all positive rational numbers, $K$, can be constructed from positive integers and can be represented as $K = \{(a, b) | a \ and \ b \ are \ positive \ integers\}$, where $a$ can be treated as the numerator and $b$, the denominator.

These set operations can be extended to a finite number of operands:

$$\bigcap_{i=1}^{n} S_i = S_1 \cap S_2 \cap \ldots \cap S_n$$

$$\bigcup_{i=1}^{n} S_i = S_1 \cup S_2 \cup \ldots \cup S_n$$

$$\prod_{i=1}^{n} S_i = S_1 \times S_2 \ldots \times S_n$$

An element in a Cartesian product of $k$ sets, denoted as $(n_1, ..., n_k)$, is called a $k$-tuple. It can be regarded as a point in the $k$-dimensional space.

The *exclusion* of $B$ from $A$, or *subtraction* of $B$ from $A$, is defined as

$$A - B = \{s | (s \in A) and (s \notin B)\}$$

which is the set of all elements in $A$ but not in $B$. As an example, the set of all positive integers $Z^+$ is $Z - \{0\}$, and the set of all positive irrational numbers is $R^+ - K$. If $U$ is the universal set, then $U - B$ is called the *complement* of $B$, denoted by $\overline{B}$.

## 7.2   Relation, Partition, Partially Ordered Set, and Lattice

Mapping $\mu$ between two sets $S_1$ and $S_2$, denoted by $\mu(S_1, S_2)$, associates each element in $S_1$ with an element or a subset in $S_2$. An example of a mapping is the procedure that takes in a photograph and outputs all colors in the photograph. In this example, $S_1$ is a set of photographs and $S_2$ is a set of colors.

A mapping that associates an element in $S_1$ with only one element in $S_2$ is called a *function*. That is, a function maps one element $S_1$ to exactly one element $S_2$. However, a function can map several elements in $S_1$ to the same element in $S_2$. Set $S_1$ is called the *domain*, and $S_2$ is called the *image* or *range* of the function. A function or mapping that maps distinct elements in $S_1$ to distinct elements in $S_2$ is called a *one-to-one function* or *one-to-one mapping*.

The *relation* $\gamma$ between $S_1$ and $S_2$ is the function $\gamma(S_1 \times S_2, \{0,1\})$. For $x \in S_1$ and $y \in S_2$, we say a relation exists between $x$ and $y$ if $\gamma(x, y) = 1$. An example of a relation is the less than relation, <, on real numbers. Relation < maps (2,3) to 1 because 2 < 3 exists. But it maps (5,1) to 0. For simplicity, we denote $x$ related to $y$ by $\gamma(x, y)$. Furthermore, for $\gamma(S \times S, \{0,1\})$, we say $\gamma$ is a relation on $S$.

A *partition* of set $S$, $\lambda = \{A_1, ..., A_n\}$, is a collection of subsets of $S$, $A_1, ..., A_n$, such that

$$\bigcup_{i=1}^{n} A_i = S$$

and

$$A_i \cap A_j = \emptyset \text{ for } i, j \in \{1, ..., n\}, \ i \neq j$$

In words, a partition of set $S$ is a division of $S$ into disjoint subsets, which together make up $S$. Subsets $A_1, ..., A_n$ are called the *elements* of the partition.

---

**Example 7.1**

Find a partition for $S = \{3, 7, 11, 17, 19, 23, 29, 31, 41\}$.

A partition is $A_1 = \{3, 7, 11\}$, $A_2 = \{17, 19\}$, and $A_3 = \{23, 29, 31, 41\}$, because these subsets are pairwise disjointed and their union is $S$. Subsets $A_1 = \{3, 17, 11\}$, $A_2 = \{7, 17, 19\}$, and $A_3 = \{3, 23, 29, 31, 41\}$ do not form a partition because $A_1$ and $A_2$ are not disjoint. For every set, there are at least two trivial partitions: the partition that consists of the set itself as the only element and the partition in which each element of the set is an element of the partition.

---

For notational convenience, let's use overlines and semicolons to denote a partition. For the partition in the previous example, we have

$$\pi = \overline{\{3, 7, 11; 17, 19; 23, 29, 31, 41\}}$$

A relation $\gamma$ on $S$ is an *equivalence relation* if it satisfies the following conditions:

1. Reflexive: $\gamma(x, x)$ for all $x \in S$.
2. Symmetric: $\gamma(x, y) \Rightarrow \gamma(y, x)$, for all $x, y \in S$.
3. Transitive: $\gamma(x, y)$ and $\gamma(y, z)$ implies $\gamma(x, z)$.

An equivalence relation is denoted by ~; for example, $x \sim y$ if $x$ is equivalent to $y$. As an example, a nonequivalent relation is relation $<$, because it is not reflexive. For example, $2 < 2$ does not exist. An example of an equivalence relation is the equality on real numbers, $=$. The main use of an equivalence relation is to partition a set to find all equivalent elements. When all equivalent elements are combined as a single element, the resulting set, called the *quotient set*, becomes smaller and easier to manipulate.

Given an equivalence relation ~, we would like to find a procedure to reduce a set based on the equivalence relation ~. Define $S_x$ as the set $\{s : s \sim x \text{ where } x \in S\}$. That is, $S_x$ is a subset of $S$, whose elements are equivalent to $x$. Element $x$ can be regarded as a representative element of $S_x$ and $S_x$, as the equivalence neighborhood of $x$. If we find all representative elements in $S$, then $S$ can be reduced to a set made of only representative elements. Theorem 7.1 guarantees this possibility.

---

**Theorem 7.1**

Given an equivalent relation ~ on $S$, the following statements are true:

1. If $x \sim y$, then $S_x = S_y$.
2. If $x$ is not equivalent to $y$, then $S_x \cap S_y = \varnothing$.
3. Let $\{x_1, .., x_n\}$ be all pairwise nonequivalent elements in $S$. In other words $x_i \sim x_j$ only if $i = j$. Then, $\lambda = \{S_{x_1}, ..., S_{x_n}\}$ is a partition of $S$.

---

Part 1 of this theorem says that any two equivalent elements must belong to the same subset. Part 2 says that any two nonequivalent elements must be in two different subsets. In other words, $S$ is broken into equivalent subsets, each of which is the maximal equivalent subset. The subsets are maximal because of the definition of $S_x$. The pairwise nonequivalent elements in part 3 can be regarded as the representatives of the equivalent subsets, each representing all the elements to which it is equivalent. Therefore, $\lambda$ is the equivalent partition with the minimum $|\lambda|$. That is, any other partition of S based on ~ has a greater or equal cardinality. $\lambda$ is the coarsest partition.

This theorem states a method of maximally partitioning a set such that all elements of any subset are equivalent. To create the partition, take any element $x$, find all its equivalent elements, and make them into a subset $S_x$. Any elements not included in $S_x$ must be not equivalent to $x$. Take any remaining elements and repeat this process. The partition ends when all elements have been included in a subset. This is summarized in the following list. Given an equivalence relation $\gamma$ on set $S$, partition $S$ according to $\gamma$.

---

**Partition S Based on an Equivalence Relation**

1. Initialize $i$ to 1.
2. While $S$ is not empty, let $x \in S$.
3. Let $A_i$ consist of all $y$ such that $y \sim x$.
4. $S = S - A_i$ and $i = i + 1$.
5. Repeat step 2.
6. The set of $A_i$ is a partition of $S$.

Note that the relation must be an equivalence relation; otherwise, the subsets formed are not a partition, as seen in Example 7.2.

---

**Example 7.2**

In this example we use two relations to partition the following set:

$S = \{33, 12, 3, 4, 21, 15, 22\}$

The first relation is an equivalence relation whereas the second one is not. The first case illustrates the partition algorithm and the second case shows the difficulties that arise if a nonequivalent relation is used.

The first relation is defined as $\gamma(x, y)$ if the sum of all digits of $x$ is equal to that of $y$. Obviously, $\gamma$ is reflexive and symmetric. It is also transitive, because if the sums of all digits of $x$ and $y$ are equal, and those of $y$ and $z$ are also equal, then the sum of digits in $x$ is equal to that of $z$. Therefore, $\gamma$ is an equivalent relation.

Following the partitioning algorithm, we set $i$ to 1. Pick $x$ to be 33. The sum of all digits of 33 is 6. In $S$, only 15 is equivalent to 33. So $A_1 = \{33, 15\}$. $S = S - A_1 = \{12, 3, 4, 21, 22\}$. Set $i$ to 2. Pick $x$ to be 12, whose sum of digits is 3. $A_2 = \{12, 3, 21\}$. $S = S - A_2 = \{4, 22\}$. Set $i$ to 3. Pick $x$ to be 4. The sum of digits is 4. $A_3 = \{4, 22\}$. Now, all elements in $S$ have been used, so the algorithm terminates. Thus, $\pi = \{\overline{33, 15}; \overline{12, 3, 21}; \overline{4, 22}\}$ is a partition of $S$.

Now, define another relation $\theta$ as $\theta(x, y)$ if $x$ and $y$ have a nontrivial common factor (a none unity factor). It can be seen that $\theta$ is reflexive and symmetric, but it is not transitive. Two and 6 have a common factor, 2. Three and 6 have a common factor, 3. However, 2 and 3 do not have a nontrivial common factor. Therefore, $\theta$ is not an equivalent relation. Nevertheless, let us attempt to partition using $\theta$ to see what problems may arise.

We see that $\theta(33, 15)$, $\theta(33, 12)$, $\theta(33, 21)$, $\theta(33, 3)$, and $\theta(33, 22)$. Let us use braces to group the elements that are related to 33 through $\theta$. Hence, $\{33, 15, 12, 21, 3, 22\}_{33}$. We also have $\{22, 33, 4\}_{22}$ and $\{12, 22, 4\}_{12}$ among other subsets. Note that 22 is in a set related to 33, and 33 is conversely in the set related to 22, but we cannot combine the two sets. It we were to combine the two sets, it would imply that 4 is related to 33, which is false. This problem is caused by the fact that $\theta$ is not transitive. In this case, 4 is related to 22 and 22 is related to 33, yet 4 is not related to 33. Therefore, we cannot group all related elements together as in the case of an equivalent relation. If we keep the subsets separate, they do not form a partition because they are not disjoint. For example, 33 belongs to more than one subset. Therefore, relation $\theta$ cannot partition $S$.

Of the three conditions for equivalent relation, if we replace the symmetric property with the asymmetric property, we have an ordering relation. An ordering relation is a generalization of the usual $\geq$ operator on real numbers. The meaning of an ordering relation extends beyond mere magnitude comparison; it is a more abstract concept of comparison. An *ordering relation*, $x \geq y$, satisfies the following three conditions:

1. Reflexive: $x \geq x$ for all $x \in S$.
2. Asymmetric: $(x \geq y, y \geq x) \Rightarrow x = y$, for all $x, y \in S$.
3. Transitive: $x \geq y$ and $y \geq z$ implies $x \geq z$.

---

**Example 7.3**

---

Determine which of the following relations are ordering relations.

1. $\alpha(x, y)$, if $x$ has more or the same number of digits than $y$, where $x$ and $y$ are positive integers.
2. $\beta(s, r)$, if $r$ is a substring of $s$, where $s$ and $r$ are strings of English letters.
3. $\gamma(m, n)$, if the determinant of m is greater than or equal to that of $n$, where $m$ and $n$ are $4 \times 4$ matrices with entries that are positive real numbers.

   a. $\alpha(x, y)$ is not an ordering relation because it does not satisfy the asymmetric condition. It is possible that two different integers have the same number of digits.
   b. A string is a concatenation of letters. For example, bceiasdfcer is string, which has eiasd as a substring. Let us examine the three conditions one by one. It is reflexive, because a string itself can be considered a substring of itself. It is asymmetric. We show this by contradiction. Assume that there exist two different strings, $p$ and $q$, such that $p \geq q, q \geq p$. Without loss of generality, we assume $p$ is of greater length than $q$. But then $p$ cannot be a substring of $q$, violating the assumption. Thus, $p$ and $q$ must have the same length. However, two equal-length strings must be identical if one is a substring of the other. Therefore, we conclude that $p = q$. Finally, assume $\beta(s, w)$ and $\beta(w, r)$. In other words, $w$ is a substring of $s$, and $r$ is a substring of $w$. It then follows that $r$ must be a substring of $s$; thus, $\beta(s, r)$. In conclusion, $\beta(s, r)$ is an ordering relation.
   c. $\gamma(m, n)$ satisfies reflexive and transitive conditions but fails the asymmetric condition because two different matrices can have the same determinant. For example, the following matrices have the same determinant:

   $$\begin{bmatrix} 1 & 0 \\ 2 & 3 \end{bmatrix} \text{ and } \begin{bmatrix} 3 & 5 \\ 0 & 1 \end{bmatrix}$$

The mathematical structure made of a set and an ordering relation is called a *partially ordered set*, or *poset*. The ordering relation operates on the elements of the set and arranges them according to the relation. It is called "partially ordered" because not all pairs of elements have a relation. Using the $\beta(s, r)$ relation from Example 7.3, $s \geq r$ if $r$ is a substring of $s$. In other words, $s$ contains or subsumes $r$. However, no order exists for these two strings $s = abc$ and $r = xy$, because we can write neither $s \geq r$ nor $r \geq s$. That is, $\beta$ cannot order $s$ and $r$.

One way to visualize ordering of a poset is to use a *Hasse diagram*. For two elements, $x$ and $y$, we say $x$ *covers minimally* $y$ if $x \geq y$, and for all $z \geq y$, $z \geq x$. That is, $x$ is the smaller element that is greater than $y$. If $x$ covers minimally $y$, we draw a downward line from node $x$ to node $y$. It may be clearer to draw an arrow from $x$ to $y$, but a Hasse diagram traditionally does not rely on arrows but north–south direction.

---

**Example 7.4**

Draw a Hasse diagram for the following partially ordered set

$$S = \left\{ \begin{bmatrix} a_{11} & a_{12} \\ a_{21} & a_{22} \end{bmatrix} \right.$$

with at least two entries equal to 1, where $a_{ij} \in \{0, 1\}$.

We define $s_1 \geq s_2$ if $a_{ij}$ in $s_1$ is 1, then $a_{ij}$ in $s_2$ is also 1. For example,

$$\begin{bmatrix} 1 & 1 \\ 0 & 1 \end{bmatrix} \geq \begin{bmatrix} 1 & 0 \\ 0 & 1 \end{bmatrix}, \begin{bmatrix} 1 & 1 \\ 0 & 1 \end{bmatrix}$$

also covers $\begin{bmatrix} 1 & 0 \\ 0 & 1 \end{bmatrix}$

There are eleven such matrices and their relations are shown in Figure 7.1.

Because of the transitive property, we see that if there is a descending path from $p$ to $q$, then $p \geq q$. If there is no descending path between the two nodes, no relation exists between them.

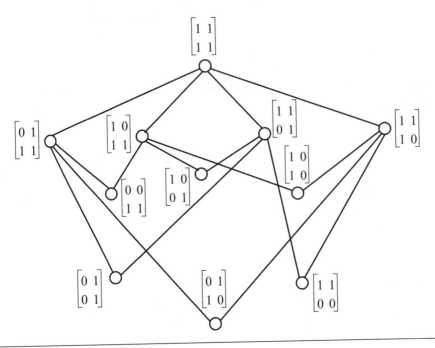

**Figure 7.1** Hasse diagram for 2 × 2 Boolean matrices

Between two elements in a partially ordered set, we can define meet and join. A *meet z* between elements x and y is the greatest lower bound of x and y. That is, $x \geq z$ and $y \geq z$. If there is an element w, such that $x \geq w$ and $y \geq w$, then $z \geq w$. We denote a meet between x and y by $x \cdot y$. In Figure 7.1, the meet of

$$\begin{bmatrix} 0 & 1 \\ 1 & 1 \end{bmatrix} \text{ and } \begin{bmatrix} 1 & 1 \\ 1 & 0 \end{bmatrix} \text{ is } \begin{bmatrix} 0 & 1 \\ 1 & 0 \end{bmatrix}.$$

Note that there may not be a meet between any two elements. For instance,

$$\begin{bmatrix} 0 & 1 \\ 1 & 1 \end{bmatrix} \text{ and } \begin{bmatrix} 1 & 0 \\ 0 & 1 \end{bmatrix} \text{ do not have a meet.}$$

Similarly, a *join* between two elements x and y, denoted as $x + y$, is the least upper bound of x and y. That is, join z of x and y is the smallest element such that $z \geq x$ and $z \geq y$. In

Figure 7.1, the join between $\begin{bmatrix} 0 & 1 \\ 0 & 1 \end{bmatrix}$ and $\begin{bmatrix} 1 & 1 \\ 0 & 0 \end{bmatrix}$ is $\begin{bmatrix} 1 & 1 \\ 0 & 1 \end{bmatrix}$. Note that $\begin{bmatrix} 1 & 1 \\ 1 & 1 \end{bmatrix}$ is not a join because

$\begin{bmatrix} 1 & 1 \\ 0 & 1 \end{bmatrix}$ is smaller than $\begin{bmatrix} 1 & 1 \\ 1 & 1 \end{bmatrix}$ and is greater than both $\begin{bmatrix} 0 & 1 \\ 0 & 1 \end{bmatrix}$ and $\begin{bmatrix} 1 & 1 \\ 0 & 0 \end{bmatrix}$.

Understanding meet and join as the greatest low bound and the least upper bound, it is easy to see that Theorem 7.2 is intuitively reasonable.

---

### Theorem 7.2

For element x and y in a partially ordered set, the following hold:

**1.** The meet $x \cdot y$ and join $x + y$ are unique if they exist.
**2.** $x \geq x \cdot y$ and $x \leq (x + y)$.
**3.** $x \leq y$ if and only if $x \cdot y = x$ and $x + y = y$.

---

As we have seen, not all pairs of elements have a meet or a join. However, when every pair of elements in a partially ordered set have a meet and a join, we call this mathematical structure (the set together with the meet and join operator) a *lattice*.

The set of partitions form a lattice if the meet and join are defined carefully. We define the meet of two partitions $\pi_1 \cdot \pi_2$ to be a partition formed by the intersection of the elements in $\pi_1$ and $\pi_2$. That is, an element in the meet partition is an intersection of an element in $\pi_1$ and an element in $\pi_2$. We define the join, $\pi_1 + \pi_2$, to be a partition formed by the union of all chain-connected elements in $\pi_1$ and $\pi_2$. Two sets, $S_1$ and $S_n$, are *chain connected* if there exists a sequence of sets, $S_2, ..., S_{n-1}$, such that $S_i$ and $S_{i+1}$ have a nonempty intersection for $i = 1, ..., n - 1$. It can be proved that with this definition of meet and join, the set of partitions constitutes a lattice.

---

### Example 7.5

Find the meet and join of the following partitions

$\pi_1 = \{\overline{a,b,c}; \overline{d,e}; \overline{f}; \overline{g,h,i}; \overline{j,k}\}$

$\pi_2 = \{\overline{a,b}; \overline{c,g,h}; \overline{d,e,f}; \overline{i,j,k}\}$

---

*continues*

---

**Example 7.5    (Continued)**

The meet partition is obtained by intersecting the subsets of the two partitions. For example, subset {a,b,c} of $\pi_1$ intersects {a,b} of $\pi_2$ to give subset {a,b} in $\pi_1 \cdot \pi_2$. Performing intersections between the subsets, we get $\pi_1 \cdot \pi_2 = \{\overline{a,b}; \overline{c}; \overline{d,e}; \overline{f}; \overline{g,h}; \overline{i}; \overline{j,k}\}$.

To obtain the join partition, we grow its elements gradually, starting with two overlapping subsets, one from $\pi_1$ and another from $\pi_2$. For example, subset {a,b,c} of $\pi_1$ and subset {a,b} of $\pi_2$ have a nonempty intersection; thus, they are chain connected. Now element c connects to other subsets. Specifically, subset {a,b,c} of $\pi_1$ and subset {c,g,h} of $\pi_2$ have a nonempty intersection, so they are chain connected. So far these subsets {a,b}, {a,b,c}, and {c,g,h} are chain connected. Then elements g and h connect more elements from $\pi_1$ (namely i) which connects j and k in $\pi_2$. So, subsets {a,b}, {a,b,c}, {c,g,h}, {g,h,i}, and {i,j,k} are chain connected. This chain connection stops at subset {j,k} of $\pi_1$ because it does not connect any more new elements. Therefore, the first subset in the join partition is the union of previous chain-connected subsets, giving {a,b,c,g,h,i,j,k}. The second subset starts with a subset not yet included, say {d,e} of $\pi_1$. {d,e} connects with {d,e,f} of $\pi_2$ and f is a subset by itself. Thus, the second subset of the join is {d,e,f}. The result is $\pi_1 + \pi_2 = \{\overline{a,b,c,g,h,i,j,k}; \overline{d,e,f}\}$.

## 7.3    Boolean Functions and Representations

Let B denote the set {0, 1}. A Boolean function of $n$ variables is a mapping from $B^n$ to B, written as $f: B^n \to B$. An example is $f(a, b, c) = ab + \overline{a}c + \overline{b}c$.

The set of variables in a Boolean function is called the *support*, denoted by *supp(f)*. Thus, for the previous example function, *supp(f)* = {a, b, c}. Even when a variable is redundant in a function (that is, it does not affect the result of the function), the variable is still included in the support. For instance, variable r in $g(a, b, r) = a\overline{r} + ar + \overline{a}b$ is redundant, but *supp(g)* = {a, b, r}. If only nonredundant variables are required, the term *minimum support* is used to include only the nonredundant variables.

A Boolean function can be represented in many ways. Let's briefly review some common forms of representation. A sum of products representation is the ORing of all AND terms. An AND term, a product term, is also called a *cube*. A sum of products is written as

$$f = \sum_{i, j, ..., n} x_i x_j ... x_n$$

where $x_i, x_j, ..., x_n$ are literals. A *literal* is either a variable or the complement of the variable. Each product term or cube $x_i x_j ... x_n$ need not have all variables present. If a product term has all variables, it is called a *minterm* or *vertex*. The sum of products representation is also called *disjunctive normal form* (DNF).

For example, consider function $f(a, b, c) = a\overline{b} + \overline{a}c + ab\overline{c}$. It is a sum of products representation. It has three cubes. There are six literals—namely, $a, \overline{b}, \overline{a}, c, b,$ and $\overline{c}$. The last cube, $ab\overline{c}$, is a minterm.

Similarly, a Boolean function can be represented as a product of sums, or *conjunctive normal form* (CNF), as written here:

$$f = \prod_{i, j, ..., n} (x_i + x_j + ... + x_n)$$

where $x_i, x_j, ..., x_n$ are literals. Each sum term need not have all variables present. A sum term, the counterpart of a cube, is called a *clause*. If all variables are present in a clause, the term is called a *maxterm*.

Finally, a Boolean function can be written in a form mixed with sums of products and products of sums. This form is generically called *multilevel representation*. A sum of products can be implemented as a two-level circuit, with the first level being AND gates and the second, an OR gate. Similarly, a product of sums can be implemented as a circuit with a layer of ORs followed by an AND gate. Therefore, a mixed representation calls for multiple levels of gates to implement, hence the name. An example is $f = \overline{a}b + c\overline{d}(a + b(c + b\overline{d})) + \overline{a}(b + \overline{c}d)$.

---

**Example 7.6**

Convert the following CNF to DNF and vice versa:

$f = (\overline{a} + b + \overline{c})(a + \overline{b} + c)(\overline{a} + \overline{b} + c)(a + b + c)$
$g = a\overline{b}c + b\overline{c} + \overline{a}b\overline{d} + \overline{a}cd$

First, we convert $f$ to DNF. This can be done by simply multiplying the terms. After multiplying the four clauses and simplifying, we get

$f = \overline{a}c + bc + a\overline{b}\overline{c}$

---

*continues*

**Example 7.6    (Continued)**

To convert $g$ from DNF to CNF, we first use De Morgan's law to obtain $\bar{g}$ in CNF and to multiply the clauses to get the CNF of $\bar{g}$. Finally, we obtain the CNF of $g$ by complementing $\bar{g}$ using De Morgan's law:

$$\bar{g} = (\bar{a} + b + \bar{c})(\bar{b} + c)(a + \bar{b} + d)(\bar{a} + c + \bar{d})$$

Multiplying the clauses, we have

$$\bar{g} = (\overline{ab} + \bar{a}c + bc + \overline{bc})(ac + a\bar{d} + \overline{ab} + \bar{b}c + \overline{bd} + \bar{a}d + cd)$$

$$= \overline{ab} + \bar{a}cd + abc + bcd + \overline{bcd}$$

Now, complement $g$ and use De Morgan's law to obtain DNF:

$$g = (a + b)(a + \bar{c} + \bar{d})(\bar{a} + \bar{b} + c)(\bar{b} + \bar{c} + \bar{d})(b + c + d).$$

Another way of obtaining DNF from CNF is to use the second form of the *distributive laws* of Boolean algebra stated here:

**1.** $a \cdot (b + c) = a \cdot b + a \cdot c$
**2.** $a + (b \cdot c) = (a + b) \cdot (a + c)$

To many people, the first form is probably more familiar than the second. However, they are related through the duality principle of Boolean. The *duality principle of Boolean Algebra* states that by replacing AND with OR, OR with AND, 1 with 0, and 0 with 1 in a Boolean identity, we arrive at another Boolean identity. Suppose we start with identity

$$a \cdot (b + c) = a \cdot b + a \cdot c$$

and apply the exchange operations of duality. We get

$$a + (b \cdot c) = (a + b) \cdot (a + c)$$

which is the second form of the distributive law.

---

**Example 7.7**

---

Let us use the second form of distributive law to convert $f = a\bar{b} + \bar{a}c + bc$ to CNF.

$$f = ((\bar{a}c + bc) + a)((\bar{a}c + bc) + \bar{b})$$
$$= (\bar{a} + bc + a)(c + bc + a)(\bar{a} + bc + \bar{b})(c + bc + \bar{b})$$
$$= ((a + c) + b)((a + c) + c)((\bar{a} + \bar{b}) + b)((\bar{a} + \bar{b}) + c)((\bar{b} + c) + b)((\bar{b} + c) + c)$$
$$= (a + b + c)(a + c)(\bar{a} + \bar{b} + c)(\bar{b} + c)$$
$$= (a + c)(\bar{b} + c)$$

To verify that our conversion is correct

$$f = a\bar{b} + \bar{a}c + bc$$

$$= a\bar{b} + \bar{a}c + (bc + abc)$$

$$= a\,(\bar{b} + bc) + \bar{a}c + bc$$

$$= a\,(\bar{b} + c) + \bar{a}c + bc \text{ (second form of distributive law)}$$

$$= a\bar{b} + ac + \bar{a}c + bc$$

$$= a\bar{b} + c + bc$$

$$= a\bar{b} + c$$

$$= (a + c)(\bar{b} + c) \text{ (second form of distributive law)}$$

---

As far as representation size is concerned, it is difficult to know when CNF or DNF is more compact. It is known that there are functions with a CNF that is of polynomial size in number of variables expands to a DNF of exponential size, and vice versa. A research result relating the size of CNF to DNF has the following bounds[1]. If function $f$ has $n$ variables and its CNF size is less than $m$, then its DNF size, $\|f(n, m)\|_{dnf}$, is bound by the following expression, where $c_1$ and $c_2$ are constants:

$$2^{n - \left(c_1 \cdot \frac{n}{\log \frac{m}{n}}\right)} \leq \|f(n, m)\|_{dnf} \leq 2^{n - \left(c_2 \cdot \frac{n}{\log \frac{m}{n}}\right)}$$

Given a function $f$ in DNF, the problem of satisfying $f$ is to determine whether there is an assignment of variables for which $f$ evaluates to 0 (that is, whether $f$ has a nonempty

---

1.  Miltersen, et al, "On Converging CNF to DNF," Mathematical Foundations of Computer Science: 28th International Symposium, Bratislava, Slovakia, August 25–29, 2003, Lecture Notes in Computer Science 2747, 612–21.

offset). Similarly, if $f$ is in CNF, the *satisfiability* problem is to determine whether $f$ can evaluate to 1 (in other words, $f$ has a nonempty on set).

A brute-force method to solve the satisfiability problem is to expand the function into minterms or maxterms. This approach, in the worst case, encounters exponential numbers of terms and is practically infeasible, because the satisfiability problem is NP complete. We return to this topic in a later chapter.

## 7.3.1  Symmetric Boolean Functions

Symmetric Boolean functions have some interesting properties and lend themselves to computational efficiency. A Boolean function can be regarded as a function of literals. For instance, $f(a,\bar{b},c) = abc + \overline{ab}c + a\overline{bc} + \overline{a}bc$ is a function of literals $a$, $\bar{b}$, and $c$, as opposed to a function of variables $a$, $b$, and $c$. With this clarification, we say that a Boolean function $f(x_1,...,x_n)$ is *symmetric* with respect to a set of literals, $x_1,...,x_n$, if the function remains unchanged by any permutation of the literals. That is, $f(x_1,...,x_n) = f(x_{\sigma(1)},...,x_{\sigma(n)})$, where $\sigma$ is a permutation map, which is a one-to-one mapping from $\{1,...,n\}$ to $\{1,...,n\}$. An example of $\sigma$ is the following: 1 maps to 2, 2 to 3, n to 1. In this case, $f(x_{\sigma(1)},...,x_{\sigma(n)}) = f(x_2,x_3,...,x_n,x_1)$. It should be emphasized that $x_i$ is a literal, not a variable. In the previous example function, $x_1 = a$, $x_2 = \bar{b}$, and $x_3 = c$.

It can be cumbersome to use this intuitive definition to determine whether a function is symmetric. Theorem 7.3 alleviates this task.

---

**Theorem 7.3**

Function $f(x_1,...,x_n)$ is symmetric if and only if there exists a set of integers $I = \{a_i : i = 0,$ $..., m \leq n\}$, such that $f(x_1,...,x_n)$ is 1 when exactly $a_i$ literals of symmetry are 1.

---

For example, $f(a,\bar{b},c) = abc + \overline{ab}c + a\overline{bc} + \overline{a}\overline{b}\overline{c}$ takes the value of 1 if and only if zero or two of the literals $a$, $\bar{b}$, or $c$ have the value of 1, no matter which two. Therefore, this function is symmetric and $I = \{0, 2\}$. Consider function $f(a,b,c) = a\overline{b}c + \overline{a}bc + \overline{ab}c$. When $a = 0$, $b = 1$, and $c = 1$, the function evaluates to 1. However, when $a = 1$, $b = 1$, and $c = 0$, which also has two literals being 1, the function fails to evaluate to 1. Therefore, evaluation of this function does depend on which literals have the value of 1. Thus, we conclude, according to Theorem 7.3, that this function is not symmetric with respect to literals $a$, $b$, and $c$.

Interestingly, if we rewrite the function as a function of literals $a$, $b$, and $\bar{c}$, and replace $\bar{c}$ with $d$, we have $f(a,b,d) = a\overline{bd} + \overline{a}b\overline{d} + \overline{ab}d$, which is symmetric with respect to $a$, $b$, and $d$. That is, the function is symmetric with respect to $a$, $b$, $\bar{c}$, but not with respect to $a$, $b$, $c$. This example shows why we need to treat symmetric functions with respect to literals, instead of variables.

Theorem 7.3 makes sense in that if a function is symmetric, then only the number of literals, not the specific literals, matters in determining the function's output value, because a symmetric function cannot distinguish literals. Because the number of literals being 1 uniquely determines the function's value, symmetric functions can be represented succinctly by the set I—in other words, $S^n_I(x_1,...,x_n)$, where $n$ denotes the number of variables in the function. For example, symmetric function $f(a,\overline{b},c) = abc + \overline{a}bc + a\overline{bc} + \overline{ab}c$ is represented as $S^3_{0,2}(a,b,c)$, and $f(a,b,\overline{c}) = a\overline{bc} + \overline{a}bc + \overline{abc}$, by $S^3_1(a,b,\overline{c})$.

With these notations for symmetric functions, manipulation of symmetric functions becomes easy with Theorem 7.4. Logical operations become set operations on the indices.

---

**Theorem 7.4**

Given symmetric functions $S^n_I(x_1, ..., x_n)$ and $S^n_J(x_1, ..., x_n)$, the following holds:

**1.** $S^n_I(x_1, ..., x_n) + S^n_J(x_1, ..., x_n) = S^n_{I \cup J}(x_1, ..., x_n)$

**2.** $S^n_I(x_1, ..., x_n) \cdot S^n_J(x_1, ..., x_n) = S^n_{I \cap J}(x_1, ..., x_n)$

**3.** $\overline{S^n_I(x_1, ..., x_n)} = S^n_{\neg I}(x_1, ..., x_n)$

**4.** $S^n_I(x_1, ..., x_n) = S^n_{n-I}(\overline{x}_1, ..., \overline{x}_n)$

---

Parts 1 and 2 state that ORing and ANDing of two symmetric functions amount to union and intersection of their respective index sets. The ORed result is 1 when at least one of its operands evaluates to 1. This means that the number of literals being 1 in the ORed result is in either or both index sets, giving rise to the union of the index sets. Similar reasoning applies to ANDing. In part 3, $\neg I$ denotes the complement of the index set I: $\neg I = \{1, ..., n\} - I$. So, complementing a function amounts to complementing the index set. The complement of a function being 1 means the function itself is 0, meaning the number of literals being 1 is not in the index set. In part 4, $n - I = \{j : j = n - i, i \in I\}$, and it states that complementing the literals translates to complementing the individual indices.

---

**Example 7.8**

---

Determine whether the following function is symmetric. Also, compute the complement of the function:

$$f(a, b, c, d) = \bar{a}b\bar{c}\bar{d} + a b\bar{c}\bar{d} + \bar{a}\bar{b}\bar{c}d + \bar{a}b\bar{c}d + \bar{a}b\bar{c}d$$

We need to know whether the number of literals being 1 uniquely determines the function's value. To start, we represent the function in a truth table (Table 7.1). Only the values giving the function a value of 1 are shown.

From Table 7.1 we see that the function's value is not solely determined by the number of 1 in literals $a$, $b$, $c$, and $d$. For example, `0100` gives output `1` but not `1000`. Therefore, the function is not symmetric with respect to literals $a$, $b$, $c$, and $d$.

However, it is possible that for other literals, the function is symmetric. We can invert some of the literals, change the values in the table, and determine whether the function becomes symmetric. By complementing variables $a$ and $c$, we get the symmetric functions seen in Table 7.2.

From Table 7.2 we see that as long as three or four literals are 1, the function is 1. Therefore, the function is symmetric with respect to literals $\bar{a}, b, \bar{c},$ and $d$. In terms of these literals,

$$f(\bar{a}, b, \bar{c}, d) = abc\bar{d} + \bar{a}bcd + a\bar{b}cd + abc\bar{d} + abcd$$

The function can be represented as $f_{3,4}{}^4(\bar{a}, b, \bar{c}, d)$. Once the function is symmetric, its complement can be computed readily, using Theorem 7.4. $\bar{f} = f^4{}_{0,1,2}(\bar{a}, b, \bar{c}, d)$. Expanding this to the sum of products, we have

$$\overline{f(a, b, c, d)} = \bar{a}\bar{b}\bar{c}\bar{d} + \bar{a}\bar{b}\bar{c}d + a\bar{b}\bar{c}\bar{d} + \bar{a}\bar{b}c\bar{d} + \bar{a}b\bar{c}\bar{d} + a\bar{b}\bar{c}\bar{d}$$
$$+ a\bar{b}\bar{c}d + a\bar{b}c\bar{d} + \bar{a}\bar{b}c\bar{d} + \bar{a}b\bar{c}d + abcd$$

**Table 7.1**    Determining Functional Symmetry

| (a,b,c,d) |
|:---:|
| 0100 |
| 1101 |
| 0001 |
| 0111 |
| 0101 |

**Table 7.2**    Symmetric Function

| $(\bar{a},b,\bar{c},d)$ |
|:---:|
| 1110 |
| 0111 |
| 1011 |
| 1101 |
| 1111 |

## 7.3.2  Incompletely Specified Boolean Functions

A Boolean function can also be characterized by its on set, off set, and don't-care set. An *on set* is the set of vertices that make the function 1; and an *off set*, 0. A *don't-care set* is the set of vertices on which the function is not specified or a don't-care. The union of on set, off set, and don't-care set makes up the universe. If a Boolean function is *completely specified*, then the don't-care set is empty; otherwise, the function is *incompletely specified*.

**Example 7.9**

The on set, off set, and don't-care set representation can be displayed spatially for functions with a smaller number of variables. Consider function

$$f(a, b, c) = \overline{a}b + a\overline{c} + \overline{b}c$$

Refer to Figure 7.2. Three axes represent variables $a$, $b$, and $c$ respectively. Each axis has only two values: 0 at the original and 1 at the far end. The coordinates are the values of the variables. For instance, 101 means a = 1, b = 0, and c = 1, and hence represents minterm $a\overline{b}c$.

The function is completely specified and its on set is all vertices in the expression. To see the vertices in the on set, we expand the three cubes into minterms. Cube $\overline{a}b$ is equal to $\overline{a}bc + \overline{a}b\overline{c}$. Therefore, $f(a, b, c) = \overline{a}bc + \overline{a}b\overline{c} + ab\overline{c} + a\overline{b}\overline{c} + a\overline{b}c + \overline{a}\overline{b}c$.

Each minterm is a vertex in the on set. Mark the vertices on the cube as small solid circles. Because the function is completely specified, the remaining vertices are in the off set, marked as small squares.

The vertices, if grouped as shown, and each group of two vertices corresponds to a cube, produce the three cubes $a\overline{b}$, $\overline{a}c$, and $b\overline{c}$. Therefore, the function is also equal to

$$f(a, b, c) = a\overline{b} + \overline{a}c + b\overline{c}$$

which is obvious from the spatial representation but not so in the sum-of-products form.

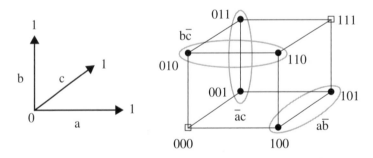

**Figure 7.2**   A spatial representation of a Boolean function

### 7.3.3   Characteristic Functions

Representing incompletely specified Boolean functions becomes more complicated because a single incompletely specified Boolean function requires at least two Boolean expressions to represent: one for the on set and the other for the off set or don't-care set. The matter becomes more so in multiple-output Boolean functions. An example of a multiple-output Boolean function is the next-state function of a finite-state machine with two D-type FFs. The next-state function has two components: $f(v) = (f_1(v), f_2(v))$, $f_1$ for the first FF and $f_2$ for the second. A multiple-output Boolean function is also called a *vector Boolean function*.

To deal with multiple-output Boolean functions, one may suggest treating each component as a function and representing each output function with its on set, off set, and don't-care set. A difficulty arises when the output of a multiple-output Boolean function takes on more than one value at a given input value. For example, when $v = 1011$, $f(v)$ can be either 11 or 00. It may be tempting to deduce that 1011 represents a don't-care point in each of the output functions. The erroneous reasoning is that $f_1(1011)$ is either 1 or 0; thus, $f_1(1011)$ is a don't-care. The same reasoning applies to $f_2(1011)$. Therefore, $f(1011) = (*, *)$, where * represents a don't-care. The fallacy is that by assuming that $f_1(1011)$ and $f_2(1011)$ are don't-cares, the output don't-care points include not just 11 and 00, but also 10 and 01, because $(*, *)$ means either component can be any value. However, the problem states that for 1011, the output can be only 11 or 00. The root cause of this difficult is that in this case the multiple-output Boolean function is no longer a function but a relation, because it maps an input value to more than one output value. In the single-output situation, we simply use a don't-care set to solve this multivalue output problem, but this technique no longer works for the multioutput situation.

To represent a Boolean relation, characteristic function plays a vital role. A *characteristic function* $\lambda(S)$ representing set S is defined as follows:

$$\lambda(r) = \begin{cases} 1, r \in S \\ 0, r \notin S \end{cases}$$

For a Boolean relation with input vector $v$ and output vector $w$, the characteristic function is $\lambda(v, w)$ and the set S is all I/O pairs describing the Boolean relation. In other words, $\lambda(v, w)$ is 1 if $w$ is an output for input $v$, and it is 0 otherwise.

**Example 7.10**

Use characteristic functions to represent the following Boolean relations.

1. A completely specified Boolean function, $f(a, b, c) = a\overline{b} + \overline{a}c + b\overline{c}$.
2. An incompletely specified Boolean function $f(a, b, c)$, with on set equal to {110, 010, 001} and don't-care set, {111, 000}.
3. An incompletely specified Boolean relation $f(v) = (f1(v), f2(v))$, $v = (a,b,c)$, described by Table 7.3 on page 353.

  a. Let $\lambda(y, a, b, c)$ be the characteristic function, where variable $y$ is the output value. The input-output pairs are {$(y, a, b, c): y = 1$ if $f(a,b,c) = 1$ and $y = 0$ if $f(a,b,c) = 0$}. Therefore, the characteristic function is defined as follows: $\lambda(y, a, b, c) = 1$ if $y = f(a,b,c)$, or $\lambda(y, a, b, c) = \overline{y \oplus f(a, b, c)}$. Substituting the equation for $f$, we get:

$$\lambda(y, a, b, c) = y \cdot f(a, b, c) + \overline{y} \cdot \overline{f(a, b, c)}$$
$$= y(a\overline{b} + \overline{a}c + b\overline{c}) + \overline{y}\overline{(a\overline{b} + \overline{a}c + b\overline{c})}$$
$$= y(a\overline{b} + \overline{a}c + b\overline{c}) + \overline{y}(abc + \overline{a}\overline{b}\overline{c})$$

As check, we know that f(101) = 1. Substitute y = 1, a = 1, b = 0, and c = 1 into $\lambda(y, a, b, c)$, and we get 1, as expected. Suppose we substitute y = 0, a = 1, b = 0, and c = 1 into $\lambda(y, a, b, c)$. We should get 0, because this output value is not produced by the input vector. A little calculation shows that $\lambda(1, 1, 0, 1) = 0$.

  b. The only difference from (a) is the don't-care points. A don't-care point simply implies that variable $y$ can be either 1 or 0. Therefore, the characteristic function $\lambda(y, a, b, c)$ is

$$\lambda(y, a, b, c) = y \cdot (ab\overline{c} + \overline{a}b\overline{c} + \overline{a}\overline{b}c) + \overline{y} \cdot (\overline{a}bc + a\overline{b}c + a\overline{b}c) + (abc + \overline{a}\overline{b}\overline{c})$$

The first group of terms represents the on set; the second, the off set; and the third, the don't-care set. We obtained the off set by subtracting the on set and the don't-care set from the universe.

  c. Let $y$ and $z$ be the variables for output $f_1$ and $f_2$ respectively. Let's go through Table 7.3 one row at a time. At each row we add a term to the characteristic function. The term describes, in Boolean form, the relation between the I/O values. For example, the first row gives $y\overline{z}\overline{a}\overline{b}c$ because output is 10 when input is 001. Then the characteristic function is

$$\lambda(y, z, a, b, c) = y\overline{z} \cdot \overline{a}\overline{b}c + \overline{y}z \cdot \overline{a}b\overline{c} + (yz + \overline{y}\overline{z}) \cdot a\overline{b}\overline{c} + (\overline{y}\overline{z} + yz) \cdot ab\overline{c} + \overline{y}\overline{z} \cdot$$
$$(\overline{a}\overline{b}\overline{c} + \overline{a}b\overline{c} + a\overline{b}\overline{c} + abc)$$

> **Example 7.10   (Continued)**
>
> The first two cubes come from the first two entries of the table. The next two groups of terms are from the third and the fourth entries, each of which have two output values. The two output values are ORed because either output value is possible. The last group of terms comes from the last entry of the table, which consists of all remaining minterms.

**Table 7.3**   Specification of a Vector Boolean Relation

| v = (a,b,c) | f = ($f_1, f_2$) |
|---|---|
| 001 | (1,0) |
| 011 | (0,1) |
| 101 | (1,1) or (0,0) |
| 110 | (1,0) or (1,1) |
| all other values | (0,0) |

## 7.4   Boolean Functional Operators

In this section, we will study some common operators encountered in formal verification. Cofactor operation on function $f$ with respect to a variable $x$ has two phases—a positive and negative phase—denoted by $f_x$ and $f_{\bar{x}}$ respectively. Cofactor $f_x$ is computed by setting variable $x$ to 1, whereas $f_{\bar{x}}$ is computed by setting $x$ to 0. Therefore, $f_x$ and $f_{\bar{x}}$ are functions without variable $x$. Functions $f_x$ and $f_{\bar{x}}$ can be further cofactored with respect to other variables. For example, cofactoring with respect to variable $y$ gives $(f_x)_y$, which is written as $f_{xy}$. It can be shown that $f_{xy} = f_{yx}$. Function $f$ is related to its cofactors through the following Shannon theorem (Theorem 7.5).

**Theorem 7.5**   Shannon Expansion Theorem

Any Boolean function $f$ can be expressed in the following forms, where $x$ is a variable of $f$:

**1.** $f = x \cdot f_x + \bar{x} \cdot f_{\bar{x}}$

**2.** $f = (\bar{x} + f_x) \cdot (x + f_{\bar{x}})$

**3.** $f = x \cdot f_x \oplus \bar{x} \cdot f_{\bar{x}}$

The Shannon cofactor can be applied successively to all variables in a function. For example, a function with two variables, $x$ and $y$, can be expressed as

$$f(x, y) = x \cdot f_x + \bar{x} \cdot f_{\bar{x}}$$
$$= x(y \cdot f_{xy} + \bar{y} \cdot f_{x\bar{y}}) + \bar{x} \cdot (y \cdot f_{\bar{x}y} + \bar{y} \cdot f_{\bar{x}\bar{y}})$$
$$= xy \cdot f_{xy} + x\bar{y} \cdot f_{x\bar{y}} + \bar{x}y \cdot f_{\bar{x}y} + \bar{x}\bar{y} \cdot f_{\bar{x}\bar{y}}$$

This last equation is a minterm expansion of the function. Because the function has only two variables, the cofactors, $f_{xy}$, $f_{x\bar{y}}$, $f_{\bar{x}y}$, and $f_{\bar{x}\bar{y}}$, are either 1 or 0. Therefore, Shannon expansion can be regarded as a partial evaluation.

**Example 7.11**

In this example, let's compute the Shannon cofactors and apply the Shannon expansion to derive various representations.

**1.** Compute the Shannon cofactor with respect to $x$ for $f(x, y, z) = x\bar{y} + y(x\bar{z} + z)(\bar{x} + \bar{y}z) + z$.

**2.** Rewrite function $f(x, y, z) = (x + \bar{y})(\bar{y} + z) + xy + \bar{x}z$ in the generalized Reed–Muller form. A Reed–Muller form is an XOR sum of cubes, in which each variable is either a positive phase or a negative phase.

a. The cofactors for $f(x, y, z) = x\bar{y} + y(x\bar{z} + z)(\bar{x} + \bar{y}z) + z$ are computed as follows:

$$f_x = f(1, y, z) = \bar{y} + y(\bar{z} + z)(\bar{y}z) + z = \bar{y} + z$$
$$f_{\bar{x}} = f(0, y, z) = y(\bar{z} + z)(1 + \bar{y}z) + z = y + z$$

Therefore, according to Theorem 7.5, $f = x(\bar{y} + z) + \bar{x}(y + z) = x\bar{y} + \bar{x}y + z$ or $f = (x + \bar{y} + z)(x + y + z)$.

**Example 7.11   (Continued)**

b. Apply the XOR version of the Shannon expansion to $f(x, y, z) = (x + \bar{y})(\bar{y} + z) + xy + \overline{xz}$ successively until it is in Reed–Muller form. First apply it to variable $x$. We get

$$f = xf_x \oplus \bar{x}f_{\bar{x}}$$
$$= x(\bar{y} + z + y) \oplus \bar{x}(\bar{y}(\bar{y} + z) + \bar{z})$$
$$= x \oplus \bar{x}(\bar{y} + \bar{z})$$

Now let $g$ be $(\bar{y} + \bar{z})$, and cofactor with respect to $y$. We obtain the required Reed–Muller form:

$$f = x \oplus \bar{x}(yg_y \oplus \bar{y}g_{\bar{y}})$$
$$= x \oplus \bar{x}(y\bar{z} \oplus \bar{y})$$
$$= x \oplus \bar{x}y\bar{z} \oplus \overline{xy}$$

You should verify that this Reed–Muller form is indeed functionally identical to the given function.

This cofactor definition, sometimes called the *classic cofactor*, is based on a cube; that is, the $c$ in cofactor $f_c$ is cube. The *generalized cofactor* function of a vector Boolean function $f$, $f_g$, allows $g$ to be any Boolean function and is defined as

$$f_g = \begin{cases} (* , \ldots , *), \ if\, g = 0 \\ f, \ if\, g = 1 \end{cases}$$

where * denotes a don't-care value. It is tempting to assume that $f_g$ is equal to $fg$, which gives $f$ when $g = 1$. The following exercise shows that they are not the same.

---

### Example 7.12

Express the generalized cofactor function using a characteristic function. From the definition, the characteristic function is

$$\lambda(y, x_1, ..., x_n) = \overline{(y \oplus f)} \cdot g + \overline{g},$$

where variable $y$ is the value of the cofactor, and $x_1$ through $x_n$ are the variables in $f$ and $g$. The function is 1 if $y$ has the same value as $f$ when $g$ is 1 or $y$ can be any value (such as don't-care), when $g$ is 0, which is exactly the definition of generalized cofactor.

Let's apply this technique to computing the generalized cofactor $f_g$, where $f = a\overline{b} + (c + \overline{d})$ $(\overline{ab} + a\overline{b})$ and $g = \overline{ab} + cd$. Using a characteristic function to represent $f_g$, we have

$$\lambda(y, a, b, c, d) = \overline{(y \oplus (a\overline{b} + (c + \overline{d})(\overline{ab} + a\overline{b})))} \cdot (\overline{ab} + cd) + \overline{(\overline{ab} + cd)}$$
$$= \overline{y}(\overline{ab} + abcd) + y(a \oplus b) \cdot cd + (a + b) \cdot (\overline{c} + \overline{d})$$

To check, let $a = b = c = d = 0$. Then, $g = 1$ and $f = 0$. $\lambda(y, 0, 0, 0, 0) = \overline{y}$, which means the value of the cofactor is 0, consistent with the value of $f$ because $g = 1$. Let $a = 1$ and $c = 0$. Then $g = 0$. So the cofactor is a don't-care, meaning the characteristic function should be 1 independent of $y$. $\lambda(y, 1, b, 0, d) = 1$, as expected.

---

The *Boolean difference* of function $f$ with respect to variable $x$, $\dfrac{\partial f}{\partial x}$, is defined to be

$$\frac{\partial f}{\partial x} = f_x \oplus f_{\overline{x}}$$

If the Boolean difference of a function with respect to variable $x$ is identically equal to 0, then the function is independent of $x$, because the cofactors with respect to $x$ are equal, meaning the function remains the same whether $x$ takes on the value of 1 or 0. Therefore, $x$ is in the minimum support of a function if and only if its Boolean difference is not identically equal to 0.

The concept of Boolean difference is used widely in practice. For example, to determine whether a circuit can detect a stuck-at fault at node $x$, the decision procedure is equivalent to determining whether there is an input vector $v$ such that

$$\frac{\partial}{\partial x} f(x, v) = 1$$

where $f$ is the function of the circuit with the fault site $x$ made as a primary input.

The problem of deciding whether a value $v$ exists such that function $f(v, w)$ evaluates to a specific function $g(w)$ is called the *existential problem*, and is often denoted by the *existential operation* as $\exists v f(v, w) \equiv g(w)$. To compute an existential problem, we simply enumerate all possible values of input $v$, evaluate the function on each of them, and OR together all outputs. If the result is equal to $g(w)$, then such an input value exists. This procedure of computing existential operation in mathematical form is

$$\exists v f(v, w) \Leftrightarrow \sum_{m \text{ is a minterm of } v} f_m(v, w)$$

where $f_m(v, w)$ is a cofactor with respect to minterm $m$. For instance, if the existential operator is on two variables, $x$ and $y$, then all minterms are $\{xy, \bar{x}y, x\bar{y}, \bar{x}\bar{y}\}$. This summation can be regarded as a Boolean operation for evaluating the existential operation. This operator, called the *smooth operator*, is

$$S_v(f) = \sum_{\text{all minterms of } v} f_m$$

A special case is when the existential operator is on a single variable, say $x$. Then the equation becomes $S_x(f) = f_x + f_{\bar{x}}$.

The counteroperator of the existential operator is the *universal operator*. An example of an application of a universal operator is deciding whether function $f(v, w)$ is always equal to some specific function $g(w)$ for all values of $v$. This is denoted by $\forall v f(v, w) \equiv g(w)$. To compute this universal operation, we can evaluate $f(v, w)$ for each of possible values of $v$ and AND the outputs. If the result is $g(w)$, the answer is yes. Mathematically, the computation is

$$\forall v f(v, w) \Leftrightarrow \prod_{m \text{ is a minterm of } v} f(m, w)$$

If the function has a single variable $x$, $\forall x f(x) = f_x \cdot f_{\bar{x}}$.

---

**Example 7.13**

**1.** Show that function $f$ can be expressed as

$$f = x \cdot \frac{\partial f}{\partial x} \oplus f_{\bar{x}}.$$

**2.** In a finite-state machine whose next-state function is given by $(f_1, f_2) = (a\bar{r} + \bar{b}s, (a + b)(rs))$, where $a$ and $b$ are input bits, and $r$ and $s$ are present state bits. $(s, r)$, find an expression for all present states whose next state can be either $(1, 0)$ or $(0, 1)$ under some input.

a. Let's start with the XOR version of the Shannon expansion theorem.

$$f = xf_x \oplus \bar{x}f_{\bar{x}} = xf_x \oplus (1 \oplus x)f_{\bar{x}} = xf_x \oplus (f_{\bar{x}} \oplus xf_{\bar{x}})$$

$$= (xf_x \oplus xf_{\bar{x}}) \oplus f_{\bar{x}} = x(f_x \oplus f_{\bar{x}}) \oplus f_{\bar{x}} = x\frac{\partial f}{\partial x} \oplus f_{\bar{x}}$$

b. Let's formulate the problem using the existential operator. The question in English is to find all present states such that for some input $(a, b)$, the output of the next-state function is either $(1, 0)$ or $(0, 1)$. Translating it into a mathematical expression, we get

$$\exists(a, b)(f_1 \cdot \bar{f_2} + \bar{f_1} \cdot f_2) = S_{a, b}(f_1 \oplus f_2)$$

$$= (f_1 \oplus f_2)_{ab} + (f_1 \oplus f_2)_{a\bar{b}} + (f_1 \oplus f_2)_{\bar{a}b} + (f_1 \oplus f_2)_{\overline{ab}}$$

Evaluating each cofactor gives

$$(f_1 \oplus f_2)_{ab} + (f_1 \oplus f_2)_{a\bar{b}} + (f_1 \oplus f_2)_{\bar{a}b} + (f_1 \oplus f_2)_{\overline{ab}}$$

$$= (\bar{r} + s) + \bar{r} + (rs) + s$$

$$= s + \bar{r}$$

Therefore, the set of present states that can be driven to 10 or 01 by an input in one cycle is $s + \bar{r}$, which represents present states $(s, r) = \{(1,0), (1,1), (0,0)\}$.

Let us check our result. If the present state is $(1,0)$, the next state is $(a + \bar{b}, 0)$, which becomes $(1, 0)$ if input $a$ is 1 or $b$ is 0. If the present state is $(1,1)$, the next state is $(\bar{b}, a + b)$, which is $(1,0)$ if $a = b = 0$, or is $(0,1)$ if $b = 1$. Finally, if the present state is $(0, 0)$, the next state is $(a, 0)$, which is $(1,0)$ if $a = 1$. The expression states that if the present state is $(0,1)$, no input can drive it to $(1,0)$ or $(0,1)$ in one cycle. Suppose the present state is $(0,1)$, the next state is always $(0,0)$, as expected.

## 7.5   Finite-State Automata and Languages

In computer science literature, finite-state automata are used in the place of finite-state machines, with minor differences. A finite-state automaton accepts input symbols and transits just like a finite-state machine, but it does not produce any output as in a finite-state machine. By adding an output function to an automaton, we have a finite-state machine or a transducer. Furthermore, a finite-state automaton has a set of states designated as acceptance states that do not have a counterpart in finite-state machines.

A *finite-state automaton* is a mathematical object with five components, or a quintuple, $(Q, \Sigma, \delta, q_0, F)$, where $Q$ is the set of states, $\Sigma$ is the set of input symbols, $\delta$ is the transition function with $\delta : Q \times \Sigma \to Q$, $q_0$ is an initial state, and $F$ is a set of states designated as acceptance or final states. The transition function $\delta$ is based on the current state and an input symbol, and it produces the next state. Another way to represent $\delta$ is $q^{i+1} = \delta(q^i, a)$, where $q^i$ is the present state at cycle i, $q^{i+1}$ is the next state, and $a \in \Sigma$ is an input symbol.

A *transducer* or *finite-state machine* is a 6-tuple $(Q, \Sigma, \Lambda, \delta, \mu, q_0)$, where $\Lambda$ is the set of output symbols and $\mu$ is the output function with $\mu : Q \times \Sigma \to \Lambda$. The output function here assumes the Mealy machine model in that the output is a function of both the current state and the input symbol. For a Moore machine, the output function is simply $\mu : Q \to \Lambda$.

A finite-state automaton can be depicted with a state diagram. A state diagram is a directed graph with nodes and directed edges connecting the nodes. A node is a state, and an edge is a transition. An input symbol that activates a transition is placed on an edge. An initial state is marked with a wedge, and an acceptance state is twice-circled node. In a finite-state machine, the output symbol is placed on a node if it depends only on the state (Moore machine); otherwise, it is placed over an edge (a Mealy machine).

---

### Example 7.14

The state diagram in Figure 7.3 represents a finite-state automaton and a Mealy finite-state machine based on the automaton. The automaton has five states: $q_0$, $q_1$, $q_2$, $q_3$, and $q_4$. The initial state, $q_0$, is marked with a wedge. The acceptance state, $q_2$, has a double circle. Note that the finite-state machine does not have an acceptance state. The input symbol set

*continues*

**Example 7.14    (Continued)**

consists of *a*, *b*, and *c*. Using the mathematical notations, this automaton has an input symbol, states, initial state, acceptance state, and transition function as follows:

$Q = \{q_0, q_1, q_2, q_3, q_4\}$

$\Sigma = \{a, b, c\}$

$q_0 = \{q_0\}$

$F = \{q_2\}$

The transition function is presented in Table 7.4. The first column in the table lists the present states. The remaining three columns are the next states on each of the three input symbols.

The finite-state machine based on this automaton has an output symbol set consisting of *x* and *y*. The output symbols are placed over the edges to denote the fact that the output depends on both the present state and the current input symbol. In mathematical notation, this state machine has the same state set, initial state, and transition function. It also has $\Lambda = \{x, y\}$. The output function is presented in Table 7.5. The first column is the present state, and the other three columns are outputs based on the present state and the present input symbol.

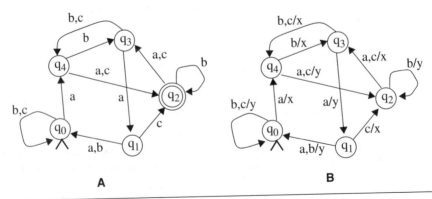

**Figure 7.3**    (A) A finite-state automaton (B) A finite-state machine

**Table 7.4**  Automaton Transition Function

| $\delta$ | a | b | c |
|:---:|:---:|:---:|:---:|
| $q_0$ | $q_4$ | $q_0$ | $q_0$ |
| $q_1$ | $q_0$ | $q_0$ | $q_2$ |
| $q_2$ | $q_3$ | $q_2$ | $q_3$ |
| $q_3$ | $q_1$ | $q_4$ | $q_4$ |
| $q_4$ | $q_2$ | $q_3$ | $q_2$ |

**Table 7.5**  Output Function for Finite State Machine

| $\mu$ | a | b | c |
|:---:|:---:|:---:|:---:|
| $q_0$ | $x$ | $y$ | $y$ |
| $q_1$ | $y$ | $y$ | $x$ |
| $q_2$ | $x$ | $y$ | $x$ |
| $q_3$ | $y$ | $x$ | $x$ |
| $q_4$ | $y$ | $x$ | $y$ |

So far, the automata and state machines discussed are deterministic, because their states over time can be predicted or determined. A *nondeterministic finite-state automaton* or finite-state machine, as an extension of the previous definitions, can have more than one initial state, or more than one next state on an input symbol, or a transition on no input symbol. It can be shown that nondeterministic finite-state automata can be transformed into a deterministic finite-state automaton with enlarged state spaces.

A finite-state automaton is *completely specified* if the next state is specified for every input symbol and every present state. Otherwise, it is *incompletely specified*. For finite-state machines, the additional requirement is that the output symbol is specified for every state in a Moore machine and for every state and input symbol in a Mealy machine.

Figure 7.4 shows a nondeterministic finite-state automaton and an incompletely specified finite-state machine. In Figure 7.4A, the automaton has two initial states: $q_0$ and $q_1$. Furthermore, state $q_1$ transits to states $q_0$ and $q_2$ on input symbol $b$. State $q_2$ can nondeterministically transit to $q_3$, even though no input symbol is present. In Figure 7.4B, ? denotes unspecified output. Furthermore, state $q_0$ does not specify the next state on input symbol $c$.

An input symbol makes an automaton transit from one state to another. Thus, a string of input symbols causes an automaton or a state machine to make a sequence of state transitions. A string of finite length is *accepted* if the automaton terminates at a final or acceptance state $q_f$ (in other words, $q_f \in F$). The set of all accepted strings of an automaton is the *language* of the automaton.

Using ending states as a condition for acceptance is just one of many possible ways of defining string acceptance and hence the language of an automaton. In fact, different kinds of acceptance conditions give rise to different types of automata. The previous definition of acceptance will not work for defining acceptance of strings of infinite length. As an example of a different kind of acceptance condition, consider the following acceptance condition of infinite strings. An infinite string is accepted if the set of states visited infinitely often is a subset of the final state set $F$. A state is *visited infinitely often* if for every N, there is M > N such that the state is visited at cycle M. Automata with this acceptance condition are called

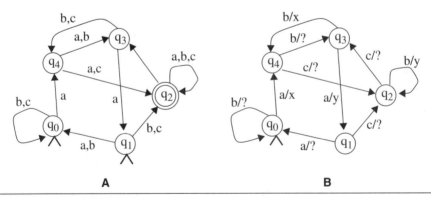

**Figure 7.4**   (A) A nondeterministic finite-state automaton. (B) An incompletely specified finite-state machine.

*Buchi automata.* Symbolically, a Buchi automaton is a quintuple, $(Q, \Sigma, \delta, q_0, F)$, with the acceptance condition that $\{q|\ q\ is\ visited\ infinitely\ often\} \subseteq F$.

Sometimes, one may be curious about situations when infinite strings may arise in practice. Indeed, all hardware machines are designed to function continuously over time, and therefore it is only realistic, although a bit counterintuitive, to assume that all input strings are of infinite length.

## 7.5.1 Product Automata and Machines

A system can consist of more than one finite-state machine that interact with each other. An example is a traffic light controller at a street intersection. There are four controllers required, one for each direction. These four controllers do not operate independently but take into consideration inputs from other controllers. Cars sensed waiting on the north–south-bound intersection prompt the east–west controller to move to red in the next cycle. Sometimes one must model a system of multiple interacting finite-state machines as one single finite-state machine. The resulting overall finite-state machine is called the *product finite-state machine*.

Given two finite-state machines, $M_1 = (Q_1, \Sigma_1, \Lambda_1, \delta_1, \mu_1, q^1{}_0)$ and $M_2 = (Q_2, \Sigma_2, \Lambda_2, \delta_2, \mu_2, q^2{}_0)$, the product machine $M = M_1 \times M_2$ is defined as $(Q, \Sigma, \Lambda, \delta, \mu, q_0)$, where

$$Q = Q_1 \times Q_2$$

$$\Sigma = \Sigma_1 \times \Sigma_2$$

$$\Lambda = \Lambda_1 \times \Lambda_2$$

$$\delta = (\delta_1, \delta_2)$$

$$\mu = (\mu_1, \mu_2)$$

$$q_0 = (q_0^1, q_0^2)$$

$M_1$ and $M_2$ are called the *component machines* of $M$. If $\Sigma_1$ and $\Sigma_2$ are identical, then $\Sigma$ is either $\Sigma_1$ or $\Sigma_2$ instead of a Cartesian product of the two.

**Example 7.15**

Let's compute the product machine of the following two finite-state machines shown in Figure 7.5. Symbolically,

$$M_1 = (\{s_0, s_1, s_2\}, \{a, b\}, \{0, 1\}, \delta_1, \mu_1, s_0)$$
$$M_2 = (\{r_0, r_1, r_2, r_3\}, \{a, b\}, \{0, 1\}, \delta_2, \mu_2, r_0)$$

The product machine is

$$\begin{aligned} Q = \{&(s_0, r_0), (s_0, r_1), (s_0, r_2), (s_0, r_3), (s_1, r_0), (s_1, r_1), (s_1, r_2), (s_1, r_3), \\ &(s_2, r_0), (s_2, r_1), (s_2, r_2), (s_2, r_3)\} \\ = \{&q_{00}, q_{01}, q_{02}, q_{03}, q_{10}, q_{11}, q_{12}, q_{13}, q_{20}, q_{21}, q_{22}, q_{23}\} \end{aligned}$$

$$\Sigma = \{a, b\}$$

$$\Lambda = \{0, 1\}$$

$$q_0 = \{(s_0, r_0)\}$$

The transition and output function for the product machine are computed component-wise as follows. For present state $(s_i, r_j)$, the next state on symbol $x$ is $(\delta_1(s_i, x), \delta_2(r_j, x))$. For instance, $(s_0, r_2)$ on input $b$ transits to $(s_1, r_1)$, and the output for the transition is *01*. State $q_{13}$ and $q_{20}$ are unreachable states. The state diagram of the product machine is shown in Figure 7.6.

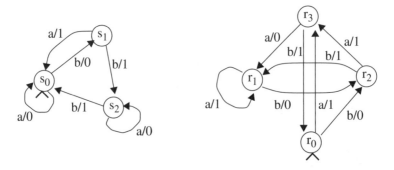

**Figure 7.5**   Component finite-state machines

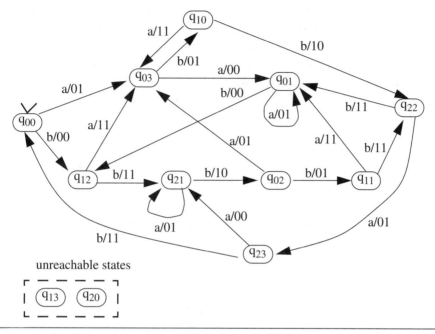

**Figure 7.6**   Product finite-state machine

In a product machine, the size of the state space is the product of the component machines' state spaces. If $M = M_1 \times \ldots \times M_n$, then

$$|Q| = \prod_{i=1}^{n} |Q_i|$$

If the component machines have approximately the same state space size, say $|Q_1|$, then $|Q| = |Q_1|^n$, which grows rather rapidly. This phenomenon is called *state explosion*.

### 7.5.2   State Equivalence and Machine Minimization

In designing a finite-state machine, it is possible that there are equivalent states (in other words, that states are indistinguishable by an external observer who examines only the input strings fed to the machine and the output strings produced by it). These equivalent states can be combined to give a smaller, reachable state space. In this section we will study a method to identify all equivalent states and to produce minimal finite-state machines.

Two states in a finite-state machine are equivalent if for any input string the output strings produced starting from each of the two states are identical. If two states are not equivalent, there must be an input string that drives the machine to produce different output strings starting from either of the two states. For brevity, we say that the input string causes the two states to produce different output strings. Let's look at an equivalent-state identification algorithm on Mealy machines. With minor modification, the algorithm can be applied to Moore machines.

The algorithm starts out by assuming all states are in one equivalent class, and then successively refines the equivalence class. Let's use the state machine in Table 7.6 to illustrate the algorithm. The first column is the present state, and the second and third columns are the next states for input $x$ equal to $a$ and $b$ respectively. The outputs are either 0 or 1 and are given after the next states. Initially, all states are assumed to be equivalent. Let's group them into an equivalent class:

$$(S_1, S_2, S_3, S_4, S_5)$$

From the definition of state equivalence, two states that produce different outputs on the same input are not equivalent. From Table 7.6, we see that on input $x = a$, $S_1$ and $S_2$ produce different outputs, $S_1$ giving 1 and $S_2$ giving 0. They must not be equivalent. Examining all states for $x = a$, states $S_1$, $S_3$, and $S_5$ could still belong to the same group because they produce the same output. Similarly, $S_2$ and $S_4$ belong to the same group. Therefore, we can split the initial single group into two groups such that the states within the same group produce the same outputs. We then have

$$(S_1, S_3, S_5) \ (S_2, S_4)$$

**Table 7.6**  State Machine for Equivalent-State Identification

| PS | NS, x = a | NS, x = b |
|----|-----------|-----------|
| $S_1$ | $S_5/1$ | $S_4/1$ |
| $S_2$ | $S_5/0$ | $S_4/0$ |
| $S_3$ | $S_5/1$ | $S_2/0$ |
| $S_4$ | $S_5/0$ | $S_2/0$ |
| $S_5$ | $S_3/1$ | $S_5/1$ |

Now we examine the other input value, $x = b$. States $S_2$, $S_3$, and $S_4$ produce output 0 whereas $S_1$ and $S_5$ produce 1. Thus, $S_3$ and $S_1$ are not equivalent. So, we can split the first group again:

$$(S_1, S_5)\ (S_3)\ (S_2, S_4)$$

If two states $S_i$ and $S_j$ in the same group, on the same input symbol (say, $y$) transit to two states $R_i$ and $R_j$ that are in two different groups, then $S_i$ and $S_j$ cannot be equivalent. The reasoning is as follows. Because $R_i$ and $R_j$ are not equivalent, there is an input string (say, $z$) that causes $R_i$ and $R_j$ to produce different output strings. The string $yz$ (symbol $y$ followed by string $z$) will cause $S_i$ and $S_j$ to produce different output strings.

Examine the state transition table for the first group, $(S_1, S_5)$. On input $x = a$, $S_1$ goes to $S_5$ and $S_5$ goes to $S_3$. Because $S_5$ and $S_3$ are in different groups, based on our reasoning we can conclude that $S_1$ and $S_5$ are not equivalent. So we can split the first group again:

$$(S_1)\ (S_5)\ (S_3)\ (S_2, S_4)$$

The only group with more than one state is $(S_2, S_4)$. On $x = a$, $S_2$ and $S_4$ both go to $S_5$. On $x = b$, $S_2$ goes to $S_4$ whereas $S_4$ goes to $S_2$. The next states, $S_4$ and $S_2$, are in the same group; thus, no splitting is needed. Up to this point, all groups have been examined and we stop. Because $S_2$ and $S_4$ are in the same group, they are equivalent. Because there are no input strings that can distinguish between $S_2$ and $S_4$, we might as well represent both states with a single state, say $S_2$. Therefore, we simplify the state transition table by replacing $S_4$ with $S_2$. The minimal-state machine is presented in Table 7.7.

**Table 7.7** Minimized Finite-State Machine

| PS | NS, x = a | NS, x = b |
|----|-----------|-----------|
| $S_1$ | $S_5/1$ | $S_2/1$ |
| $S_2$ | $S_5/0$ | $S_2/0$ |
| $S_3$ | $S_5/1$ | $S_2/0$ |
| $S_5$ | $S_3/1$ | $S_5/1$ |

The state equivalence identification and state minimization algorithm is summarized here:

---

### State Equivalence and Minimization Algorithm

1. For each input symbol, create groups of states such that states within the same group produce the same output.
2. For each group, examine the states on each input symbol. Two states whose next states are in different groups of the partition from the previous step must be split.
3. Repeat step 2 until there is no more group splitting.

---

We need to understand why states in the same group at the termination of the algorithm are equivalent. Suppose that states $S$ and $R$ are in the same group but are not equivalent. Then there must be an input string that causes $S$ and $R$ to produce different output strings. Let $T$ be the first time the two output strings differ. Let $S_T$ be the present state at time $T$ for the run of the machine starting from $S$. Similarly, let $R_T$ be the present state at time $T$ for the run starting from $R$. Because $S_T$ and $R_T$ produce different outputs, the algorithm guarantees that they be in different groups. Step 2 of the algorithm also guarantees that the state pair one cycle before $S_T$ and $R_T$ belongs to different groups. Again, the state pair two cycles before $S_T$ and $R_T$ also belongs to different groups. Going backward, we will eventually arrive at $S$ and $R$, and will conclude that $S$ and $R$ must belong to different groups, which is a contradiction.

A rigorous proof is based on mathematical induction on the length of the input string. When the length is 1, step 1 of the algorithm guarantees equivalence of states. Assume that the states produced by the algorithm are equivalent for any string of length less than $n$. For a string of length $n$, feed the first symbol to the machine. Two states in the same group will transit to the same group, as ensured by step 2 of the algorithm. Now we have an input string of length less than $n$. The induction hypothesis guarantees the two states be equivalent. I'll leave the details of a proof based on mathematical induction to you to determine.

Furthermore, the algorithm finds all equivalent states. If two states are equivalent, then they must belong to the same group at the end of the algorithm. To see why this is so, we notice that initially all states are grouped by their outputs. So these two states must be grouped together in step 1. Because they are equivalent, they will not transit to two different groups; hence, step 2 will not split them. Therefore, they must remain in the same group at the termination of the algorithm.

This grouping of equivalent states can be regarded as a state partition. The state partition produced by the algorithm is unique. That is, if there is another partition of the states, then that partition must be identical to the one produced by the algorithm. Suppose this were not true. There must be two states, $S$ and $R$, that belong to one group in one partition and belong to two different groups in the other partition. Belonging to two different groups means that they are not equivalent. However, on the other hand, they belong to the same group, meaning they are equivalent—a contradiction. Therefore, the state partition given by the algorithm is unique.

So far, the machine minimizaton algorithm has been applied only to finite-state machines. It can also be applied to finite-state automata. Because automata do not produce output, the meaning of state equivalence must be redefined. We define two states to be equivalent if the following condition is satisfied: whether the automaton accepts or rejects an input string is independent of which of the two states it starts at. In other words, an accepted (rejected) string by the machine starting at one state is also accepted (rejected) by the machine starting at the other state.

With this definition of state equivalence, the first step of the algorithm is to group final states as one group and the remaining states as another, because states in one group are accepted whereas the others are not. Thus, states in these two groups are not equivalent. Then we examine the next state of the states in each group, on each of the input symbols. Two states transiting to two different groups on the same input symbol are not equivalent, and thus the group must be split. This process continues until no group splitting exists. Just like the algorithm for finite-state machines, the reduced automata produced by combining the equivalent states indicated by this algorithm are minimal and unique.

---

### Example 7.16

Consider the finite-state automaton described by the state diagram in Figure 7.7. The first partition of states is to group final states as a group: states $a$ and $g$, and the remaining states as another group. We have $(a, g) (b, c, d, e, f)$.

Examine the states in each group for each input symbol. On input 0, $a$ and $g$ go to $b$ and $e$ respectively. Both are in the same group, so no splitting is required. In the second group, $b$ goes to $c$, and $c$ goes to $a$. Because these next states are in different groups, $b$ and $c$ belong to different groups. A similar analysis produces the following: $(b, c, d, e, f)$ goes to $(c, a, c, f, g)$. Therefore, $(b, c, d, e, f)$ splits to $(c, f) (b, d, e)$. States $c$ and $f$ belong to the same group because their next states on input 0 are $a$ and $g$, which are in the same group. At this point we have $(a, g) (c, f) (b, d, e)$.

*continues*

---

**Example 7.16    (Continued)**

---

On input 1, *a* and *g* go to *d* and *e* respectively, which are in the same group. States *c* and *f* go to *f* and *c* respectively, which are in the same group. States *b*, *d*, and *e* go to *a*, *g*, and *g* respectively, which are in the same group. At this point, there is no splitting. Now on input 0, *(a, g)* goes to *(b, e)*, *(c, f)* goes to *(a, g)*, and *(b, d, e)* goes to *(c, c, f)*. There is no splitting on input 0. Therefore the process terminates. The final partition is *(a, g)* *(c, f)* *(b, d, e)*. States *a* and *g* are equivalent, as are *c* and *f*, and *b*, *d*, and *e*. By combining the equivalent states, we have the reduced automaton shown in Figure 7.7B.

## 7.5.3    Finite-State Machine Equivalence

During the course of a design project, it is sometimes required to determine whether two state machines are equivalent. Two finite-state machines are equivalent if their input and output behavior is identical. Comparing state machines ensures that a state machine, after undergoing a timing enhancement, is functionally equivalent to the original machine. For smaller, deterministic state machines, the following procedure can be used. First, minimize both machines using the previous algorithm. If these two reduced machines are indeed equivalent, there is a one-to-one correspondence between the states in the machines. That is, for every state in the first machine, there is a corresponding state in the second machine such that they play the same role in the machines except for possibly their names (in other words, they are isomorphic). To determine this one-to-one correspondence, assume their initial states correspond to each other. Then for each input symbol, make the next states from the respective initial state correspond to each other if they produce the same output.

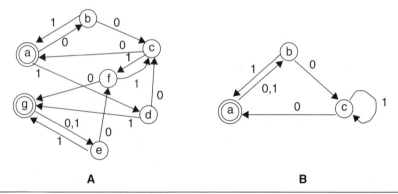

**A**                                    **B**

**Figure  7.7**  Finding equivalent states and minimizing a finite-state automaton. (A) An automaton with equivalent states (B) A reduced automaton

When all next states of the initial state have been mapped, continue the same process on the next states of the respective mapped states. If, at the end, all states have been mapped, the two machines are equivalent; otherwise, they are not.

The state mapping step is difficult if the machines are nondeterministic, because there could be more than one initial state or next state, so that the correspondence between states is difficult to determine.

---

### Example 7.17

Consider the two minimal state machines in Figure 7.8. The first machine, in Figure 7.8A, has initial state $a$, and the second machine has initial state $z$. Map state $a$ to $z$. On input 0, the next state of the first machine is $c$ with output 1, and that of the second machine is $s$ with output 1. Because the outputs are identical, map state $c$ to $s$. On input 1, the next states of the first and the second machine are $b$ and $y$ respectively, with identical output. So map state $b$ to state $y$. Choose mapped states $c$ and $s$ for the next iteration of the mapping process. On input 0, the next states are $b$ and $y$ with identical output, which is consistent with the existing mapping. On input 1, the next states are $e$ and $x$ respectively. Because the outputs are equal, map state $e$ to state $x$. Repeat until all states and transitions have been visited. At the end, we find that all states have been mapped, and we therefore conclude that the two machines are equivalent. The final state mapping is

$$a \leftrightarrow z$$
$$b \leftrightarrow y$$
$$c \leftrightarrow s$$
$$d \leftrightarrow r$$
$$e \leftrightarrow x$$

Note the increased difficulty if there is more than one initial state or next state.

---

## 7.5.4 Graph Algorithms

Because finite-state automata or machines can be represented by state diagrams, algorithms on automata and state machines often take the form of graph algorithms. For example, to inquire whether a state can be reached from the initial state, it is a question of whether the state vertex in the state diagram can be reached from the initial state vertex. In this section we will study several graph algorithms commonly encountered during formal verification procedures.

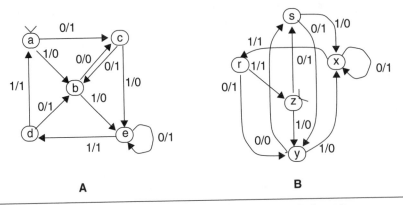

**Figure 7.8**    Determine state machine equivalence between the machines in A and B.

A *directed graph* or *digraph* consists of a set of vertices $V$ and a set of directed edges $E$—in other words, $G(V, E)$. An *edge* from vertex $n$ to vertex $m$ is represented by *(n, m)*. In the context of a state diagram, the vertices represent states and the edges represent transitions. A vertex has a fanout if there is an edge from the vertex to another vertex, which is called the *fanout vertex*. A *path* from vertex $n$ to vertex $m$ is a sequence of alternating vertices and edges, with the first vertex being $n$ and the last vertex $m$ such that the $i$th edge goes from the $i$th node to the $i + 1st$ node. Vertex $n$ can be reached from vertex $m$ if there is a path from $m$ to $n$. Given a node $n$, the set of all nodes that can be reached from $n$ is called the *transitive closure* of $n$.

---

### Example 7.18

In this example we look at representing a graph as a Boolean relation. Representing a graph in a Boolean domain translates graphical manipulations into Boolean operations. Consider the graph in Figure 7.9. Two things need to be represented: vertices and edges. We first encode the vertices. There are four vertices; thus, 2 bits are required to encode the vertices. Vertices $a, b, c,$ and $d$ are encoded as 00, 01, 10, and 11.

An edge is identified by its head and its tail, which are vertices; therefore, 4 bits are required to encode an edge. Let us choose variables $r, s, v,$ and $w$. For example, edge *(a, c)* is encoded as 0010. The first 2 bits are the code of the edge's head, vertex $a$, and the last 2 bits are the code of the tail, vertex $c$.

**Example 7.18    (Continued)**

We can define a relation on vertex pairs $(e_1, e_2)$. The relation is true, if edge $(e_1, e_2)$ exists in the graph. For instance, $(b, d)$ is true but $(c, a)$ is false. This relation is represented by a characteristic function. Therefore, the characteristic function $\lambda(r, s, v, w)$ is 1 for the vertex pairs: $(a, b)$, $(a, c)$, $(b, b)$, $(b, d)$, $(b, c)$, $(c, d)$, $(c, c)$, and $(d, a)$. Replacing the edges by their code, we have:

$$\lambda(r, s, v, w) = \overline{r}\overline{s}\overline{v}w + \overline{r}\overline{s}v\overline{w} + \overline{r}s\overline{v}w + \overline{r}svw + \overline{r}sv\overline{w} + r\overline{s}vw + r\overline{s}v\overline{w} + r\overline{s}\overline{v}\overline{w}$$

To check, for edge $(b, c)$, the code for the edge is 0110 (in other words, $r = 0$, $s = 1$, $v = 1$, and $w = 0$). We can see that $\lambda(0, 1, 1, 0) = 1$ because the fifth term becomes 1, as expected. Consider edge $(c, a)$, which is not in the graph. The code is 1000. We calculate $\lambda(1, 0, 0, 0) = 0$, as expected.

## 7.5.5  Depth-First Search

A basic procedure on a graph is to visit, or *traverse*, the vertices in a systematic way so that all vertices are visited only once. Two popular traversal algorithms are DFS and breadth-first search (BFS). Let us study DFS first.

DFS starts from an arbitrary vertex, called the *root of DFS*, selects a fanout vertex, and visits all vertices that can be reached from the fanout vertex before repeating the traversal on the other fanout vertices. A DFS algorithm is shown here. The number N records when a

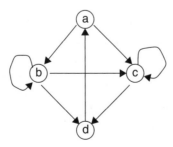

**Figure 7.9**   A graph for Boolean representation

vertex is visited. When a vertex is visited, N is stored in the entry time attribute of the vertex. When a vertex is done visiting, N is stored in the exit time attribute of the vertex. N is included for the sole purpose of showing the order in which the vertices are visited, and hence is not necessarily a part of the algorithm. In a DFS operation, unlike vertices, which are all traversed, not all edges have to be traversed. We will see that all traversed edges form a set of trees, or a *forest*. A tree is a graph with no loops. The set of traversed edges is called the *DFS tree*.

---

### DFS Algorithm

```
Input: G(V, E)
Output: a systematic traversal of vertices.
Initialization: N = 1 // used to record entry and exit time of a node

DFS (G) {
    while (vertex v in V is not marked visited) VISIT(v);
}

VISIT(v) {
    mark v visited;
    v.entry = N;
    N = N + 1; // record node entry time
    for_each (u = fanout of v)
        if ( u is not marked visited) VISIT(u);
    v.exit = N;
    N = N + 1;
}
```

---

### Example 7.19

Let us perform a DFS on the graph in Figure 7.10. We start the DFS from vertex *G*, which we mark visited and tag with $N = 1$. In Figure 7.10, the first number is entry time and the second number is exit time. Following the arrow, we arrive at B, which has two fanouts. We mark *B* visited and tag *B* with $N = 2$, and follow a fanout to *D*, which is then marked visited and is tagged with $N = 3$. Because *D* has no fanout, *D* is done visiting and we return from *D*.

---

**Example 7.19   (Continued)**

On exiting from $D$, we tag $D$ with exit time 4. Following the other fanout of $B$, we arrive at $C$, which is marked and tagged with $N = 5$. Because $C$ has no fanout, we return from $C$ and tag it with exit time $N = 6$. On returning to $B$, we find that all of its fanouts have been visited, so we exit from $B$ and tag its exit time with 7. Finally, we return to node $G$ and tag its exit time with 8. This completes the first traversal. However, there are still unvisited vertices, such as $A$. So, we start from another vertex, say, $E$. We tag $E$ with entry time 9 and follow its sole fanout to $A$. On arriving at $A$, we tag its entry time with 10. Because both fanouts of $A$ have already been marked visited, we return from $A$ and tag its exit time with $N = 11$. On returning to $E$, we tag its exit time with 12. The remaining unvisited nodes, $F$ and $H$, having no unvisited fanouts, are marked and tagged. The edges that were traversed in the DFS are in bold and form the DFS trees.

---

In a DFS tree, a vertex is called a *parent* of another vertex if there is an edge in the DFS tree that starts from the former and goes to the latter. The second vertex is called the *child* vertex. A vertex $A$ is called a *descendant* of vertex $B$ if $A$ can be reached from $B$, following directed edges of the DFS tree. Vertex $B$ is called an *ancestor* of $A$. Once a DFS tree or forest is determined, edges in the graph can be classified into the following four types: tree edges, back edges, forward edges, and cross edges. *Tree edges* are just edges of the DFS tree. *Back edges* connect descendants to ancestors. *Forward edges* connect ancestors to descendants. The remaining edges are *cross edges*.

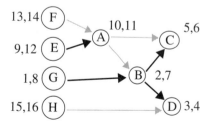

**Figure 7.10** A graph with entry and exit times from a DFS

One application of DFS is to find cycles or loops in a directed graph. Theorem 7.6 provides the necessary and sufficient condition.

---

**Theorem 7.6**

Main property of DFS trees and cycle detection

1. If edge *(n, m)* is such that the entry time of *n* is less than that of *m*, then *m* is a descendant of *n*.
2. Graph *G(V, E)* has a directed cycle, or loop, if and only if *G* contains a back edge with respect to the DFS tree of *G*.

---

**Example 7.20**

Let's apply DFS to the graph in Figure 7.11 to illustrate the previous definitions and theorem. The numbers shown are entry times. The bold edges form the DFS tree. Edges *(b, f)*, *(f, c)*, *(c, d)*, *(c, e)*, *(e, a)*, and *(a, g)* are tree edges. According to part 1 of Theorem 7.6, *g* is a descendant of *b*. So, edge *(b, g)* is a forward edge. Vertex *a* is a descendant of vertices *c* and *b* because there is a path from *c* or *b* to *a* along the DFS tree. Therefore, edges *(a, b)* and *(a, c)* are back edges. Finally, vertex *g* and vertex *d* are not related (neither is a descendant of the other); therefore, edge *(g, d)* is a cross edge. So is edge *(e, d)*. According to part 2 of Theorem 7.6, there are directed cycles or loops because of the back edges *(a, b)* and *(a, c)*. Indeed, the loops are *(a,b,f,c,e,a)* and *(a,c,e,a)*.

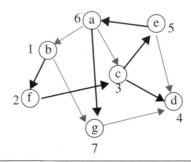

**Figure 7.11** Detect loops using DFS

Directed cycles are special instances of SCCs. An SCC of a directed graph is a maximal subgraph such that every vertex can be reached from any other of its vertices. SCCs are an expanded notion of loop. A DAC is a tree or forest, and hence is free of SCCs. It is known that any directed graph can be decomposed into a DAC and SCCs. In other words, this decomposition breaks any directed graph into looping components and straight components. Once a graph is decomposed in such a way and the SCCs are encapsulated, the resulting graph is a DAC and is loop free. SCCs can be determined by applying DFS, as follows:

**Algorithm for Finding SCCs**

Input: graph G

output: a collection of SCCs in G

1. Perform DFS on G and record the exit time for the nodes.
2. Reverse the edges of G and apply DFS to this graph, selecting the nodes in order of decreasing exit number in the `while` loop step.
3. The vertices of a DFS tree from step 2 are SCCs.

In step 2 of the algorithm, reversing an edge means making the head of the edge its tail, and vice versa. When applying DFS to this graph, in the `while` loop of the DFS algorithm, select the unvisited nodes in the order of decreasing exit numbers that were derived in step 1.

**Example 7.21**

Identify strongly connected components.

Apply the SCC-finding algorithm to the graph in Figure 7.12. In Figure 7.12A, the numbers underlined are the exit times of the nodes from the DFS in step 1. Figure 7.12B is obtained from A with edges reversed. To apply DFS to this derived graph, the first unvisited vertex is vertex *d* because it has the largest exit number. This search ends after it has visited vertices *a*, *b*, *d*, and *h*. The next unvisited vertex is *f*, because it has the largest exit number in the remaining unvisited vertices. This search concludes the DFS after it has visited *g* and *e*. The bold edges in Figure 7.12B are the DFS trees from this DFS. The vertices of a DFS tree form an SCC. Therefore, {a,b,d,h} is an SCC, as is {e,f,g}. Making an SCC into a composite vertex, the resulting graph is a DAC, which is shown in Figure 7.12C.

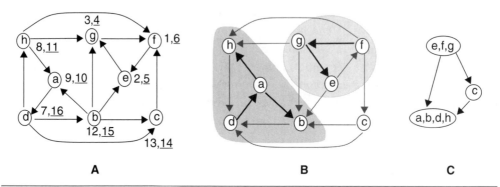

**Figure 7.12** Finding SCCs. (A) Result of a DFS. The underlined numbers are exit times. (B) DFS on a graph derived by reversing the edges. (C) Resulting DAC by treating SCCs as composite nodes.

## 7.5.6   Breadth-First Search

A BFS visits vertices level by level. First it selects a vertex, the root of the DFS, as the starting point of the search and visits all its fanout vertices before visiting other vertices. Then the fanout vertices of the recently visited fanout vertices are visited, and so on, until all vertices are visited. If we define the level number of a vertex with respect to the root to be the minimum number of edges from the root to the vertex, then vertices in a BFS are visited according to their level numbers from the root in increasing order, because the level numbers of fanout vertices of a vertex are one more than that of the vertex itself. Similar to DFS, all traversed edges form a BFS tree or forest. A BFS algorithm is as follows:

```
Input: G(V, E)
Output: a systematic traversal of vertices.
Initialization: N=1 // used to record entry time of a node

BFS (G) {
   while (vertex v in V is not marked visited) VISIT(v);
}

VISIT(v) {
   mark v visited;
   place v in a queue Q.
   while (Q is not empty) {
      remove head of Q and call it u.
      u.entry = N; N = N + 1; // record node entry time
      for_each edge (u, r) where r is not yet marked {
         mark r;
         add (u, r) to the BFS tree;
         place r in Q;
```

```
      } // end of for_each
   } // end of while
}
```

Theorem 7.7 shows two properties about BFS. Because BFS visits vertices in increasing order of level numbers, the first property, a direct result of the search procedure, says that for all edges *(n, y)* leading to *y*, the one closest to the root is selected first and thus belongs to a BFS tree. The second property is a consequence of successive manifestations of the first property. A path is made of a sequence of connecting edges. According to property 1, the edges on a BFS tree are the ones closest to the root; therefore, the shortest path from a root to a vertex is the one made of only tree edges in a BFS tree.

---

**Theorem 7.7**

Properties of BFS

1. Suppose edge *(x, y)* is a tree edge of a BFS tree and *(z, y)* is any edge leading to *y*. Then the entry number of *x* is less than that of *z*.
2. The shortest path from a root of a BFS to a vertex *x* is the path from the root to *x*, which lies in the BFS tree.

---

**Example 7.22**

Let's apply BFS to the graph in Figure 7.13, for which we performed a DFS. You may want to compare the two search methods. Let's select vertex *b* as the starting point. In routine VISIT, vertex *b* is marked and placed into the queue. It is then removed from the queue and tagged with an entry time of 1. Then its fanout vertices *f* and *g* are enumerated, marked, and placed in the queue. Now the queue contains *f* and *g*. Vertex *f* is removed. Vertex *f* is tagged with entry time 2 and its fanout is *c*, which is then marked and placed in the queue. Now the queue has *g* and *c*. Next, *g* is removed from the queue and is tagged with an entry time of 3. Its fanout is *d*, which is then marked and placed in the queue. The queue has *c* and *d*. Vertex *c* is taken off the queue and is tagged with an entry time of 4. Its fanouts are *e* and *d*, but *d* has already been marked. Thus, only *e* is marked and is placed in the queue. The queue has *d* and *e*. Vertex *d* is removed from the queue and is assigned an entry time of 5. It has no fanout. The queue has *e*. Next, vertex *e* is taken off the queue and is given an entry time of 6. Its sole fanout is vertex *a*, which is then marked and placed in the queue. The queue has *a*. Vertex *a* is removed from the queue and is assigned an entry time of 7. Its three fanouts—*b*, *g*, and *c*—are all marked, and the BFS terminates.

*continues*

---

**Example 7.22     (Continued)**

---

The graph with entry times and the BFS tree in bold are shown in Figure 7.13A. In Figure 7.13B, the vertices are rearranged according to the BFS tree. In this rearrangement, it is easy to see that BFS levels the vertices, and the shortest distance from the root (vertex $b$) to any vertex is the path that lies in the BFS tree, as stated in part 2 of Theorem 7.7. Furthermore, the nontree edges also obey the rule stated in part 1 of Theorem 7.7. For example, edges $(c, d)$ and $(e, d)$ are not tree edges, whereas $(g, d)$ is. Hence, the entry times of $c$ and $e$ are greater than that of $g$.

## 7.6    Summary

In this chapter we reviewed some of the basic mathematics encountered in formal verification. We started with set operations and then defined relations among sets. An especially interesting relation is the equivalence relation, based on which sets can be partitioned. Next we studied partially ordered sets along with meet and join, and introduced lattices.

We then moved to Boolean functions and their representations. We examined CNF, DNF, and multilevel representation. Symmetric Boolean functions have a concise way of representation, and we showed how operations on symmetric functions could be

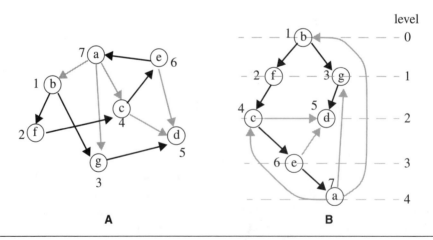

**Figure 7.13** Graph for BFS. (A) Result of a BFS run shows a BFS tree with entry numbers. (B) Rearrange vertices according to their levels.

performed easily. For incompletely specified functions, we introduced representation using on set, off set, and don't-care set. To represent Boolean relations, we used characteristic functions.

We went over several Boolean operators—namely, cofactor, which gives rise to Shannon expansion, generalized cofactor, Boolean difference, and existential and universal operators. In passing, we mentioned the generalized Reed–Muller form.

From the discussion of combinational domain we continued to sequential domain. First we defined finite-state automata, finite-state machines, completely specified, incompletely specified, deterministic, and nondeterministic. A language of an automaton is defined in terms of its acceptance condition. There are various forms of acceptance conditions, some for finite strings and some for infinite strings. Finally, we considered product machines and the state explosion phenomenon.

States may be equivalent in the sense that an external observer would not be able to distinguish the states. We provided an algorithm to determine all equivalent states and to reduce automata and machines to their minimal forms. We also looked at a procedure to determine the isomorphism of deterministic finite-state machines.

Lastly we studied some graph algorithms. We defined common graph terms, then looked at DFS and BFS in detail. We defined a DFS and BFS tree, applied DFS to finding directed cycles and SCCs, and showed how BFS levels nodes and produces the shortest path.

## 7.7 Problems

**1.** The power set of set $S$, denoted by $2^S$, is defined to be the set of all subsets of $S$.

    a. Find the power set of set $S = \{\{a, b\}, \{8, 11\}, \{\%\}\}$.
    b. For any arbitrary set $S$, relate $\left|2^S\right|$ to $|S|$.

**2.** Determine which of the following is an equivalence relation. If it is not an equivalence relation, explain why.

    a. $\gamma(x, y)$ is a relation on numbers of the form $a + b\sqrt{2}$, where $a$ and $b$ are integers. Let $x = a_1 + b_1\sqrt{2}$ and $y = a_2 + b_2\sqrt{2}$. Then $\gamma(x, y)$, if $a_1 = a_2$.
    b. Similar to (a) except that $a$ and $b$ are real numbers.
    c. Similar to (a) except that $\gamma(x, y)$ if $a_1 \cdot b_1 = a_2 \cdot b_2$.

**3.** Define two relations on the set of all regular polygons inscribed inside a unit cycle with a number of sides less than or equal to a constant $N$.

a. Relation $\alpha(x, y)$ if polygon $x$ is congruent with polygon $y$ under rotation transformation (in other words, $x$ can be rotated so that it is identical to $y$). Is $\alpha(x, y)$ an equivalence relation? If yes, describe the partition under $\alpha(x, y)$. If no, which condition is not satisfied?

b. If $\alpha(x, y)$ is not an equivalence relation, how do you modify it to be an equivalence relation? Let $\alpha(x, y)$ be the previous relation if it is an equivalence relation, or your modified version if not. Relation $\gamma$ is defined based on the elements of the partition under $\alpha(x, y)$ such that $\gamma(x, y)$ if $\|x\| \geq \|y\|$, where $\|x\|$ is the area of polygon $x$. Is $\gamma(x, y)$ an ordering relation?

c. If $\gamma(x, y)$ is an ordering relation, what are meet and join of $x$ and $y$? Is the set under $\gamma(x, y)$ a lattice? Why?

**4.** Determine support, minimum support, and Boolean difference.

a. Determine the support and minimum support of Boolean function $ab\bar{c} + (\bar{b} + \bar{c})(b + c) + \bar{a}b\bar{c}$.

b. If variable $x$ is in support of $f$ but not in minimum support of $f$, what can you say about $\dfrac{\partial f}{\partial x}$?

**5.** Convert the following DNF to CNF. What is the ratio of the number of literals in DNF to that of CNF?

$$\overline{abc} + ab + bc + ac + a\bar{b}c$$

**6.** Prove or disprove

$$\frac{\partial}{\partial x}(f \oplus g) = \frac{\partial f}{\partial x} \oplus \frac{\partial g}{\partial x}$$

**7.** Determine whether the following function is symmetric. If it is symmetric, represent it using short-hand notation. Determine the complement of the function.

$$f(a,b,c,d) = abc\bar{d} + abcd + \bar{a}bc\bar{d} + a\bar{b}c\bar{d} + abc\bar{d}.$$

8. Represent the following Boolean functions/relations using characteristic functions.

   a. The Boolean function is $f(a,b,c,d,e) = ab + bc + ad + db + cd + \overline{a}\overline{b}c$
   b. The incompletely specified Boolean function has on set $f(a,b,c)$, off set $g(a,b,c)$, and don't-care set $h(a,b,c)$, where $f$, $g$, and $h$ are Boolean functions.
   c. The vector Boolean function $f = (f_1, f_2)$ is described in Table 7.8.

9. The output of a two-input combinational circuit, $(f, g)$, is described by vector Boolean function $f(a,b,c) = \overline{a}(b + \overline{c}) + \overline{ac} + \overline{b}(\overline{ac} + \overline{cb})$ and $g(a,b,c,) = b\overline{c} + \overline{a}(b + c)$. Determine the Boolean function that represents all inputs $b$ and $c$ such that a change at input $a$ is seen at both outputs.

10. A finite-state machine's next-state function is given by $(f, g) = (as + br, b(r + as))$, where $a$ and $b$ are inputs, and $r$ and $s$ are state bits.

   a. Determine all states whose next state can be in state 11.
   b. Given the current state of 01 (= rs), find all states that can be reached in two cycles or less.

11. Derive a generalized Reed–Muller representation for the following function:

$$f(a,b,c,d) = \overline{a} + \overline{b}(c + d) + d(\overline{c + ab})$$

**Table 7.8** An Incompletely Specified Vector Boolean Function

| (a,b,c,d) | ($f_1$, $f_2$) |
|---|---|
| 0101 or 1100 | 11 or 00 |
| 10-0 | 01 |
| 00-- | 10 |
| other values | 01 or 10 |

**12.** Review generalized cofactor calculation.

  a. Calculate the generalized cofactor $f_g$, where $f = a(b + cd) + e$ and $g = b\bar{c} + \bar{a}c\bar{d}$.
  b. Calculate $f \cdot g$.
  c. Are the results from (a) and (b) equal?

**13.** Decide whether the following CNF can be satisfied (in other words, equal to 1).

$$(a + \bar{b} + c + \bar{d})(\bar{a} + \bar{b} + c + d)(\bar{a} + b + c + \bar{d})(\bar{a} + \bar{b} + \bar{c} + d)(a + \bar{b} + \bar{c} + d)$$

**14.** Represent the finite-state machine's transition function and output in Table 7.9 using characteristic functions.

**15.** Study product machines of nondeterministic and deterministic component machines.

  a. For the two finite-state machines in Figure 7.14, determine their product machine.
  b. What states cannot be reached? What is the ratio of the number of states that can be reached in the product machine to the sum of the numbers of states in the component machines?
  c. Suppose state s is also an initial state. How does it affect the size of your product machine?

**16.** For the finite-state machine represented in Table 7.10, minimize the machine.

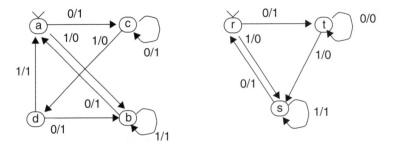

**Figure 7.14** Component machines for a product machine

**Table 7.9**    A Nondeterministic Finite-State Machine

| PS | input | NS | output |
|----|-------|-------|--------|
| **00** | 1- | 01 or 11 | 00 |
| **01** | 00 | 10 | 11 |
| **10** | 11 | 01 or 00 | 01 |
| **11** | 01 | 11 | 00 |

**Table 7.10**    A Finite-State Machine with Equivalent States

| PS | NS/output input = 1 | NS/output input = 0 |
|----|---------------------|---------------------|
| **a** | e/0 | e/0 |
| **b** | e/1 | f/0 |
| **c** | d/0 | e/0 |
| **d** | e/1 | a/0 |
| **e** | d/1 | a/0 |
| **f** | h/0 | b/0 |
| **g** | e/0 | d/0 |
| **h** | a/0 | d/0 |

**17.** Perform a DFS on the graph in Figure 7.15. Choose vertex *a* as the root of the DFS. Label the DFS tree. List the forward edges, back edges, and cross edges.

**18.** Find all SCCs in Figure 7.15.

**19.** Perform a BFS on Figure 7.15, with the BFS root being vertex *a*. Show the BFS tree and rearrange the vertices in levels according to the BFS.

**20.** Represent the graph in Figure 7.16 using a characteristic function. From the characteristic function, find a Boolean expression for all paths of length 2 between vertices *x* and *y*. Each edge is of length 1.

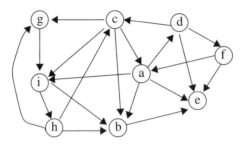

**Figure 7.15** Graph for DFS

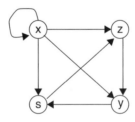

**Figure 7.16** Graph for Boolean representation and calculation

## CHAPTER 8

# Decision Diagrams, Equivalence Checking, and Symbolic Simulation

## Chapter Highlights

- Binary decision diagrams
- Decision diagram variants
- Decision diagram-based equivalence checking
- Boolean satisfiability
- Symbolic simulation

**A** fundamental problem in formal verification is determining whether two Boolean functions are functionally equivalent. A case in point is to determine whether a circuit, after timing optimization, retains the same functionality. Logical equivalence can be done on either combinational functions or sequential functions. Determining even combinational equivalence can be hard, such as deciding whether $\overline{a}c + b\overline{c} + a\overline{b}$ is equivalent to $a\overline{c} + \overline{b}c + \overline{a}b$. The second expression can be obtained from the first expression by shifting the complementation operation to the other literal of each cube. Although somewhat counterintuitive, these two expressions are functionally equivalent. Sequential equivalence is concerned with determining whether two sequential machines are functionally equivalent. One way to do this is based on combinational equivalence. First we minimize the two machines to eliminate redundant states, then we map their states based on one-to-one correspondence, and finally we decide whether their next-state functions are combinationally equivalent. The problem of combinational equivalence is a focus of this chapter.

A brute-force method to determine combinational equivalence is to expand the combinational functions in minterm form and compare them term by term. The minterm representation has the property that two equivalent Boolean functions have identical minterm expressions. This property in general is called *canonicity of representation*. A representation having the property is a canonical representation. In addition to minterm representation, maxterm and truth table are also canonical representations. One way to decide whether two functions are equivalent is to represent them in a canonical form. If their forms are identical, they are equivalent; otherwise, they are not equivalent. However, this method runs into the problem of exponential size, because the number of minterms or maxterms of a function can be exponential with respect to the number of variables.

The sum-of-products representation is more compact than minterm representation, but it does not reveal readily functional equivalence or unequivalence. As demonstrated by our previous example, the two functions have a completely different appearance, even though they are functionally equivalent. Thus, a second desirable property of a representation, besides canonicity, is compactness. Sum-of-products is more compact than truth table representation.

An ideal representation should be both canonical and compact. Because logical equivalence is an NP-complete problem, it is likely that all canonical representations are exponential in size in the worst case. However, a canonical representation is still useful if its size for many practical functions is reasonable. In this chapter we will look at BDDs as a representation of Boolean functions. Then we will examine how BDDs are used in determining functional equivalence. As an alternative to BDDs in checking equivalence, we will study SAT. Finally, we will examine symbolic simulation, to which BDDs and SAT are applicable.

## 8.1   Binary Decision Diagrams

A BDD is a directed acyclic graph (DAG) that represents a Boolean function. A node in a BDD is associated with a Boolean variable and has two outgoing edges. One edge represents the TRUE (1) value of the variable, and the other, the FALSE (0) value. The two edges point to other BDD nodes. Let's call the node pointed by the 0-edge the *0-node* and the node by the 1-edge, the *1-node*. Using this convention, the left edge represents the 1 value of the variable and the right edge represents the 0 value. The node without an incoming edge (in other words, not being pointed to) is called the *root* of the BDD. A leaf node or terminal node, denoted by a square, is not associated with a variable and is a constant node that evaluates to either 0 or 1. The size of the BDD, denoted as $|BDD|$, is the number of nodes in the BDD. To evaluate a function represented by a BDD on a given value of its inputs, we start from the root and follow the edges to trace the path determined by the variable values. When a node is encountered, whether the right or left edge is followed is determined by the value of the variable of the node. For instance, at node $x$, the right edge is followed if $x$ takes on value 0; otherwise, the left edge is followed. If the path arrives at terminal node 0, the function evaluates to 0. If the path arrives at terminal node 1, the function evaluates to 1.

## Example 8.1

An example of a BDD is shown in Figure 8.1. By looking at Figure 8.1, we can determine that node $a$ is the root of the BDD because it has no incoming edges. More than one node can represent the same variable. For example, there are two nodes representing variable $b$. The two square nodes at the bottom are terminal nodes, representing 0 and 1. For input value $a = 1$, $b = 0$, $c = 1$, we evaluate the function. Start at the root, which represents variable $a$. Because $a = 1$, we take the left edge and arrive at node $b$ on the left-hand side. Because $b = 0$, we take the right edge and arrive at node $c$. Because $c = 1$, we follow the left edge and arrive at terminal node 1. Therefore, for $a = 1$, $b = 0$, $c = 1$, the function evaluates to 1. The path traced is P1. Note that variable $d$ is not involved in this evaluation and this means the value of the function for this particular input combination is independent of variable $d$.

In this example we traced a path in a BDD from an input value. Conversely, we can select a path in a BDD and deduce from it the value of the function and the input value giving rise to the function value. For instance, consider path P2, which arrives at terminal node 0. Along the path, $a$ takes on 0 because the right edge is taken, $b$ takes on 1, and $c$ takes on 0. This path implies that the function evaluates to 0 when $a = 0$, $b = 1$, and $c = 1$. In other words, cube $\bar{a}bc$ is in the off set of the function. Therefore, any path ending at terminal 0 represents a cube in the off set. Similarly, any path ending at terminal 1 represents a cube in the on set. Thus, we can derive the function this BDD represents. In sum-of-products form, we look at all paths ending at terminal 1. There are four paths. These four paths represent cubes $\overline{ab}d$, $\bar{a}\bar{b}c$, $\bar{a}bc$, and $ab$. Because each of these cubes causes the function to be 1, the function is a sum of these cubes:

$$f = \overline{ab}d + \bar{a}\bar{b}c + \bar{a}bc + ab$$

*continues*

**Example 8.1     (Continued)**

To derive the function in product-of-sums form, we need to work on the off set. There are three paths ending at terminal 0. These paths represent cubes $\overline{abd}$, $\overline{abc}$, and $a\overline{bc}$:

$$\overline{f} = \overline{abd} + \overline{abc} + a\overline{bc}$$

Using DeMorgan's Law, we can obtain the function in product-of-sums form:

$$f = (a+b+d)(a+\overline{b}+c)(\overline{a}+b+c)$$

The BDD rooted at node $x$ represents a Boolean function $f$, and the 0-node represents the cofactor of the function, $f_x$, and the 1-node, $f_x$. This is because the function of the 0-node is derived from the root via the setting of $x$ to $0$, which is just $f_{\overline{x}}$. The same goes for the relation between the 1-node and $f_x$. Consequently, to obtain the Boolean function of a BDD, one can apply the Shannon cofactor theorem in reverse. Once the functions at the 0-node and 1-node of a node have been computed to be $g$ and $h$, then the function at the node is $xh + \overline{x}g$. This can be done from either the bottom up, starting from the constant nodes, or from the top down recursively. A top-down recursive algorithm is as follows:

**Derive Boolean Function from a BDD: BDDFunction(BDD)**

Input: a BDD

Output: the Boolean function represented by the BDD

   **1.** $x$ = root of BDD.
   **2.** If $x$ is a constant, return the constant.
   **3.** Let $y$ and $z$ be the 1-node and 0-node respectively of $x$.
   **4.** Return $x$BDDFunction($y$) + $\overline{x}$BDDFunction($z$).

You may observe that the paths in a BDD determine the cubes in the function's on set and off set, and a path (consisting of a sequence of nodes) produces a cube regardless of the order in which the nodes appear along the path. For instance, path $a$,right; $b$,right; and $c$,right gives the same cube as path $b$,right; $c$,right; $a$,right in another BDD. The two BDDs shown in Figure 8.2 represent the same function even though the two graphs do not resemble each other.  Using the previous path tracing technique, we can verify that the two

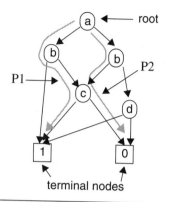

**Figure 8.1**   A sample BDD

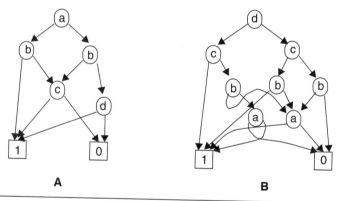

**Figure 8.2**   (A, B) BDDs of the same function have different variable ordering

BDDs represent the same function. Consider the BDD in Figure 8.2B. There are six paths that end at terminal node 1. These paths produce cubes $dc$, $d\bar{c}b\bar{a}$, $d\bar{c}\bar{b}\bar{a}$, $\bar{d}\bar{c}\bar{b}a$, $\bar{d}cb$, and $\bar{d}c\bar{b}a$. Therefore, the function is

$$f = dc + d\bar{c}b\bar{a} + d\bar{c}\bar{b}\bar{a} + \bar{d}\bar{c}\bar{b}a + \bar{d}cb + \bar{d}c\bar{b}a$$

$$= cd + ab\bar{c}d + \overline{ab}cd + a\bar{b}\bar{c}\bar{d} + ab\bar{c}\bar{d} + bc\bar{d}$$

$$= (ab + \overline{ab} + \bar{a}b + a\bar{b})cd + ab\bar{c}d + \overline{ab}cd + a\bar{b}c\bar{d} + ab\bar{c}\bar{d} + (\bar{a} + a)bc\bar{d}$$

$$= (abcd + ab\bar{c}d) + (\overline{abcd} + \overline{ab}c\overline{d}) + (a\overline{b}cd + a\overline{b}c\overline{d}) + ab\overline{cd} + (\overline{a}bcd + \overline{a}b\overline{cd}) + a\overline{b}c\overline{d}$$

$$= abd + \overline{ab}d + a\overline{b}c + ab\overline{cd} + \overline{a}bc + a\overline{b}c\overline{d}$$

$$= (ab(d + \overline{cd} + c\overline{d})) + \overline{ab}d + a\overline{b}c + \overline{a}bc$$

$$= ab(d + \overline{d}) + \overline{ab}d + a\overline{b}c + \overline{a}bc$$

$$= ab + \overline{ab}d + a\overline{b}c + \overline{a}bc$$

which is identical to the function given by the BDD in Figure 8.2A. You may note that between the two BDDs, the order of the variables along the paths is different. In the first BDD, along any path, the order of variables is $a, b, c, d$, whereas in the second BDD, the order is $d, c, b, a$. One of our objectives is to find a canonical representation. In this case, two functionally equivalent functions should have graphically identical BDDs. Therefore, to achieve canonicity, an ordering of nodes must be imposed. An ordered BDD (OBDD) is a BDD with variables that conform to an ordering. Given a variable ordering $<$, an OBDD with the ordering is a BDD with the restriction that for any path in the OBDD containing variables $x$ and $y$, node $x$ is an ancestor of node $y$ if $x < y$.

With an ordering imposed on variables, we have to consolidate many different BDDs of the same function. However, the remaining OBDDs are not yet unique. For instance, the two OBDDs in Figure 8.3 have the same variable ordering and represent the same function, but they are not graphically identical. Note that the two $c$ nodes in Figure 8.3B

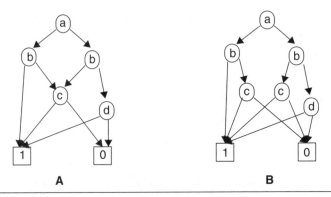

A                                              B

**Figure 8.3**   (A, B) Two OBDDs represent the same function.

can be merged because these two nodes represent the same function—namely, *c*. Once these two *c* nodes are merged, the two OBDDs are now identical. Indeed, it can be proved that if the following irredundancy conditions are satisfied on an OBDD, then the OBDD is canonical:

**1.** There are no nodes A and B such that they share the same 0-node and 1-node.
**2.** There is no node with two edges that point to the same node.

If the two nodes in condition 1 do exist, the function represented at nodes A and B is equivalent, according to the Shannon theorem, and thus the nodes should be combined. That is, if condition 1 is violated, then nodes A and B can be merged into a single node. If a node with two edges points to the same node, it means that the value of the variable at the node, whether 1 or 0, has no effect on the outcome, because the same node is encountered for either value of the variable; hence, this node can be ignored. That is, if condition 2 is violated, the node can be eliminated. The two transformations, merge and eliminate, are illustrated in Figure 8.4. With the merge transformation, when a node is merged with another, all incoming edges are redirected to the new node, and the node is deleted. In Figure 8.4A, node A and node B are to be merged; node A is picked as the new node and node B is deleted. The incoming edge of node B is redirected to node A. In eliminate transformation in Figure 8.4B, both edges of node N point to the same node M, and

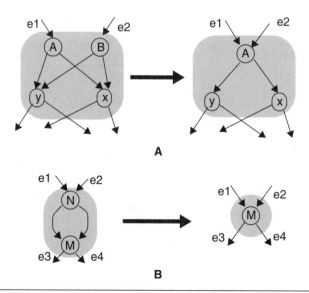

**Figure 8.4**   Transformations to reduce OBDDs. (A) Merge (B) Eliminate

hence node N is redundant. Node N is removed and its incoming edges are redirected to node M as shown.

---

**Example 8.2**

---

Reduce the OBDD in Figure 8.5.

The two $c$ nodes violate the first condition and thus are combined to give Figure 8.6A. Now the left and the middle $b$ nodes violate condition 1, and they are combined to produce Figure 8.6B. The left node $d$ violates condition 2 and is eliminated to give Figure 8.6C. At this point, both conditions are satisfied. This final BDD is canonical.

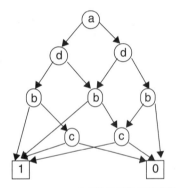

**Figure 8.5**   An OBDD for reduction

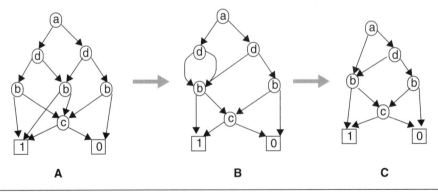

**Figure 8.6**   Reducing an OBDD. (A) Original BDD (B) Merging of two ⓑ nodes (C) Eliminating node ⓓ

An OBDD satisfying the two irredundancy conditions is called a *reduced OBDD* (ROBDD). The major property of ROBDDs is canonicity, as stated in Theorem 8.1.

---

**Theorem 8.1**

Canonicity of Reduced OBDDs

Two Boolean functions are equivalent if and only if their reduced OBDDs are identical with respect to any variable ordering.

---

## 8.1.1   Operations on BDDs

Once Boolean functions are represented using BDDs, Boolean operations are translated into BDD manipulations. In this section we will study algorithms for constructing BDDs from a Boolean function and a circuit, and for computing various Boolean operations using BDDs.

**Construction.**   Construction algorithms build a BDD for a function or circuit. The first question is whether any Boolean function has a BDD representation. The answer is affirmative because a BDD can be constructed for any function based on the Shannon cofactor theorem. According to the cofactor theorem, any function can be expressed as $f = x f_x + \bar{x} f_{\bar{x}}$. Therefore, we can create a BDD with node $x$ with a 1-edge that points to $f_x$ and a 0-edge that points to $f_{\bar{x}}$, and then recur the procedure on $f_x$ and $f_{\bar{x}}$ until the function is a constant, as demonstrated in Example 8.3.

---

**Example 8.3**

Construct a BDD for $f = ab + \bar{b}c$. Let's choose variable ordering $a < b < c$. First we compute cofactors of $f$ with respect to variable $a$. $f_a$ and $f_{\bar{a}}$ are equal to $(b + \bar{b}c)$ and $(\bar{b}c)$ respectively. Thus, $f = a(b + \bar{b}c) + \bar{a}(\bar{b}c)$. Create node $a$ and point its edges to $f_a$ and $f_{\bar{a}}$. Repeat the same procedure for the two cofactors: $b + \bar{b}c$ is cofactored as $b(1) + \bar{b}(c)$; thus, $f_{ab} = 1$ and $f_{a\bar{b}} = c$. Create node $b$ and point its two edges to $f_{ab}$ and $f_{a\bar{b}}$. The other cofactor, $\bar{b}c$, is cofactored as $\bar{b}(c)$; thus, $f_{\bar{a}\bar{b}} = c$ and $f_{\bar{a}b} = 0$. Create another node $b$ and point its edges to $f_{\bar{a}b}$ and $f_{\bar{a}\bar{b}}$. Function $c$ is $c(1) + \bar{c}(0)$. Create node $c$ and point its edges to $0$ and $1$. The BDDs for these cofactors are shown in Figure 8.7A. Connecting all these BDDs together, we have a BDD in Figure 8.7B. The cofactors are labeled in the BDD to illustrate the relationship between cofactor functions and BDD nodes. Then we apply the merge and eliminate transformations on the BDD to derive an ROBDD, as shown in Figure 8.7C.

---

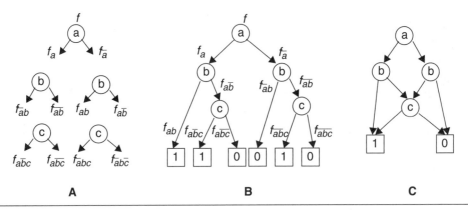

**Figure 8.7**    Constructing BDDs using Shannon cofactoring. (A) BDD nodes and Shannon cofactors.
(B) Connect BDD nodes. (C) Reduce BDD.

The major drawback of this method is the exponential number of operations. For the first variable, two cofactor functions are created. For the next variable, four cofactor functions are created, and so on. In general, $2^n$ cofactor functions are created for $n$ variables. In practice, BDDs are built from the bottom up, starting from variables or primary inputs of a circuit. As the building process progresses, a dynamic programming technique keeps track of the intermediate functions being built and returns the result if an intermediate function has already been built. This reuse strategy cuts down runtime substantially.

Let's look at the bottom-up construction process. Then we will see how dynamic programming helps to reduce runtime. From the start, BDDs are built for the variables, then for simple expressions made of the variables, followed by more complex expressions made of the simple expressions, until the function is completed.

---

**Example 8.4**

To build a BDD for $f = ((ab + c)(ad + ef) + bdf) + bce$, BDDs are built for $a$, $b$, $c$, $d$, $e$, and $f$. There are six operations in this step. Then, based on the BDDs for the variables, BDDs are constructed for these simple functions: $ab$, $ad$, $ef$, $bdf$, and $bce$. To construct these BDDs, the BDDs for the variables are ANDed together. For now, let's defer the study of ANDing and other Boolean operations of BDDs to a later section. There are seven ANDing operations in this step. Next, BDDs are created for $(ab + c)$ and $(ad + ef)$, which comprise two ORing operations of BDDs—namely, ORing of BDD of $ab$ with BDD of $c$, and ORing of BDD of $ab$ with BDD of $ef$. Finally, combining the clauses, BDDs are constructed for $((ab + c)(ad + ef) + bdf)$ and $((ab + c)(ad + ef) + bdf) + bce$. Three operations are in this final step. There are a total

---

**Example 8.4     (Continued)**

of 18 operations, and each operation is of polynomial complexity in the number of BDD nodes. Compare this number of operations with $2^6 = 64$ operations using the cofactoring method. In the section on Boolean operations, we will get into the details of using dynamic programming in computing Boolean operations of BDDs.

If BDDs are built from a circuit, BDDs for the primary inputs are constructed first, followed by the BDDs for the intermediate nodes, and finishing with BDDs for the primary outputs. The progression resembles the previous bottom-up method for Boolean functions. Consider the circuit in Figure 8.8. The sequence of BDD construction follows alphabetical order. The rule is that a circuit node is ready for BDD construction if the BDDs of all its fanins have already been constructed.

---

The complexity of BDD construction in general depends on variable ordering. However, there are functions with BDD sizes that are exponential for any variable ordering. In practice, however, many functions have polynomial sizes for some variable orderings. We will revisit the issue of variable ordering in a later section.

**Reduction.** Reduction operation transforms a BDD into a reduced BDD by applying recursively the merge and eliminate transformations on the BDD until the irredundancy conditions (see page 393) are met. The complexity of making a BDD canonical is $O(|BDD|)$, because each node is examined a fixed number of times in the transformations. To reduce a BDD with variable order $\pi$, we apply the merge and eliminate transformations to BDD nodes following the *reverse* ordering of $\pi$. Initially, we apply the emerge transformation to all constant nodes. Then we apply the merge and eliminate transformations to all BDD nodes of the last variable in $\pi$. Next, we repeat the transformations on the next-to-the-last variable and so on, until all variables are examined. At the end, the resulting BDD satisfies

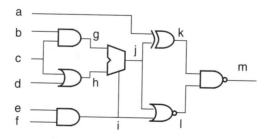

**Figure 8.8**   A circuit for illustrating the BDD building sequence

the irredundancy conditions. For instance, the BDD in Figure 8.7B does not satisfy the irredundancy conditions. We start out by combining the constant terminal nodes, which then reveal that the two $c$ nodes have their respective 1-nodes and 0-nodes pointing to the same nodes. We then apply the transformations to the c nodes and combine them. Finally, we repeat these actions on the $b$ nodes. At the end, the resulting BDD shown in Figure 8.7C satisfies the irredundancy conditions and hence is canonical.

**Restriction.** Restriction operation on a function sets certain variables to specific values. For example, restricting $x$ to 0 and $y$ to 1 in $f(r, s, x, y, z) = s\overline{r}xz + \overline{r}sy + \overline{r}sxyz + \overline{x}y\overline{z}$ produces $f(r, s, z) = srz + \overline{r}s + \overline{z}$. This operation can be easily done on BDDs. To restrict variable $v$ to 1, simply redirect all incoming edges to BDD nodes labeled $v$ to their 1-nodes. Similarly, restricting $v$ to 0 redirects all incoming edges to the 0-nodes. After the edges are redirected, it is possible that redundant nodes and equivalent nodes may appear. Thus, the reduction operation is called to reduce the BDD. The BDD nodes labeled $v$ still exist in the BDD, but they have no incoming edges. To clean up, successively remove all nodes, except the root, that have no incoming edges.

---

**Example 8.5**

Restrict the BDD in Figure 8.9A to $c = 0, d = 1$, and reduce it. Redirect all incoming edges to nodes labeled $c$ to their 0-nodes (see Figure 8.9B). The redirected edges are darker than the other edges. Now redirect all incoming edges to the $d$ node to its 1-node (see Figure 8.9C). Now, the right $b$ node and $a$ node are redundant, because both their edges point to the same node. Apply the reduction operation to remove them. Finally, all nodes but the root are removed that do not have any incoming edges; nodes $c$ and $d$ are removed. The resulting reduced BDD is shown in Figure 8.9D.

---

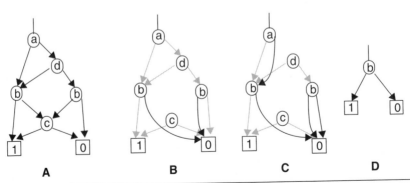

A          B          C          D

**Figure 8.9** Restriction operation on a BDD. (A) Original BDD (B) Restriction to c = 0 (C) Restriction to d = 1 (D) Reduced BDD

**Boolean operations.** When functions are represented by BDDs, operations on the functions translate into operations on BDDs. For example, $f$ AND $g$ becomes BDD($f$) AND BDD($g$). Here we examine various operations on BDDs. First we encounter the *ITE* (if–then–else) operator on Boolean functions $A$, $B$, and $C$: $ITE(A, B, C) = AB + \overline{A}C$. Note that the function represented by a BDD node, say $x$, is a special case of an *ITE* operation; the function of the node can be written as *ITE(x, function at 1-node, function at 0-node)*. The *ITE* operator notation also has the added convenience of relating to BDD nodes. For example, *ITE(x, y, z)*, where $x$, $y$, and $z$ are BDD nodes, simply represents BDD node $x$ with its 1-edge pointing to node $y$ and its 0-edge pointing to node $z$. The *ITE* operator encompasses all unary and binary operators. Table 8.1 lists some common Boolean operators and their corresponding ITE representations.

The complementation operation, $\overline{X}$, simply switches the 0 and 1 constant nodes. Composition of Boolean functions substitutes a variable in a function with another function and

**Table 8.1**    ITE Operator Representation of Boolean Operators

| Operator | ITE form |
|---|---|
| $\overline{X}$ | $ITE(X,0,1)$ |
| $XY$ | $ITE(X,Y,0)$ |
| $X+Y$ | $ITE(X,1,Y)$ |
| $X \oplus Y$ | $ITE(X,\overline{Y},Y)$ |
| $MUX(X,Y,Z)$ | $ITE(X,Y,Z)$ |
| composition, $f(x,g(x))$ | $ITE(g(x),f(x,1),f(x,0))$ |
| $\exists x f(x)$ | $ITE(f(1),1,f(0))$ |
| $\forall x f(x)$ | $ITE(f(1),f(0),0)$ |

can be computed as follows. Composing $f(x,y)$ with $y = g(x)$, giving $f(x, g(x))$, is equivalent to $f(x,1)$ when $g(x) = 1$ and $f(x,0)$ when $g(x) = 0$. Symbolically,

$$f(x, g(x)) = f(x, 1) \cdot g(x) + f(x, 0) \cdot \overline{g(x)}$$

which is $ITE(g(x), f(x,1), f(x,0))$. The existential and universal quantification operators are

$$\exists x f(x) = f(1) + f(0)$$

$$\forall x f(x) = f(1) \cdot f(0)$$

which consist of restriction, and OR and AND operations. The existential quantification $\exists x f(x) = f(1) + f(0)$ has the interpretation that there exists a value of $x$ such that $f(x)$ holds. Because $x$ is binary, either $f(0)$ or $f(1)$ must hold. In other words, $\exists x f(x) = f(1) + f(0)$. Similarly, the universal quantification means that for all values of $x$, $f(x)$ holds. That is, $f(0)$ and $f(1)$ must hold; hence, $\forall x f(x) = f(1) \cdot f(0)$.

Therefore, once we have an algorithm to perform $ITE$ operations on BDDs, we can perform any unary and binary Boolean operations on BDDs. The key to $ITE$ operations is based on the following identity:

$$
\begin{aligned}
ITE(A, B, C) &= AB + \bar{A}C \\
&= x(AB + \bar{A}C)_x + \bar{x}(AB + \bar{A}C)_{\bar{x}} \ \ // \text{ Shannon cofactor theorem} \\
&= x(A_xB_x + \bar{A}_xC_x) + \bar{x}(A_{\bar{x}}B_{\bar{x}} + \bar{A}_{\bar{x}}C_{\bar{x}}) \\
&= (x, ITE(A_x, B_x, C_x), ITE(A_{\bar{x}}, B_{\bar{x}}, C_{\bar{x}}))
\end{aligned}
$$

which shows that the problem can be reduced to a smaller problem by cofactoring with respect to a variable, and it lends itself to a recursive algorithm. When calls to $ITE(A_x, B_x, C_x)$ and $ITE(A_{\bar{x}}, B_{\bar{x}}, C_{\bar{x}})$ return, a BDD node $x$ is created and its two edges are pointed to $ITE(A_x, B_x, C_x)$ and $ITE(A_{\bar{x}}, B_{\bar{x}}, C_{\bar{x}})$. The recursion stops when it reaches a terminal case for which the $ITE$ operation is trivial. The terminal cases are the ITEs that are equal to either constants, one of its operands, or complement one of its operands. For example, these are terminal cases: $ITE(1,X,Y)$, $ITE(0,Y,X)$, $ITE(X,1,0)$, $ITE(Y,X,X)$, all of which are equal to $X$, and $ITE(1, 1, Y)$, $ITE(0, Y, 1)$, $ITE(X, 1, 1)$, all of which are equal to $1$.

This recursive operation is shown in Figure 8.10. The triangles are input BDDs of functions A, B, and C. Figure 8.10A is operation $ITE(root(A), root(B), root(C))$, where we use notation *root(A)* to emphasize that the operand is the root node of BDD A. The first variable $x$ selected for cofactoring is the earliest ordered root variable of A, B, and C. Figure 8.10B depicts the situation after one recursive call: A BDD node for variable $x$ is created and two *ITE* calls are invoked on the 1-node and 0-node of the root of BDD A, B, and C; that is, the two recursive calls are as follows, the operands of the *ITE* operator being BDD nodes: $ITE(root(A_x), root(B_x), root(C_x))$ and $ITE(root(A_{\bar{x}}), root(B_{\bar{x}}), root(C_{\bar{x}}))$.

Directly using the cofactoring identity would generate an exponential number of subfunctions. To curb the exponential growth, dynamic programming is used to remember what subfunctions have been operated on. If there is a match, the result is returned immediately. A *computed table* stores results of computed ITEs and the results are indexed by their operands. A match is found if all the operands of a pending *ITE* operation match the operands of an entry in the computed table. In this case, the pending *ITE* operation terminates and returns with the result found in the table. Let us compute how many distinct entries are possible in a computed table. Because each operand in an *ITE* is a BDD node, there are only $|A| \cdot |B| \cdot |C|$ possible combinations of operands. Therefore, at most $|A| \cdot |B| \cdot |C|$ *ITE* operations are required, compared with $2^n$, where $n$ is the number of variables. Hence, the complexity of BDD operations is $O(|A| \cdot |B| \cdot |C|)$.

The previous algorithm will create BDDs that may not be reduced. Thus, we must modify the ITE procedure to include the merge and eliminate transformation so that the resulting BDDs are canonical. To incorporate the merge transformation, a unique table remembers all unique BDDs that have already been created, which are indexed by keys consisting of the node variable and its 1-node and 0-node. When calls of $ITE(A_x, B_x, C_x)$ and $ITE(A_{\bar{x}}, B_{\bar{x}}, C_{\bar{x}})$ return, and before $ITE(A,B,C) = (x, ITE(A_x, B_x, C_x), ITE(A_{\bar{x}}, B_{\bar{x}}, C_{\bar{x}}))$ is created, we first

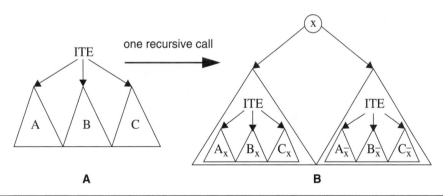

**Figure 8.10** Recursive algorithm for computing the *ITE* operator. (A) Before an ITE call (B) After an ITE call on variable $x$

check the unique table for an entry with a key of node $x$, and with a 1-node and 0-node that are $ITE(A_x, B_x, C_x)$ and $ITE(A_{\bar{x}}, B_{\bar{x}}, C_{\bar{x}})$ respectively. If such an entry exists, the node found in the table is used. This extra checking step in effect does the merge transformation.

Second, to incorporate the eliminate transformation, we need to check whether the *then* node is identical to the *else* node. For example, $ITE(A_x, B_x, C_x) = ITE(A_{\bar{x}}, B_{\bar{x}}, C_{\bar{x}})$. If they are, no new node will be created. Instead, either $ITE(A_x, B_x, C_x)$ or $ITE(A_{\bar{x}}, B_{\bar{x}}, C_{\bar{x}})$ is returned. This step eliminates redundant nodes. If neither of these two checks succeeds, a new node $x$ is created and its edges are pointed to $ITE(A_x, B_x, C_x)$ and $ITE(A_{\bar{x}}, B_{\bar{x}}, C_{\bar{x}})$. Example 8.6 illustrates the ITE operations, dynamic programming, and merge and eliminate transformation embedding.

---

### Example 8.6

In this example, we want to OR the two BDDs shown in Figure 8.11. The operation is $ITE(X, 1, Y)$. For the convenience of referring to the nodes of the BDDs, we number the nodes. To refer to a node, we use notation, *BDD_name.node_number*; for instance, node $c$ of *BDD X* is *X.4*.

To select the first variable for cofactoring, we examine the root nodes of both BDDs and select the one with an earlier ordering. In this case, both roots are the same, node $a$. Select variable $a$ as the first cofactoring variable. The cofactors of $a$ in *BDD X* are the BDDs rooted at *X.2* and *X.3*. Similarly, the cofactors of $a$ in *BDD Y* are the BDDs rooted at *Y.2* and *Y.3*. Using the ITE cofactoring identity, we have

$$ITE(X, 1, Y) = (a, ITE(X.2, 1, Y.2), ITE(X.3, 1, Y.3)).$$

The algorithm recurs on $ITE(X.3, 1, Y.3)$. Cofactoring variable $b$, $ITE(X.3, 1, Y.3) = (b, ITE(X.4, 1, Y.3), ITE(X.6, 1, Y.3))$. Here, the *Y.3* root node is $c$ and hence it remains the same in the cofactoring of $b$. At this step, because $X.6 = 0$, we have reached a terminal case—namely, $ITE(X.6, 1, Y.3) = Y.3$. Thus, the BDD rooted at *Y.3* is returned as the result of $ITE(X.6, 1, Y.3)$, and is entered into the compute and unique tables. The index for *Y.3* in the compute table is $(X.6, 1, Y.3)$, the operands of the ITE operator. Recur on the *then* component, $ITE(X.4, 1, Y.3)$.

$$ITE(X.4, 1, Y.3) = (c, ITE(X.5, 1, Y.4), ITE(X.6, 1, Y.5))$$

$$= (c, ITE(1, 1, 1), ITE(0, 1, 0))$$

$$= (c, 1, 0).$$

## Example 8.6   (Continued)

Before $(c, 1, 0)$ is created, we need to consult the unique table and find a match, which is $Y.3$. Therefore, $Y.3$ is returned as the result for $ITE(X.4, 1, Y.3)$. Now $Y.3$ indexed $(X.4, 1, Y.3)$ is entered into the compute table. At this step we return from $ITE(X.3, 1, Y.3) = (b, Y.3, Y.3)$. Because the two edges point to the same node, no BDD node is created; instead, $Y.3$ is returned as the result for $ITE(X.3, 1, Y.3)$. $Y.3$ indexed $(X.3, 1, Y.3)$ is entered into the compute table.

The other branch of recursion is $ITE(X.2, 1, Y.2)$. $ITE(X.2, 1, Y.2) = (b, ITE(X.5, 1, Y.5), ITE(X.6, 1, Y.3))$. Select $ITE(X.5, 1, Y.5)$ for recursion, which is equal to $ITE(1, 1, 0) = 1$, a terminal case. Thus, $ITE(X.5, 1, Y.5)$ returns 1 and is entered into the compute table. Next, recur on $ITE(X.6, 1, Y.3)$. Because $ITE(X.6, 1, Y.3) = ITE(0, 1, Y.3) = Y.3$, a terminal case, we consult the unique table and find $Y.3$ there. Thus, $Y.3$ is returned as the result of $ITE(X.6, 1, Y.3)$. Returning from $ITE(X.2, 1, Y.2) = (b, ITE(X.5, 1, Y.5), ITE(X.6, 1, Y.3))$, we have $ITE(b, 1, Y.3)$. Because $ITE(b, 1, Y.3)$ is not yet in the unique table, we create BBD node $b$ with its 1-node pointing to constant node 1 and its 0-node pointing to $Y.3$. This new node is entered into the unique and compute tables. Let us call this node $N1$.

Returning from the top-level recursion $ITE(X, 1, Y) = (a, ITE(X.2, 1, Y.2), ITE(X.3, 1, Y.3))$, we have $ITE(X, 1, Y) = (a, N1, Y.3)$. This is the resulting BDD. Figure 8.12A shows the ITE recursions and Figure 8.12B shows the final BDD. Notice how the embedded merge and eliminate transformations produced a reduced BDD at the end.

To verify the result, BDD $X$ represents Boolean function $ab + \bar{a}bc$, and BDD $Y$, $a\bar{b}c + \bar{a}c$. ORing the two functions gives

$$ab + \bar{a}bc + a\bar{b}c + \bar{a}c = ab + c$$

The BDD in Figure 8.12B represents $ab + c$.

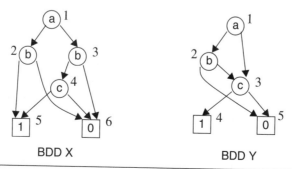

BDD X                                    BDD Y

**Figure 8.11**  Operant BDDs for ITE operation

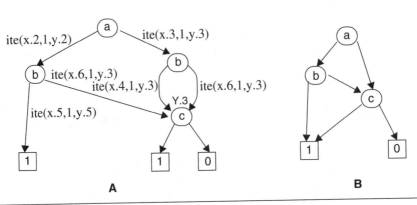

**Figure 8.12** Intermediate and final result from BDD ITE operations (A) BDD created by ITE operations (B) Reduced BDD

The ITE algorithm is summarized here:

---

**Boolean Operations on BDDs: ITE(A, B, C)**

Input: Three BDDs, A, B, and C.

Output: ROBDD representing *ITE(A, B, C)*.

1. If *ITE(A, B, C)* is a terminal case, return with result.
2. If *ITE(A, B, C)* is in the computed table, return with result.
3. Select the root variable $x$ that is ordered earliest.
4. Compute $BDD_0 = ITE(A_{\bar{x}}, B_{\bar{x}}, C_{\bar{x}})$ and $BDD_1 = ITE(A_x, B_x, C_x)$.
5. If $(BDD_0 = BDD_1)$, return $BDD_0$.
6. If $(x, \text{root of } BDD_1, \text{root of } BDD_0)$ is in the unique table, return the existing node.
7. Create BDD node $x$ with 0-edge and 1-edge pointing to $BDD_0$ and $BDD_1$ respectively.

---

## 8.1.2   Variable Ordering

So far in our discussion of BDDs, we have assumed an arbitrary variable ordering. Different variable orderings can cause drastic differences in BDD size. Example 8.7 dramatizes the effect and sheds some insight into what makes a variable ordering good.

## Example 8.7

Consider building BDD for $(a_1 \oplus a_2)(b_1 \oplus b_2)(c_1 \oplus c_2)$, first with variable ordering $a_1 < a_2 < b_1 < b_2 < c_1 < c_2$ and then with $a_1 < b_1 < c_1 < a_2 < b_2 < c_2$. The BDD with the first ordering is in Figure 8.13A and the one with the second ordering is in Figure 8.13B. The size ratio of the second BDD to the first BDD is 23:11, more than a 100% increase.

In the first ordering, $a_1 < a_2 < b_1 < b_2 < c_1 < c_2$, when $a_1 = a_2$. The function's value is completely determined to be 0 regardless of the values of the other variables. Therefore, two paths have only two variable nodes, $a_1$ and $a_2$. Similarly, $b_1 = b_2$ or $c_1 = c_2$ completely determines the value of the function. Thus, we observe that when the variables are ordered together early that completely determine the value of the function, fewer nodes appear on the paths from BDD root to constant roots, and hence a simpler BDD results.

If $a_1$ and $a_2$ are not equal, then the function is determined solely by the remaining variables without retaining any knowledge of the specific values that $a_1$ and $a_2$ have taken. In other words, $b_1 = b_2$ or $c_1 = c_2$ alone determines the value of the function for all unequal values of $a_1$ and $a_2$. Therefore, we see that variables $b_1$ and $c_1$ are being shared by the two paths that represent $a_1 = 1, a_2 = 0$ and $a_1 = 0, a_2 = 1$. We observe that the less knowledge about past variable assignment is required to determine the function's value the more nodes that are shared. More sharing gives smaller BDDs.

In the second ordering, $a_1 < b_1 < c_1 < a_2 < b_2 < c_2$, the earliest time the function's value is decided is when four variables are assigned values, which is worse than the situation from the first ordering. If the values assigned to the first four variables do not determine the function's value, some already assigned values need to be remembered to assign values to the remaining variables. Therefore, this ordering produces a larger BDD size.

The following two observations provide insight into what makes up a good variable ordering. First, the "width" of a BDD, crudely defined as the number of paths from root to the constant nodes, is determined by the "height" of the BDD, which is the average number of variable nodes along paths, because each variable node branches to two paths. Thus, the more nodes there are on a path, the more branches do the nodes produce, and the wider the BDD. The size of a BDD is proportional to the product of the width and height; hence, the height determines the BDD size. Roughly speaking, the larger the average number of

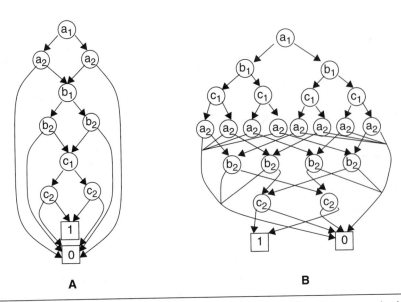

A　　　　　　　　　　　B

**Figure 8.13** Impact of variable ordering on BDD size. (A) BDD with variable ordering $a_1$, $a_2$, $b_1$, $b_2$, $c_1$, $c_2$ (B) BDD with variable ordering $a_1$, $b_1$, $c_1$, $a_2$, $b_2$, $c_2$

variable nodes along the paths, the bigger the BDD size. Now let's relate variable ordering to the number of variable nodes along paths. The variable nodes on a path represent the minimum knowledge about the variables such that the value of the function is determined. Therefore, the less knowledge required to know about variables to determine the function's value, the smaller the BDD size. Hence, a good variable ordering should have the property that, as variables are evaluated one by one in the order, the function value is decided with the fewest number of variables the sooner a function's value is decided, the fewer variable nodes are on the paths.

Second, node sharing reduces BDD size. Let node $v$ be shared by two paths. Let us call the variables ordered before $v$ the predecessors, and the variables ordered after $v$ the successors. The portions of the two paths from the root to $v$ correspond to two sets of values assigned to the predecessors. The fact that the two paths share node $v$ implies that the valuation of the function by variable $v$ and its successors is independent of the two assigned values to the predecessors. Hence, the more independent the successors and predecessors are in a variable ordering, the more sharing that occurs and thus the smaller the resulting BDD. In other words, a good variable ordering groups interdependent variables closer together.

Deriving a variable ordering from circuits also follows these two principles but manifests slightly differently and uses the circuit structure to select the variables. The key is to arrange the variables such that as evaluation proceeds in the order, the function simplifies as quickly as possible and is as much independent of the already assigned values as possible. Many heuristics exist. An example heuristic is to level the gates from the primary output (the level of a gate being the shortest distance to the output and the distance being the number of gates in between) and order the inputs in increasing level number. The rationale is that variables closer to the output have a better chance of deciding the value of the circuit and hence create BDD paths with fewer nodes. In addition, order the primary input variables such that the intermediate node values are determined as soon as possible. The idea lies in the hope that when intermediate node values are determined, the circuit simplifies.

---

**Example 8.8**

Consider the circuit in Figure 8.14. The numbers are the level numbers. Input $a$ has a level number of 1; $b$, 4; $c$, 3; $d$, 3; and $e$, 3. An order would be $a, c, d, e, b$. Variable $a$ is closest to the output and makes the output 0 when it is 0. Variables $c$, $d$, and $e$ are next closest to the output. Because variables $c$ and $d$ alone can force the output value of the NAND and the AND gate, whereas $e$ cannot, $c$ and $d$ are ordered before $e$. The last variable is $b$. As with any heuristic, there is no guarantee of any optimality in the result, but a good heuristic should yield near-optimal or just good results for many practical situations.

---

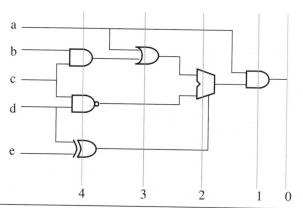

**Figure 8.14** A sample circuit for variable ordering

When composing or computing an operation on several Boolean functions, each of which has its own good ordering, a question arises about how to derive a good ordering for the resulting BDD based on the orderings of the operands. A heuristic is to interleave variable orderings to form an initial overall ordering. For example, if operand 1 has variable ordering $a$, $b$, $c$, $d$, and operand 2 has $x$, $y$, $z$, then an interleaved ordering is $a$, $x$, $b$, $y$, $c$, $z$, $d$. How variables are interleaved also depends on the specific operators. After an initial overall ordering is obtained, dynamic algorithms can be used throughout the computing process to adjust the variable ordering to minimize the sizes of intermediate BDDs. Dynamic variable ordering is discussed next.

**Dynamic Variable Ordering.** So far, the discussion on ordering assumes that an ordering algorithm calculates an ordering of the variables involved in a Boolean computation beforehand and the ordering is used throughout the computation of the functions. Such an ordering algorithm is called *static*. When a function $f$ is composed with another function $g$ and the result is then operated with yet another function $h$, a method based on static ordering would first calculate a variable ordering for the variables in $f$, $g$, and $h$, and then use the ordering throughout these two operations. However, a better strategy is to choose a good variable ordering for each of the operations, as opposed to one ordering for all operations. Even during the process of computing a single operation, oftentimes BDD size varies over the entire computational process and peaks in the middle of the process. Therefore, it is difficult to predict an optimal or good ordering for the entire computing process. In fact, an optimal algorithm may have to change variable ordering during the computing process. A dynamic variable ordering algorithm changes the variable ordering as the BDD size approaches a threshold, and usually makes local and incremental adjustments. A typical application of a dynamic algorithm is to build a BDD with a variable ordering derived from a static algorithm, then use a dynamic algorithm to modify the ordering to improve the BDD size during subsequent computations. A requirement for dynamic algorithms is that they must have minimal computational cost so that their use, instead of finding a new ordering using a static algorithm, is justified.

A simple dynamic ordering algorithm is based on the repetitive application of the swap operation, which exchanges the position of two adjacent variables. Referring to Figure 8.15A, variable $x$ is ordered ahead of $y$ and we want to reverse their order. To swap variable $x$ with $y$, the labels of the nodes are exchanged and the inner two nodes, $b$ and $c$, are swapped. This swap operation preserves the functionality of the BDD, because for all possible values of $x$ and $y$, the same child nodes ($a$, $b$, $c$, $d$) are reached in each case. For example, $x = 1$, $y = 0$ reaches node $b$ before and after the swap. After swapping, a reduce

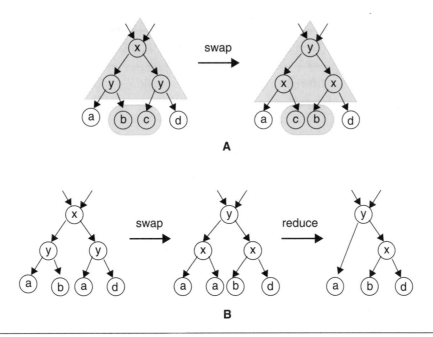

**Figure 8.15** Swap two adjacent BDD variables. (A) Swap operation. (B) Swap operation followed by a reduce operation.

operation may be required to maintain *local* canonicity, as is the case in Figure 8.15B. Therefore, we conclude that a swap operation has a low computational cost and can be done locally.

Any variable ordering can be obtained from an initial ordering through multiple swap operations, and thus, in theory, the optimal ordering can be achieved from an initial ordering using only swap operations. However, in practice, there is no known guidance for selecting variables for swapping for optimal ordering, except for exploring all permutations. Therefore, a greedy algorithm is often used with the swap operation. A sifting algorithm moves a selected variable to all possible positions and chooses the one with the smallest BDD size. For example, suppose the current variable ordering is $a < b < c < d < e$. Sifting variable $d$ produces the five orderings shown in Table 8.2. In this example, the ordering $a < d < b < c < e$ gives the smallest BDD and thus is kept. Sifting a single variable may be regarded as searching along the variable's axis. Sifting multiple variables simultaneously explores more space and may produce a better result.

**Table 8.2**    Sifting Variable $d$

| Operation | Variable Ordering | BDD Size |
|---|---|---|
| **Sift right** | a,b,c,e,**d** | 121 |
| *Original* | *a,b,c,**d**,e* | 133 |
| **Sift left** | a,b,**d**,c,e | 118 |
| **Sift left** | a,**d**,b,c,e | 117 |
| **Sift left** | **d**,a,b,c,d | 127 |

**Functions and BDD Sizes.** Variable ordering algorithms are only useful for the functions that have a good ordering. There are functions that always have BDD sizes exponential in the number of input variables for any variable ordering. A well-known such function is the multiplication function. Specifically, let $x_1, ..., x_n$ and $y_1, ..., y_n$ be the Boolean variables of two multiplicands, and $z_1, ..., z_{2n}$, those of the result—that is, $(z_{2n}, ..., z_1) = multiply(x_n, ... x_1, y_n, ..., y_1)$. The output functions of 3-bit multiplication are illustrated in Figure 8.16. Variable $c_i$ is the carry from output bits before the $i$th bit. For example, $c_3$ is 1 only if the two product terms in $z_2$ are both 1. In other words, $c_3 = x_1x_2y_1y_2$ and $z_3 = x_1y_3 + x_2y_2 + x_3y_1 + c_3$. Then, at least one output bit function, $z_i$, has a BDD of size at least $2^{n/8}$ for *any* ordering of $x_1, ..., x_n, y_1, ..., y_n$.

At the other extreme, there are functions with BDDs that are always of polynomial size with respect to any variable ordering. Because symmetric functions are invariant to variable interchanges, their BDD sizes are independent of variable ordering. Recall that a symmetric function is 1 if and only if there is a set of integers, $\{a_1, ..., a_k\}$, such that the number of 1s in the inputs is one of the integers. A symmetric function is completely specified by the set of integers. This fact leads to a universal implementation of symmetric functions. This implementation first adds up all input bits. The result is the number of 1s in the inputs. This result is connected to a multiplexor select line, which selects a 1 if the select value is one of the integers in the symmetric function. Figure 8.17 shows the universal implementation of symmetric function with integer set $\{a_1, ..., a_k\}$. This configuration of

| | | | $x_3$ | $x_2$ | $x_1$ |
|---|---|---|---|---|---|
| | | X | $y_3$ | $y_2$ | $y_1$ |
| | | | $x_3y_1$ | $x_2y_1$ | $x_1y_1$ |
| | | $x_3y_2$ | $x_2y_2$ | $x_1y_2$ | |
| | $x_3y_3$ | $x_2y_3$ | $x_1y_3$ | | |
| $c_6$ | $x_3y_3+c_5$ | $x_2y_3+x_3y_2+c_4$ | $x_1y_3+x_2y_2+x_3y_1+c_3$ | $x_2y_1+x_1y_2$ | $x_1y_1$ |
| $\uparrow$ | $\uparrow$ | $\uparrow$ | $\uparrow$ | $\uparrow$ | $\uparrow$ |
| $z_6$ | $z_5$ | $z_4$ | $z_3$ | $z_2$ | $z_1$ |

**Figure 8.16** Output functions of a 3-bit multiplier

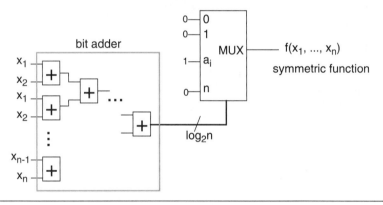

**Figure 8.17** A universal implementation of symmetric functions

adder and multiplexor has a polynomial-size BDD. Therefore, all symmetric functions have polynomial BDD sizes for any ordering.

Unlike these two extreme cases, many functions are sensitive to variable ordering, and their BDDs can change from polynomial size to exponential size and vice versa. A practical and interesting property is how easy it is to find a variable ordering that yields a BDD of reasonable size.

## 8.2   Decision Diagram Variants

In this section we will study several variants of decision diagrams, all of which are canonical and have efficient manipulation algorithms, such as construction and Boolean operations. Each variant is targeted toward a specific application domain. Therefore, as the application of decision diagrams become more widespread, more variants are sure to come.

### 8.2.1   Shared BDDs (SBDDs)

Strictly speaking, SBDDs are not a variant of BDDs; rather, they are an application of a BDD to vector Boolean functions. A vector Boolean function has multiple outputs. An example is the next-state function of a finite-state machine having more than one FF. When constructing BDDs for a number of Boolean functions, instead of having a number of isolated BDDs, one for each function, the BDD nodes representing the same functionality are shared. This overall BDD having multiple roots each representing a function is called a *shared BDD* (or SBDD). The main advantage of an SBDD is compactness through node sharing. An SBDD preserves all the properties of a BDD.

---

**Example 8.9**

Construct an SBDD for the following functions:

$$f = a\overline{b}c + \overline{a}b\overline{c} + ab\overline{c}$$

$$g = \overline{a}(b\overline{c} + \overline{b}c) + a\overline{b}\overline{c}$$

For illustration purposes let's first construct a BDD for each function and then merge their nodes. In practice, BDDs for $f$ and $g$ are constructed simultaneously, and node sharing is a part of the construction process. Choose variable ordering $a < b < c$. To merge nodes of the same functionality, we start from the leaf nodes and move toward the root. The BDDs for $f$ and $g$ and the SBDD are shown in Figure 8.18. The ratio of BDD sizes before and after sharing is 13:9.

---

### 8.2.2   Edge-Attributed BDDs

It is simple operation to complement a function: by exchanging the constant nodes in the function's BDD. That is, except for the two constant nodes, the BDD of the function and that of its complement are identical. Being able to share the common structures between the function and its complement can have a drastic effect on overall size, especially for a function with many subfunctions that are complements of each other. One solution to this

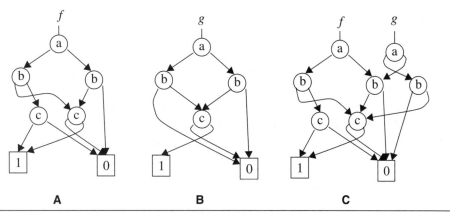

**Figure 8.18** (A) BDD for *f*. (B) BDD for *g* (C) SBDD for *f* and *g*

problem allows BDD edges to carry an attribute that can be an inversion or no inversion. When an edge's attribute is inversion, the function seen at the tail of the edge is the complement of the function at the head of the edge. To denote an inversion attribute, we place a small dot on the edge.

Allowing unconditional placement of inversion attributes on an edge violates canonicity. For instance, the pairs of BDDs in Figure 8.19 have equivalent functionality, even though they are structurally different. Therefore, we must restrict which edges are allowed to be complemented. For the two equivalent configurations in each pair, we only allow the one with no dot on the 1-edge (in other words, the first configuration). With this restriction, we restore canonicity.

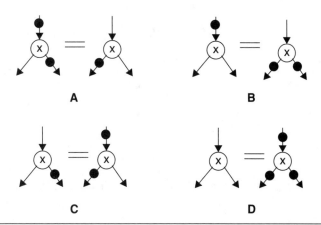

**Figure 8.19** (A–D) Equivalent-edge attributes

**Example 8.10**

As an example, let's apply edge attributes to the SBDD in Figure 8.18C. Recognizing that the two *c* nodes are complements of each other, we choose to eliminate the *c* node on the right and redirect its incoming edges to the other *c* node. Add the inversion attribute to each of the redirected edges (see Figure 8.20B). Redistribute the dots to avoid complementation on the 1-edge to preserve canonicity. The resulting BDD is shown in Figure 8.20C.

### 8.2.3   Zero-Suppressed BDDs (ZBDDs)

The two reduction rules for BDD are removing nodes with two edges that point to the same node and merging nodes that represent the same functionality. In ZBDDs, the first reduction rule is replaced by removing nodes with a 1-node that is the constant node 0. As illustrated in Figure 8.21, the two transformations for the ZBDD are the following:

**1.** Remove the node with a 1-edge that points to constant node 0 and redirect all incoming edges to its 0-node.
**2.** Merge nodes that represent the same functionality.

It can be shown that ZBDDs are canonical. However, ZBDD nodes may exist with both edges that point to the same node. Consequently, a ZBDD has the following interpretation.

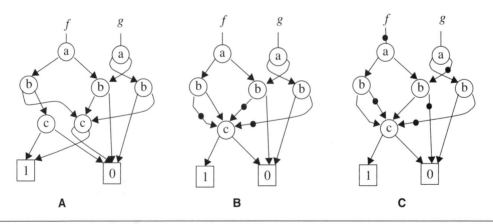

**Figure 8.20** SBDD with edge attributes. (A) SBDD without edge attributes (B) Add edge attributes (C) Canonical form with edge attributes

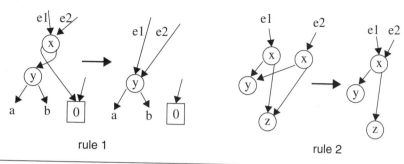

**Figure 8.21** ZBDD reduction rules

A path from the root to the constant node 1 represents a cube. If a variable is missing in the path, the variable node's 1-node is constant 0 and therefore the variable is a negative literal in the cube.

Conceptually, a ZBDD can be obtained from a completely unreduced decision diagram by applying the two previous reduction rules. A completely unreduced decision diagram is a graphical representation of the truth table of a function, and hence has exactly $2^n$ paths and $2^n$ leaf nodes, where $n$ is the number of variables, and it can be constructed by applying successive cofactor operations for all variables in the function. Each path represents a minterm, and the value of the leaf node is the value of the function at the minterm. In practice, ZBDDs are not constructed directly from unreduced decision diagrams, but from a procedure that applies the two reduction rules throughout the whole construction process. For more details, please refer to the bibliography.

---

**Example 8.11**

Figure 8.22A shows a completely unreduced decision diagram. From this decision diagram, we can apply the ZBDD reduction rules. First, remove all the $c$ nodes with a 1-edge that points to constant node 0, and redirect their incoming edges to their respective 0-nodes, as in Figure 8.22B. Now the left $b$ node has its 1-node as constant 0 and is thus removed, as shown in Figure 8.22C. At this point, node $a$ becomes eligible for removal. Getting rid of node $a$ and sharing the constant node 1, we have a reduced ZBDD in Figure 8.22C. An ROBDD for the same function is shown in Figure 8.23A. Note that the reduced BDD has more nodes and edges. Also note that variable $a$ does not even appear in the ZBDD. Let us try to read the function off the ZBDD. There are two paths to constant node 1. The first path consists of $b$ and $c$.

---

*continues*

**Example 8.11    (Continued)**

Because variable *a* is missing in the path, it is a negative literal in the cube. The cube indicated by the first path is $\overline{a}bc$. The second path has only variable *b*; therefore, the two missing variables show up in the cube as negative literals. This cube is $\overline{abc}$. The function is the sum of the two cubes: $\overline{a}bc + \overline{abc}$. The same function can be derived from the BDD.

Not all functions have ZBDDs smaller than BDDs. Apart from sharing nodes, a ZBDD gets reduced in size by removing nodes with 1-edges that point to 0, whereas BDD reduces from removing nodes with both edges that point to the same node. Therefore, a ZBDD has a smaller size than the BDD of the function if there are more such 1-edge-to-0 nodes than the nodes with both edges that point to the same node, and vice versa. A path in a decision diagram represents a cube. Thus, a path having a 1-edge-to-0 node means the cube has a negative phase of the variable. Therefore, intuitively, a ZBDD is more compact if the

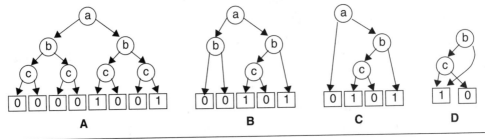

**Figure 8.22** (A) Unreduced decision diagram (B, C) Nodes eliminated via reduction (D) Reduced ZBDD

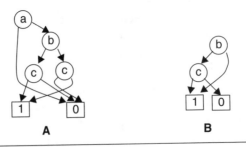

**Figure 8.23** (A, B) Compare ROBDD (A) with ZBDD (B)

function to be represented has a lot more negative literals than positive literals in the minterms of its on set. A minterm of $n$ literals can be interpreted as a selection among the $n$ objects: A positive literal at position $i$ means the $i$th object is selected; a negative literal, the $i$th object is not selected. Therefore, a function of $n$ variables, expressed as a disjunction of minterms, represents all acceptable selections. A more compact ZBDD results if there are more negative literals, implying that few objects are selected. With this interpretation, sometimes ZBDDs are said to be compact for sparse representation (in other words, the objects selected are sparse).

## 8.2.4   Ordered Functional Decision Diagrams (OFDDs)

A BDD is based on the first form of the Shannon cofactoring expansion. It is possible to derive decision diagrams based other expansions. OFDDs are based on positive Davio expansion. Positive and negative Davio expansions are

$$\text{Positive: } f = f_{\bar{x}} \oplus \left( x\frac{\partial f}{\partial x} \right)$$

$$\text{Negative: } f = f_x \oplus \left( \bar{x}\frac{\partial f}{\partial x} \right).$$

We will focus our discussion on positive expansion. A function, when expanded with respect to variable $x$, generates two subfunctions: $f_{\bar{x}}$ and $\frac{\partial f}{\partial x}$, each of which is independent of $x$. Expansion continues on the subfunctions until all variables are exhausted. The resulting expansion is called the Reed Muller form, which is an XOR sum-of-products term. This expansion of a function can be represented graphically, and the resulting graph is called an OFDD. When expanding about variable $x$, a node labeled $x$ is created and the 1-edge is pointed to the node of the Boolean difference whereas the 0-edge is pointed to the node of the cofactor, as shown in Figure 8.24A. If the cofactor or the Boolean difference is not a constant, the expansion continues.

Conversely, given an OFDD, the represented function is derived by ANDing the node variable with its 1-node and XORing with its 0-node, as in Figure 8.24B, starting from the leaf nodes and ending at the root. Each path from the root to a leaf node represents a cube. Based on the semantics of an OFDD node, if a path takes the 1-edge of a node, the node variable is included in the cube. If the 0-edge is taken, the variable is not included. Therefore, to obtain a function from an OFDD, enumerate all paths from the root to the constant 1-node. Then generate the cubes for the paths. Finally, XOR all the cubes.

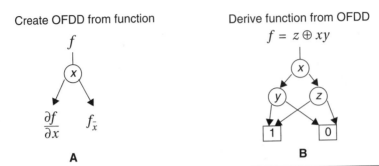

**Figure 8.24** OFDD from Davio expansion. (A) Edge semantics of an OFDD node (B) An OFDD for $f = z \oplus xy$

Two reduction rules for OFDDs are the same as those for ZBDDs:

1. If a node's 1-edge points to constant 0, remove the node and redirect its incoming edges to its 0-node.
2. Merge nodes that have the same child nodes.

Rule 1 follows directly the expansion $f = g \oplus xh$; if the 1-node is constant 0 (in other words, $h = 0$, then $f = g$), meaning that the incoming edges directly point to the 0-node. Rule 2 combines isomorphic nodes.

---

**Example 8.12**

Expand the following function using positive Davio expansion and create an OFDD:

$$f(a,b,c,d) = \overline{a}bc + \overline{b}cd + \overline{a}cd$$

Choose variable ordering $a < b < c < d$.

$$f = g \oplus ah = f_{\overline{a}} \oplus a\frac{\partial f}{\partial a} = (\overline{b}c + cd) \oplus a(\overline{b}cd \oplus (\overline{b}c + cd)) = (\overline{b}c + cd) \oplus a(\overline{b}c \oplus cd)$$

$$g = \overline{b}c + cd = p \oplus bq = g_{\overline{b}} \oplus b\frac{\partial g}{\partial b} = (c) \oplus b(c\overline{d}) = c \oplus bc\overline{d}$$

$$h = \overline{b}c + cd = r \oplus bs = h_{\overline{b}} \oplus b\frac{\partial h}{\partial b} = 0 \oplus b(0 \oplus cd) = bcd$$

$$q = c\overline{d} = q_{\overline{c}} \oplus c\frac{\partial q}{\partial c} = 0 \oplus c(\overline{d})$$

$$\overline{d} = 1 \oplus d(1)$$

$$c = 0 \oplus c(1)$$

**Example 8.12   (Continued)**

Now the remaining subfunctions are constants, so we stop. The resulting expansion is as follows and the OFDD is shown in Figure 8.25A:

$$f = g \oplus ah = (c \oplus bc\bar{d}) \oplus a(c\bar{d} \oplus bc) = c \oplus bc\bar{d} \oplus ac\bar{d} \oplus abc$$

Now apply the reduction rules to simplify the OFDD. The reduced OFDD is shown in Figure 8.25B. Note that the $d$ node has both edges pointing to the same node, yet the node remains in the OFDD.

As a check, let us derive the Reed Muller form from the OFDD. There are six paths from the root to constant node 1. For each path, the node variable is included in the cube if the path takes the 1-edge of the node. For the path shown Figure 8.25B, because only the 1-edge of node $c$ is taken, the path gives cube $c$. The other five paths give cubes $abc$, $acd$, $ac$, $bc$, and $bcd$. The function is the XORing of the cubes

$$\begin{aligned}
f &= c \oplus bc \oplus bcd \oplus ac \oplus acd \oplus abc \\
&= c \oplus (bc \oplus bcd) \oplus (ac \oplus acd) \oplus abc \\
&= c \oplus bc\bar{d} \oplus ac\bar{d} \oplus abc
\end{aligned}$$

which is equal to the original function, as expected.

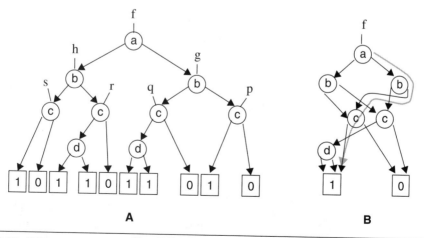

**Figure 8.25** OFDD of $f(a, b, c, d) = a\bar{b}c + \bar{b}c\bar{d} + \bar{a}cd$ (A) An OFDD (B) The reduced OFDD

## 8.2.5  Pseudo Boolean Functions and Decision Diagrams

Thus far, the Boolean domain is implied in our study. It is interesting to explore decision diagrams in other domains to see whether advantages such as representation size can be achieved. In this section we will briefly study several decision diagrams for functions that have either or both non-Boolean domain or range. These decision diagrams are extensions from the binary domain to the integer/real domain, and they have found application in specific design and analysis areas.

A generalization of a BDD is to relax the values a variable can take from binary to any set of numbers. Thus, a node can have more than two outgoing edges and each edge is labeled with the value the variable assumes. This decision diagram is called a *multivalue decision diagram* (MDD). An application of MDDs is to represent a design at an early stage when signals are in symbolic form as opposed to encoded binary form. For instance, the input variables to the function can be control signals with permissible values WAIT, SYNC, ACK, ABORT, ERR, and REQ.

For a multivalue function $f: M^k \text{->} M$, where $M = \{1, ..., n\}$ (in other words, $f$ has $k$ variables each of which can have a value between 1 and $n$), an MDD for $f$ is a DAG with $n$ leaf nodes. The leaf nodes are labeled as $1, ..., n$. Each internal node is labeled with a variable and has $n$ outgoing edges. Each edge is labeled with a value between 1 and $n$. Along any path from the root to a leaf node, a variable is traversed (at most) once. The semantics of an MDD is that a path from the root to a leaf node represents an evaluation of the function. Along the path, if a node takes on value $v$, the edge labeled $v$ is followed to the next node. When a leaf node labeled $i$ is reached, the function evaluates to $i$.

An ordered MDD conforms to a variable ordering such that variables along any path are consistent with the ordering. A reduced MDD satisfies the following two conditions: (1) no node exists with all of its outgoing edges pointing to the same node, and (2) no isomorphic subgraphs exist. Like a BDD, a reduced ordered MDD is canonical and has a suite of manipulation operations associated with it. Furthermore, several multivalue functions can be merged and represented as a shared MDD.

---

**Example 8.13**

---

Consider a robot controller equipped with three one-dimensional coordinate sensors. Each sensor has three possible outputs: 1, 2, and 3. The output of the controller is 1, 2, and 3, denoting stop, continue, and turn respectively. Let $x$, $y$, and $z$ denote the output of the three sensors respectively. The controller's behavior is described in Table 8.3. A reduced ordered MDD for the controller is shown in Figure 8.26.

**Table 8.3** A Multivalue Function $f: M^3 -> M$

| (x,y,z) | f(x,y,z) |
|---------|----------|
| (1, 1, -) | 1 |
| (1, 2, z) | z |
| (1, 3, -) | 2 |
| (2, 1, -) | 1 |
| (2, 2, -) | 2 |
| (2, 3, z) | z |
| (3, -, z) | z |

An algebraic decision diagram (ADD) represents a function with a domain that is Boolean but with a range that is real: $f: B^n -> R$. An ADD differs from a BDD only in the number of leaf nodes. An ADD has a number of leaf nodes equal to the number of possible output values of the function it represents. A reduced ordered ADD has the same variable ordering and reduction rules as in a BDD. The semantics of ADD nodes are the same as those of BDD nodes. One application of an ADD is in the area of timing analysis. For instance, the function gives the power consumption or delay of a circuit based the circuit input values. Example 8.14 illustrates such an application.

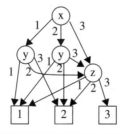

**Figure 8.26** An example MDD

**Example 8.14**

In this example we will look at an application of an ADD in representing static leakage power of circuit. Assume that the leakage power of a gate is one unit if the gate's output is 0, and is 2 otherwise. We can use an ADD to represent all leakage power behavior of the circuit in Figure 8.27. Let's choose variable ordering $a < b < c$. First we calculate the power for all inputs. The results are tabulated in Table 8.4. The result is translated into an ADD (Figure 8.27B), and a reduced ADD is shown in Figure 8.27C. Maximum power occurs when the input vectors are 101 or 001. Minimum power is consumed when the input is 01-.

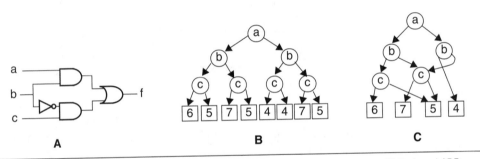

A                              B                              C

**Figure 8.27** (A) A circuit for leakage power representation (B) ADD of leakage power (C) Reduced ADD

**Table 8.4**    Power Leakage Results

| Input (abc) | Power | Input (abc) | Power |
|-------------|-------|-------------|-------|
| 000         | 5     | 001         | 7     |
| 010         | 4     | 011         | 4     |
| 100         | 5     | 101         | 7     |
| 110         | 5     | 111         | 6     |

## 8.2.6   Binary Moment Diagram (BMD)

An interesting way to look at Boolean variables is to treat them as integer variables restricted to 0 and 1. TRUTH value is denoted by $x$ and FALSE is denoted by $1 - x$. When applying BMD to a function mapping from $B^n$ to R, the inputs to the function can be conveniently expressed as "cubes." For instance, referring to Table 8.4, the first entry, 000 giving 5, can be expressed as $5(1 - a)(1 - b)(1 - c)$, which is 5 only if $a = b = c = 0$ as required. Using this technique and by adding up the "cubes," a function from $B^n$ to R can be expressed as a polynomial. The function in Example 8.14 has polynomial

$$f(a, b, c) = 5(1 - a)(1 - b)(1 - c) + 4(1 - a)b(1 - c) + 5a(1 - b)(1 - c) + 5ab(1 - c)$$
$$+ 7(1 - a)(1 - b)c + 4(1 - a)bc + 7a(1 - b)c + 6abc$$
$$= -b + ab + 2c - 2bc + abc + 5$$

A BMD represents the coefficients of the polynomial in leaf nodes. Nonterminal nodes have the interpretation that if the 1-edge (left) is taken, the variable of the node is included; otherwise, the variable is excluded. The path from the root to a leaf node represents a term in the polynomial by multiplying the value of the leaf node and all included variables along the path. Then the function is the sum of the terms represented by all paths. This interpretation is similar to that of OFDDs, which, instead of summation, XORs all cubes. A BMD for the leakage power function is in Figure 8.28. The two paths shown produce the terms $ab$ and $-b$. The left path takes 1-edge of node $a$ and node $b$, and 0-edge of node $c$; thus, variables $a$ and $b$ are included. Furthermore, the leaf node value is 1. Multiplying the variables with the leaf node value gives the term $ab$. Similarly, the right path

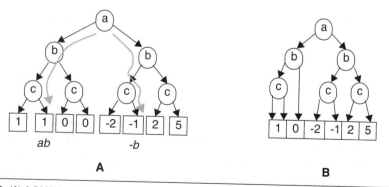

**Figure 8.28**  (A) A BMD for leakage power function (B) A reduced BMD

takes the 0-edge of node *a* and node *c*, and the 1-edge of node *b*. The leaf node value is -1. Therefore, the term is *-b*.

The two reduction rules for BMDs are the same as those for OFDDs. That is, first, if the 1-edge of a node points to 0, the node is removed and its incoming edges are redirected to its 0-node. And second, all isomorphic subgraphs are shared. The first rule follows from the multiplicative action of the node variable on the 1-edge. If the 1-edge points to 0, the node variable is zeroed and only the 0-node is left. Therefore, all incoming edges can be redirected to the 0-node. Applying these reduction rules to Figure 8.28A, we obtain the reduced BMD in Figure 8.28B.

A further generalization of BMDs allows edges to carry weight. As a path is traversed, the included variables, the weights on the traversed edges, and the leaf node value are multiplied to produce a term. This generalization gives more freedom to distribute the coefficients, and results in more isomorphic subgraphs for sharing, producing more compact representations. With proper restrictions on subgraph transformations, these generalized BMDs can be reduced to be canonical. For more information, consult the citations in the bibliography.

## 8.3 Decision Diagram-Based Equivalence Checking

In this section we will study the use of BDDs in determining the functional equivalence of two Boolean functions or circuits, known as *equivalence checking*. The ideas presented here are applicable not only to BDDs but also to any canonical decision diagrams. We use BDDs to illustrate the ideas because of their widespread use. A situation in which equivalence checking is called for is when a designer needs to determine whether an RTL description of a circuit is functionally equivalent to its gate-level model or a timing optimized version of the same design. In determining functional equivalence, ROBDDs for the two functions are built and compared structurally. Because of the canonicity of ROBDDs, the two functions are equivalent if and only if their ROBDDs are isomorphic.

Let's first consider determining the functional equivalence of combinational circuits and then sequential circuits. For combinational circuits, the shared ROBDDs of the outputs of circuits are built and compared. To compare two ROBDDs, one can compare them graphically—that is, map the nodes and edges between the two ROBDDs. We start by identifying the corresponding pairs of ROBDD roots and then follow the edges of the roots. If the 1-node (0-node) of one root is the same variable as the 1-node (0-node) of the other root, the mapping process recurs on the 1-nodes and 0-nodes. Otherwise, a mismatch occurs and the two ROBDDs are not isomorphic. The two ROBDDs are isomorphic if the procedure terminates at the leaf nodes without encountering any mismatched nodes.

Another way to determine whether two shared ROBDDs are isomorphic is to perform an XOR operation on the ROBDDs. If the resulting ROBDD is a constant node 0, the two ROBDDs are isomorphic; otherwise, they are not. This method is also applicable to checking circuit equivalence. To compare two circuits for equivalence, pairs of corresponding outputs are identified. Then these pairs are XOR, and the XOR gate outputs are ORed to produce a single output, as shown in Figure 8.29. If any pair of the outputs is not equivalent, the XOR output of the pair becomes 1, thus producing a 1 at the output of the OR gate. Next, an ROBDD is built for output differ of this overall circuit. Therefore, the circuits are equivalent if the ROBDD is equal to constant node 0. If the ROBDD is not constant 0, any path in the ROBDD from root to constant 1 is an input vector that demonstrates the difference in output response from the two circuits.

Checking the equivalence of sequential circuits is simplified when correspondence of state bits between the two circuits can be identified. For instance, the $i$th FF in circuit A is identified with the $j^{th}$ FF in circuit B. If such a one-to-one state correspondence is known, equivalence of the two circuits reduces to checking equivalence of the next-state functions, which are combinational circuits. A next-state function has as input the primary inputs and outputs of the FFs, and has as outputs the primary outputs and inputs of the FFs. If a one-to-one mapping between the state bits is not available, the two sequential circuits can be connected, as shown in Figure 8.29. Techniques in model checking are used to search the state space to prove or disprove that output differ is identically equal to 0. We discuss modeling checking in Chapter 9 (see "Equivalence Checking" on page 502).

## 8.3.1   Node Mapping and Constraining

In practice, it is rare that these methods work outright without modification. The main problems are running out of memory space or runtime taking too long. To mitigate these problems, two common techniques are used. The first technique identifies as many possible pairs of internal or I/O nodes that are required or known to be equivalent. These nodes

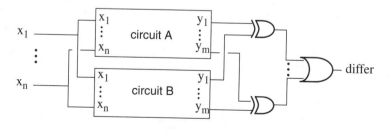

**Figure 8.29** Auxiliary circuit for checking circuit equivalence

are called *cut points*. This can be done through user-defined mapping between pairs of nodes in the two circuits or by assuming that nodes with the same name correspond to each other. Equating nodes with the same name is based on the rationale that one circuit is often a synthesized version or slightly modified version of the other, and therefore most node names are preserved between the two circuits. Once such a mapping is obtained, the mapped nodes are broken so that the fanins to the nodes are considered as outputs, and the fanouts are considered inputs. The two circuits with these newly formed inputs and outputs, together with the original primary inputs and outputs, are checked for equivalence. The advantage of breaking the known equivalent points and creating new inputs and outputs is that, although there are more outputs to check, the BDDs for the outputs have become smaller, as illustrated in Example 8.15.

---

### Example 8.15

Let's assume we want to check the equivalence of the circuit shown in Figure 8.30A, which is a gate-level model and the RTL model of the same circuit in Figure 8.30B. The first method simply builds ROBDDs for the two circuits and compares the BDDs. The second method identifies three cut points—namely $x$, $y$, and $z$ in the gate-level model and $x'$, $y'$, and $z'$ in the RTL model. That is, node $x$ should be functionally equivalent to node $x'$, $y$ to $y'$, and $z$ to $z'$. This knowledge can be supplied by the designer. In this example, let's compare the largest BDD sizes in the two methods. For simplicity, we will use only the BDDs of the gate-level model in this exercise.

With the first method there is only one output, $f$, and five inputs: $a$, $b$, $c$, $d$, and $e$. So two BDDs are built, one for the gate model and the other for the RTL model. The BDD for the gate model is shown in Figure 8.31A, which has 13 nodes.

With the second method, the three pairs of internal nodes are identified to be equivalent and the cut points are $x$, $y$, and $z$. We then cut the circuit at these points and create three inputs and outputs, as shown in Figure 8.32. Now, instead of just one BDD, we have to construct four BDDs for the four outputs $f$, $x$, $y$, and $z$. Then we construct ROBDDs for the four pairs of nodes—$(x,x')$, $(y,y')$, $(z,z')$, and $(f,f')$—and compare the respective BDD pairs. If any one of these BDD pairs is not equivalent, the circuits are not equivalent. In practice, if these BDD pairs are not equivalent, the first thing the designer needs to do is to reexamine the assumption that these internal node pairs are indeed equivalent. The BDDs for the internal nodes are shown in Figures 8.31B, C, and D. Because the internal nodes are also inputs, output $f$ can be expressed in terms of the internal nodes ($f = x + y + \overline{zce}$), the BDD of which is shown in Figure 8.31E. The largest BDD size of the second method is the BDD size of node $x$, which has eight nodes. As illustrated in this example, we can see that the largest BDD size has been reduced greatly by using cut points.

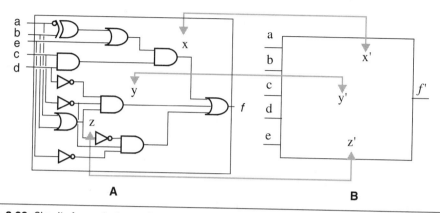

**Figure 8.30** Circuits for equivalence checking (A) Gate-level model (B) RTL model

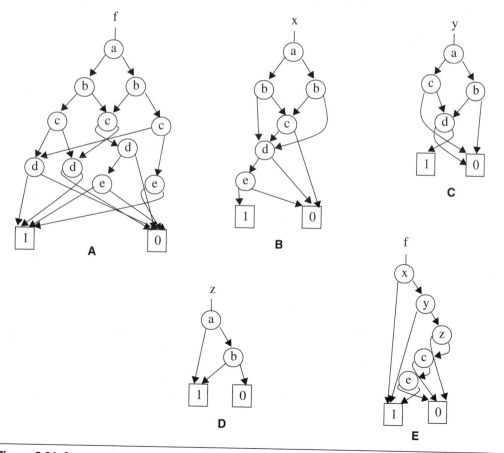

**Figure 8.31** Compare the largest BDD sizes with and without node mapping. (A) BDD for node f (B) BDD for node x (C) BDD for node y (D) BDD for node z (E) BDD for node f in terms of new variables

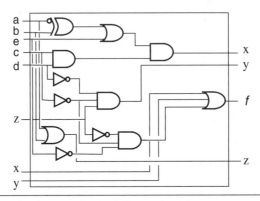

**Figure 8.32** Create intermediate inputs and outputs from cut points

Thus far, we have assumed that the inputs of the circuits undergoing equivalence checking are unconstrained (that is, the inputs can take on any values). When two circuits are equivalent only in a restricted input or state space, then the input or state space must be constrained. An example of equivalence under restricted space is comparing the next-state function of two finite-state machines. To perform equivalence checking in this situation, the FFs are removed and the next-state functions are compared as stand-alone combinational circuits. Because the next-state functions are now stand-alone functions, their inputs can take on any value. If the finite-state machines do not reach the entire state space, then the inputs to the next-state function must be constrained to the reachable state space of the finite-state machines; otherwise, it is possible that the next-state functions are not equivalent outside the reachable state space. Constraining eliminates the unreachable portion of search space from giving a false negative. Besides limiting input space, constraining can also restrict the structure of a circuit for equivalence checking (for example, by deactivating the scanning portion of a circuit by setting the scan mode to false). The following are some common applications for which constraining plays an important role:

1. Compare next-state functions. Constrain the state space to be the reachable state space of the finite-state machines.
2. Compare prescan and postscan circuits. Constrain to eliminate the scan logic so that only the main logic portions of the circuits are checked.
3. Reduce runtime and memory use. Constrain the input search space so that an equivalence checking operation finishes within a specific time and space limit.

**4.** Compare a unit of two circuits. Constrain the peripheral circuitry surrounding the unit so that the environment of the unit is the same as that inside the circuit. This is a more general situation than the reachable state space constraining method mentioned earlier. Figure 8.33 shows the problem and the idea of input constraining, where the constraining function maps the entire input space to the smaller realizable input space that unit A expects to see when embedded within the original circuit.

Constraining can be done directly by mapping an unconstrained input space into a permissible input space, as illustrated in Figure 8.33, or indirectly by creating an exclusion function that defaults the two circuits to be equivalent when it detects an input that lies outside the permissible region, as illustrated in Figure 8.34. The entire input space, permission and forbidden, is applied to the circuits under comparison. The exclusion function outputs 1 when an input is forbidden, which then forces the comparison to be equal. Therefore, for the output of this comparison structure always to be 1, the circuits must be equivalent. This indirect method of constraining is easy to implement, as illustrated in Example 8.16.

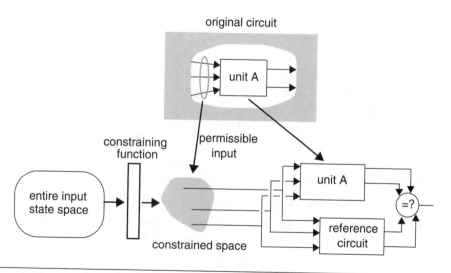

**Figure 8.33** Input constraining for checking embedded circuitry

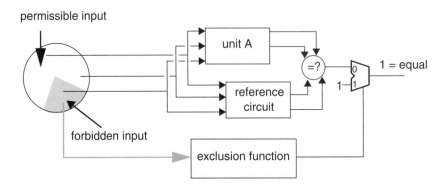

**Figure 8.34** Constrain by forcing output to be equal at forbidden inputs

---

**Example 8.16**

Suppose that a finite-state machine never reaches states 1001 and 1100. Let $a, b, c,$ and $d$ be the state bits. To check equivalence of this machine's next-state function, we can either prevent these two states from being fed to the comparison structure or allow them to be input, but force the result to be equal. Let's choose the latter. To exclude these two states, we define an exclusion function to be $g = a\bar{b}\bar{c}d + ab\bar{c}\bar{d}$. If $f$ and $h$ are the outputs of the next-state function and a reference circuit, respectively, the output of the comparison structure is $d = g + \overline{(f \oplus h)}$. If $f$ and $h$ are not equivalent, there exists an input in the permissible region, $g = 0$, such that $\overline{(f \oplus h)} = 0$, and hence $d = 0$. Therefore, output $d$ is *always* 1 if and only if $f$ and $h$ are equivalent. So, we build a BDD for $d$ and determine whether this BDD reduces to the constant 1.

---

## 8.4   Boolean Satisfiability

An alternative to BDDs in checking equivalence uses techniques from determining Boolean satisfiability called SAT. The problem of Boolean satisfiability decides whether a Boolean expression or formula is identically equal to 1. That is, it is satisfiable if there is an assignment to the variables of the expression such that the expression evaluates to 1. Equivalence checking is translated into the SAT problem by XORing the two functions: $d = f \oplus g$. If expression $d$ is satisfiable, then $f$ is not equivalent to $g$. Otherwise, $f$ is equivalent to $g$. The expression in Boolean satisfiability is in a conjunctive form—the product of sums. A sum is called a *clause*. If each clause of the conjunctive form has at most two variables, then the problem of Boolean satisfiability is called 2-SAT. Similarly, if each clause has at most three variables, it is called 3-SAT. 2-SAT can be solved in polynomial time, whereas 3-SAT is NP complete. Furthermore, satisfiability of any Boolean expression can be polynomially reduced to 3-SAT. In other words, if 3-SAT can be solved in polynomial time, so can any SAT problem. Hence, in the following discussion, we will not distinguish 3-SAT from a general SAT problem. We will just use the term *SAT*.

---

### Example 8.17

Determine whether the following expression is satisfiable:

$$f(a, b, c) = (a + b + c)(\bar{a} + b + \bar{c})(a + \bar{b} + \bar{c})(\bar{a} + \bar{b} + c)(\bar{a} + \bar{b} + c)$$

For this expression to evaluate to 1, all clauses must evaluate to 1 for some variable assignment. Let us solve this problem by trial and error. Variable assignment $a = 1$ makes the first clause 1, $b = 1$ makes the second clause 1, and $c = 0$ makes the fourth clause 1, but the fifth clause evaluates to 0. Try again. Pick $a = 1$ and $c = 0$ to make the first four clauses 1. The last clause is 1 if $b = 0$. Thus, for variable assignment $a = 1, b = 0, c = 0$, the expression is satisfied. In the worst case, we may have to try all $2^3$ (or eight) variable assignments to reach the conclusion.

---

In the following discussion, we will look at two classes of algorithms to solve SAT problems. We will first discuss a mathematically elegant method (a resolvent method) and then a more efficient method based on search.

### 8.4.1  Resolvent Algorithm

This method is based on the following resolvent and consensus identity:

$$f = (x + A)(\bar{x} + B) = (x + A)(\bar{x} + B)(A + B)$$

and

$$f = xC + \bar{x}D = xC + \bar{x}D + CD$$

where $A$ and $B$ are sums of literals, and $C$ and $D$ are product of literals. $A + B$ is called the *resolvent* of $(x + A)$ and $(\bar{x} + B)$. $CD$ is called the *consensus* of $xC$ and $\bar{x}D$. The resolvent identity is applicable to the conjunctive form whereas the consensus identity is applicable to the disjunctive form. In the following discussion we will use the resolvent identity on conjunctive forms.

Now, we want to show that $A + B$ is satisfiable if and only if $(x + A)(\bar{x} + B)$ is satisfiable. According to the resolvent identity, if $A + B$ is satisfiable, then $A$, $B$, or both is satisfiable. Let us assume $A$ is satisfiable. Then, when $x = 0$, $(x + A)(\bar{x} + B)$ becomes $A$ and thus is satisfiable. Similarly, if $B$ is satisfiable, then when $x = 1$, $(x + A)(\bar{x} + B)$ become $B$ and thus is satisfiable. On the other hand, if $(x + A)(\bar{x} + B)$ is satisfiable, then $(x + A)(\bar{x} + B)$ is satisfiable when $x = 0$ or $x = 1$. Let us consider both cases. At $x = 0$, $(x + A)(\bar{x} + B)$ becomes $A$ and therefore $A$ must be satisfiable. At $x = 1$, $(x + A)(\bar{x} + B)$ becomes $B$ and hence $B$ must be satisfiable. Therefore, either $A$ or $B$ must be satisfiable if $(x + A)(\bar{x} + B)$ is satisfiable. In conclusion, $(x + A)(\bar{x} + B)$ is satisfiable if and only if $A + B$ is satisfiable. Note that the variable assignments satisfying $A + B$ do not necessarily satisfy $(x + A)(\bar{x} + B)$. For a counterexample, see Example 8.18 on page 433. By translating the satisfiability problem of $(x + A)(\bar{x} + B)$ to that of $A + B$, the problem complexity is reduced by one variable— namely, variable $x$. Extending from two clauses to more clauses, consider $(x + A)(\bar{x} + B)$ $(x + C)(\bar{x} + D)$. We form the resolvent for all pairs of clauses containing both phases of $x$. There are a total of four pairs: $(x + A)(\bar{x} + B)$, $(x + A)(\bar{x} + D)$, $(x + C)(\bar{x} + B)$, and $(x + C)(\bar{x} + D)$. For each pair, we obtain its resolvent. Combining all resolvents, we conclude that $(x + A)(\bar{x} + B)(x + C)(\bar{x} + D)$ is satisfiable if and only if $(A + B)(A + D)(C + B)$ $(C + D)$ is satisfiable. Replacing clauses containing variable $x$ or $\bar{x}$ by their resolvents is called *resolving* variable $x$.

After a variable has been resolved, there are straightforward cases that other variable assignments can be easily inferred. If these straightforward cases exist, as many as possible variable assignments are inferred before resolving another variable. Here we will examine some of straightforward cases. First, if a resolvent contains both phases of a variable, such as the resolvent $y + \bar{y}$ in resolving $(x + y)(\bar{x} + \bar{y})$, then the resolvent, being 1, is removed.

If both phases of a variable are in a formula, the variable is called *binate*. If only one phase of the variable appears in the formula, the variable is *unate*. For example, in $(a + \bar{b} + c)(a +$ $b + \bar{c})(a + c)$, $a$ is unate, and $b$ and $c$ are binate. For a unate variable, we simply assign the variable to a value so that the clauses having the variable become 1. Consequently, the resulting formula formed by replacing the clauses containing the unate variable with 1 is satisfiable if and only if the original formula is satisfiable. This replacement operation is called the *unate variable rule* or the *pure literal rule*. Therefore, only binate variables need to be resolved.

Another straightforward case is when a clause contains only one literal. For instance, in $(a + b + c)(\bar{b})(b + \bar{c} + d)$, the second clause contains only one literal, $\bar{b}$. Then the variable must be assigned to the value at which the clause is satisfied, which is $b = 0$ in this case. This rule is called the *unit clause rule* and the literal in the clause is called the *unit literal*.

We define an empty clause to be a clause that contains no literals. The resolvent of $(x)(\bar{x})$ is an empty clause. A formula containing an empty clause is unsatisfiable. With this definition, the resolvent algorithm is as follows:

---

**Resolvent Algorithm for Satisfiability**

**1.** Select a variable $x$ in the expression.
**2.** If $x$ is unate, apply the unate variable rule.
**3.** If $x$ is a unit literal, apply the unit clause rule.
**4.** If $x$ is binate, resolve $x$.
**5.** Repeat these steps until all variables are resolved.
**6.** If the final expression is 1, the expression is satisfiable; otherwise, it is not (it is an empty clause).

---

**Example 8.18**

Let us use the resolvent algorithm to determine the satisfiability of

$$f(a, b, c) = (a + b + c)(\bar{a} + b + \bar{c})(a + \bar{b} + c)(\bar{a} + \bar{b} + \bar{c})(\bar{a} + \bar{b} + c)$$

Select variable $a$ for resolution. For each pair of clauses containing both phases of $a$, form its resolvent. Clause pairs and their resolvents are shown in Table 8.5. The first column lists the clause pairs containing both phases of $a$. The second column is the resolvents. The third column is the simplified resolvents. Therefore,

$$f(a, b, c) = (a + b + c)(\bar{a} + b + \bar{c})(a + \bar{b} + c)(\bar{a} + \bar{b} + \bar{c})(\bar{a} + \bar{b} + c)$$

is satisfiable if and only if $\bar{b} + \bar{c}$ is satisfiable. Because both variables, $b$ and $c$, are unate, we apply the unate rule. So $\bar{b} + \bar{c}$ reduces to 1. Therefore, the original expression is satisfiable.

Note that the resolvent algorithm does *not* imply that variable assignments satisfying $\bar{b} + \bar{c}$ also satisfy the original expression. As a simple counterexample, $a = b = c = 0$ satisfies $\bar{b} + \bar{c}$, but does not satisfy the original expression.

**Table 8.5**    Binate Clauses and Resolvents

| Binate Clauses | Resolvent | Simplifies To |
|---|---|---|
| $a + b + c, \bar{a} + b + \bar{c}$ | $b + c + b + \bar{c}$ | 1 |
| $a + b + c, \bar{a} + \bar{b} + \bar{c}$ | $b + c + \bar{b} + \bar{c}$ | 1 |
| $a + b + c, \bar{a} + \bar{b} + c$ | $b + c + \bar{b} + c$ | 1 |
| $\bar{a} + b + \bar{c}, a + \bar{b} + \bar{c}$ | $b + \bar{c} + \bar{b} + \bar{c}$ | 1 |
| $\bar{a} + \bar{b} + \bar{c}, a + \bar{b} + \bar{c}$ | $\bar{b} + \bar{c} + \bar{b} + \bar{c}$ | $\bar{b} + \bar{c}$ |
| $\bar{a} + \bar{b} + c, a + \bar{b} + \bar{c}$ | $\bar{b} + c + \bar{b} + \bar{c}$ | 1 |

As demonstrated by Example 8.18, a resolvent is produced for every pair of clauses containing both phases of a variable. In the worst case, an exponential number of resolvents is computed in solving a SAT problem. Therefore, although mathematically elegant, the resolvent algorithm is reserved for small SAT problems. To deal with large problems, a more efficient search algorithm is required.

## 8.4.2   Search-Based Algorithm

A search-based algorithm enumerates all input assignments and evaluates them to decide whether a formula is satisfiable. The set of all possible assignments of the variables in a formula forms the search space. The search space can be represented by a binary tree. A node in the tree represents a variable and has two edges. The left edge represents the variable taking value 1; and the right edge, value 0. A path from the root to a leaf node is a variable assignment, and the value of the leaf node is the value of the formula under the variable assignment. There are a total of $2^n$ leaf nodes for $n$ variables. A partial path gives a partial variable assignment. If the value of the formula evaluates to 1 at a leaf node, the formula is satisfiable. Therefore, a formula is satisfiable if there is a leaf node of value 1.

Although, in the worst case, a search-based algorithm has to traverse the entire binary tree, oftentimes pruning techniques are used to cut down the number of paths with values that can be deduced. Figure 8.35 illustrates the concept of the search strategy and pruning technique. Suppose we start the search following path P1. Path P1 is a variable assignment of $a = 0, c = 1, d = 1$, and $b = 1$. The leaf node for the path is 0. Thus, for this variable assignment, the formula is not satisfied. Let's backtrack to the last decision node $b$ and take the

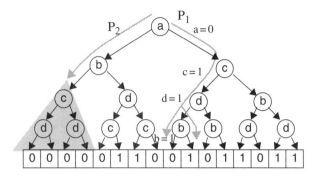

**Figure 8.35** A search strategy, search path, and pruning by deduction

other choice, $b = 0$. Backtracking to a previous decision variable and selecting the other value for the variable is called *flipping* the variable. With this new assignment, the formula is satisfied. In general, backtracking does not have to return to the last decision node; it can return to any prior decision node (for example, decision node $c$).

Next let's consider a situation in which pruning eliminates the necessity of visiting all paths. Partial path P2 represents the partial variable assignment $a = 1$, $b = 1$. Note that the all leaf nodes following this partial path have value 0, meaning that the formula cannot be satisfied if $a = 1$ and $b = 1$. Therefore, no further assignments need to be considered; that is, the tree rooted at the $c$ node can be pruned. An intelligent pruning algorithm has to detect this type of subtree at the earliest time.

A search-based SAT solver finds a satisfying variable assignment by making one variable assignment at a time. The candidate variable for assignment is selected based on some heuristic that determines the variable with the "best" chance of making the formula satisfied. Then, after the assignment of the variable is done, the solver attempts to infer as many possible assignments to other variables that must be made to satisfy the formula. After the variable assignment and consequent inferred assignments are made, the formula is evaluated for the partial assignment and is simplified. For example, if $x$ in $(x + \bar{y} + z)(\bar{x} + y + z)(\bar{x} + \bar{y} + z)(x + y + \bar{z})(x + y + z)$ is selected and assigned 0, the SAT solver substitutes 0 into $x$ and evaluates it to be $(\bar{y} + z)(\bar{y} + z)(y + z)$.

During the evaluation process, three cases are possible for a clause: First, the clause is resolved (evaluates to 1); second, the clause is conflicted (evaluates to 0); and third, the clause does not evaluate to a constant (is indeterminate). If the result of an evaluation of the formula turns out to be *0* and there are still unenumerated variable assignments, some prior variable assignments must be reversed; that is, the solver needs to backtrack. For the other two cases, the search continues. If an evaluation produces a result of *1* for the formula, the formula is satisfiable. On the other hand, if all assignments produce a result of *0*, the formula is unsatisfiable.

An efficient search-based algorithm should not enumerate all assignments one by one, but should prune as many assignments as possible that would not satisfy the formula. An outline of the structure of a generic search-based algorithm is shown here. Almost all search-based SAT solvers follow this algorithmic structure and differ in the implementation specifics in routines `select_branch()` and `backtrack()`. Routine `infer()` deduces variable assignments based on the partial assignment made so far, and consists of a collection of special cases that imply other variable assignments:

---

**Search-Based SAT Algorithm**

SolveSAT()

input: a formula

output: SAT or UNSAT

```
      forever {
          state = select_branch(); // choose and assign a variable
          if (state == EXHAUSTED) return UNSAT;

          result = infer(); // infer variable values
          if ( result == SAT)
              return SAT;
          else if (result == UNSAT)
              backtrack(); // backtrack to a prior decision
          else // result == INDETERMINATE
              continue; // need further assignment
      }
```

---

Routine `select_branch()` selects a variable that has not yet been assigned, called a *free variable*, and assigns a value to it. Besides, this routine keeps track of previously visited paths and prunes paths based on the values of already assigned variables. If all assignments have been enumerated, it returns a state EXHAUSTED, which means the formula is unsatisfiable. To select a variable that may satisfy the formula as early as possible, `select_branch()` relies on some measure of predicting the quality of a new variable assignment. An assignment to a variable is of good quality if the assignment has a high probability of being part of a satisfying assignment or reveals that many paths can be pruned. A reliable quality measure is essential to the performance of a SAT solver. In addition to prediction accuracy, a good assignment quality measure must be easy to compute so that it does not become the bottleneck of the solver. Many heuristics exist for measuring assignment quality. An example predictive measure is the number of literals of a variable in the remaining unsatisfied clauses, the rationale being that an assignment to such a variable will satisfy most clauses. Using this quality measure, `select_branch()` returns an assignment to the variable that satisfies most remaining clauses. For example, if the remaining clauses of a formula are $(a + b + \bar{c} + d)(\bar{a} + \bar{b} + \bar{c} + e)(b + c + e + \bar{f})(b + d + f + \bar{g})$, then variable $b$ assigned to 1 satisfies three clauses, whereas other variables satisfy fewer clauses. Therefore, `select_branch()` selects variable $b$ and assigns 1 to it.

Routine `infer()` looks for special cases in the formula that imply immediate assignments for the remaining unassigned variables. These special cases are the same as those discussed in our study of resolvent algorithms: pure literal rule and unit clause rule. In the pure literal case, there is a variable that appears in only one phase in the formula. The assignment for the variable is then inferred such that the literal becomes 1. For example, a negative-phase literal $\bar{x}$ implies $x = 0$. In the unit clause case, there is a clause that contains only one literal, so the literal is assigned to 1. The inferred assignments are then substituted into the formula, and the satisfied clauses are removed. This propagation of assignments is called *Boolean constraint propagation* (BCP). An important requirement in implementing BCP is that it must save a history of previous assignments so that, at the request of routine `backtrack()`, it can return a previously unassigned value. As an example of the operation of `infer()`, suppose that the current partial assignment is $a = 0, b = 1$ in satisfying $(a + \bar{b} + \bar{c})(\bar{b} + c + d)(\bar{a} + b + d)$. After BCP, the formula becomes $(\bar{c})(c + d)$. Hence it can be inferred that the first clause being a unit clause, $c = 0$. After a BCP for $c = 0$, it infers from the second clause that $d = 1$. In this case, the result is SAT. If the result of `infer()` is SAT, the formula is satisfiable. If the result is UNSAT, it has to backtrack to a prior decision.

When a partial assignment fails to satisfy the formula, some of the previous assignments must be reversed. Determining the variables to reverse their values is the task for routine `backtrack()`. Many heuristics exist for selecting variables. The simplest one is to flip the most recently assigned free variable selected by `select_branch()` that has not been flipped. Note that a variable most recently assigned by `infer()` must not be chosen for flipping, because such an assignment was forced by another assignment. Therefore, a candidate variable for flipping must be a variable that was assigned by `select_branch()`. This strategy of reversing the last decision is called *chronological backtracking*. On the other hand, *nonchronological backtracking*, when choosing variables to be flipped, does not necessarily flip the last decision variable. It first analyzes the reasons for a conflict, determines the variable assignments that are probable causes of the conflict, and then reverses the root, causing assignment.

If the current partial assignment is not sufficient to determine satisfiability of the formula, the result from `infer()` is UNDETERMINATE, and it calls `select_branch()` for the next branch.

---

### Example 8.19

Determine satisfiability of the following formula:

$$(a + b)(\overline{a} + b)(a + \overline{b} + c)(\overline{a} + \overline{c} + d)(\overline{a} + \overline{b} + c)(\overline{b} + c + \overline{d})(a + \overline{d})(\overline{d} + e)(\overline{d} + e + f)(a + \overline{e} + f)$$

With the `select_branch()` routine, we use the heuristic that chooses the literal with the highest count. In the formula, $\overline{d}$ and $a$ each appear four times. To break the tie, let's arbitrarily select $\overline{d}$. Propagating $d = 0$ into the formula, we obtain

$$(a + b)(\overline{a} + b)(a + \overline{b} + c)(\overline{a} + \overline{c})(\overline{a} + \overline{b} + c)(a + \overline{e} + f)$$

Now literals $a$ and $\overline{a}$ have the highest count of three each. Arbitrarily selecting $a$ and propagating a = 1, we get $(b)(\overline{c})(\overline{b} + c)$.

From this, the first two clauses are unit clauses. So, we infer that $b$ must be 1 and $c$ must be 0. However, when we apply BCP to this expression, it becomes 0. Therefore, we backtrack chronologically to the last decision, $a = 1$. Reversing the decision, we assign 0 to $a$. Propagating this assignment to $(a + b)(\overline{a} + b)(a + \overline{b} + c)(\overline{a} + \overline{c})(\overline{a} + \overline{b} + c)(a + \overline{e} + f)$, we get $(b)(\overline{b} + c)(\overline{e} + f)$.

We infer that $b$ must be 1. After applying the unit clause rule to remove literal $b$, we have $(c)(\overline{e} + f)$, from which we further infer that $c$ must be 1. Again, using the unit clause rule to remove literal $c$, we arrive at $(\overline{e} + f)$, to which we apply the pure literal rule to deduce $e = 0$, and conclude that the formula has been satisfied. Therefore, the formula is satisfiable and a satisfying assignment is $d = 0$, $a = 0$, $b = 1$, $c = 1$, and $e = 0$.

---

## 8.4.3 Implication Graph and Learning

When an assignment is searched to satisfy a formula, some free variables are selected to take on certain values. The values form a branch in the decision diagram. The selected variables are called *decision variables*. If the assigned values cause a conflict, other values are assigned. When a decision variable is selected, we associate a decision-level number with it. The decision-level number keeps track of how deep the assignment is from the top of the search and it facilitates backtracking. Variables with values that are forced by the current decision variable, called *implied variables*, are consequences of the decision variables and are thus assigned the current decision-level number. The decision number starts with 1 when the first decision variable is selected.

To facilitate backtracking, we need to understand the causes of the conflict. The cause of a conflict is an assignment of decision variables that eventually leads to the conflict. To express causes of conflicts, implication graphs are useful. An implication graph is a DAG with vertices that represent assignments to variables and edges that represent consequences of variable assignments. A vertex is labeled with the variable name along with its decision level and is assigned a value. A conflict occurs if both 1 and 0 are in the implication graph. The variable of the conflicting vertices is the conflicting variable. Vertices without incident edges are decision variables, because they are not implied by values of other variables.

---

**Example 8.20**

Figure 8.36 shows an implication graph. A vertex label has a variable name, its decision level, and the value it is assigned. The value assigned is 0 if there is a minus sign in front of the label; otherwise, it is 1. For instance, vertex $b(4)$ represents variable $b$ assigned to 1 at decision level 4; vertex $-a(3)$ represents variable $a$ assigned to 0 at decision level 3. Decision vertices are solid circles and implied variables are unfilled circles. In Figure 8.36, the decision vertices are $-a(3)$, $b(4)$, $j(2)$, and $k(1)$. The rest of the vertices are implied vertices. According to Figure 8.36, at decision level 1, variable $k$ is assigned to 1. This assignment affects variable $g$, but is not sufficient to imply $g$'s value. At decision level 2, variable $j$ is assigned 1. At this point, variable $g$ is forced to take on value 1; thus, the decision level of variable $g$ is also 2. At decision level 3, variable $a$ is selected to be 0, which implies variable $f$ to be 0, which in turn implies variable $d$ to be 1. At decision level 4, variable $b$ is assigned 1, which determines the values of variables $n$, $e$, $h$, and $c$. At this point, variable $x$ is implied to be both 0 and 1, which is a conflict. Therefore, the assignment has to backtrack.

---

With an implication graph defined, some concepts of an implication graph must be discussed. Vertex $v$ is said to *dominate* vertex $u$ if any path from a decision variable at the decision level of $v$ to $u$ has to go through $v$. Intuitively, vertex $v$ determines vertex $u$. For example, in Figure 8.36, vertex $e$ dominates vertices $c$ and $h$, because any path from decision vertex $b$ to vertices $c$ or $h$ has to go through $e$. A *unique implication point* (UIP) is a

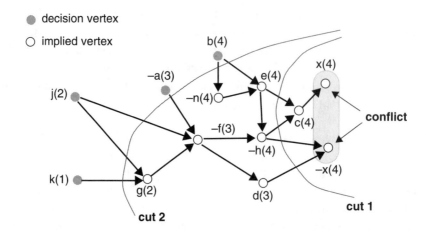

**Figure 8.36** An implication graph showing the relationship between decision and implied variables

vertex at the current decision level that dominates both vertices of a conflict. For example, the vertices of conflict are *x(4)* and *–x(4)*, and vertex *e(4)* dominates these vertices; therefore, *e(4)* is a UIP. The concept of a UIP is used to identify variables to be reversed to resolve a conflict.

When a conflict exists, a cause of the conflict can be obtained by forming a cut on the implication graph such that the graph is partitioned into two parts: reason and conflict. The conflict part contains the conflicting vertices and possibly other implied variables. Note that there is no decision variable in the conflict part. The other part is called the *reason part* and it consists of all other vertices. Two cuts are shown in Figure 8.36.

The decision variables in the reason part that have an edge to the conflict part contribute to a cause of the conflict. The edges of decision variables crossing a cut represent assignments leading to the conflict and hence must be avoided.

When a search encounters a conflict, the path leading to the conflict should be avoided in any future search. This path of conflict can be learned by including a clause in the formula so that this particular variable assignment can be avoided early during the search. To do so, a cut of a conflict is first determined. Then, the edges of the decision variables crossing the cut are identified. These edges form a clause, which when satisfied avoids the conflict, resulting in an assignment. For example, cut 1 in Figure 8.36 corresponds to variable assignment *e = 1, h = 0, d = 1*, which results in a conflict. To avoid this partial assignment, any future assignment must satisfy one of the following: *e = 0* or *h = 1*, or *d = 0*. In other words, clause $\bar{e} + h + \bar{d}$ must be satisfied. This new clause can be added to the original formula. Note that the added clause is redundant in the sense that it does not change the satisfiability of the formula. Its only function is to speed up decision making by preventing a repeat of known conflicting assignments.

There can be more than one cut per conflict. Different cuts give different learning clauses. A preferred cut is one that contains only one decision variable assigned at the current level and others assigned at lower levels, because this is the variable that has the most immediate impact on the conflict, and its value should be reversed in backtracking. Cuts with this property are called *asserting cuts*. At a given decision level, an asserting cut can always be found because the decision variable at the current level is always a UIP. Different cuts represent different learning schemes. For more in-depth information, consult the bibliography.

---

**Example 8.21**

---

Consider the satisfiability of

$$e(\bar{a} + \bar{c})(a + b + c + d)(\bar{a} + \bar{b} + c + \bar{d})(\bar{a} + \bar{b} + \bar{c} + d)(a + b + c + \bar{d})(\bar{a} + \bar{b} + c + d)$$

To satisfy the formula, variable $e$ must be assigned 1. Because there is no decision variable chosen yet, variable $e$ has decision level 0. At level 1, we choose variable $a$ as a decision variable and assign it to 1. This selection reduces the formula to

$$(\bar{c})(\bar{b} + c + \bar{d})(\bar{b} + \bar{c} + d)(\bar{b} + c + d)$$

which implies that variable $c$ must be 0. Thus, the formula further simplifies to $(\bar{b} + \bar{d})(\bar{b} + d)$. The next decision variable selected is $b$. Suppose we assign it to 1. This assignment forces variable $d$ in the first clause to be 0, and variable $d$ in the second clause to be 1—a conflict. An implication graph for the process so far is shown in Figure 8.37. The cut shown gives clause $\bar{a} + \bar{b} + c$ which is ANDed with the original formula to prevent future decisions from arriving at this conflict. Therefore, the formula with this learned clause is

$$(\bar{c})(\bar{b} + c + \bar{d})(\bar{b} + \bar{c} + d)(\bar{b} + c + d)(\bar{a} + \bar{b} + c)$$

Now we can either decide this formula or backtrack. To continue the decision process, we infer that $c = 0$, which implies (through the learned clause) $b = 0$. This assignment satisfies the formula. If we backtrack, we first find a cut. The cut shown is an asserting cut, and the variable to flip is $b$. Assigning $b$ to 0 satisfies the formula.

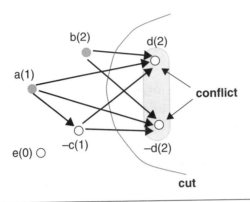

**Figure 8.37** Implication result from a satisfiability process

## 8.5   Symbolic Simulation

With traditional simulation, stimuli as well as responses consist of bits of binary constants. In a combinational circuit with $n$ input bits, the input space has $2^n$ vectors, or points. A traditional simulator simulates one point at a time. On the contrary, symbolic simulation takes in variables, called *free variables*, as input and produces output expressions in terms of the variables. An input at each cycle is a free variable. Because a variable represents both 0 and 1, a symbolic simulator simulates simultaneously the entire set of the points that the input variables can take on. With traditional simulation, constants 0 and 1 are sufficient to represent the simulation values at any time, whereas (because a set of points is computed) a symbolic simulator uses functions to represent sets of points. The functions in a symbolic simulator can be represented using BDDs.

### Example 8.22

Simulate the circuit in Figure 8.38 using traditional and symbolic simulation for two cycles. For traditional simulation, assume the input vector sequence is (abcd) = (0011) at cycle 1 and (1101) at cycle 2 and the initial state of the FF is 0. Now do the following: (1) compare the output values at the end of cycle 2 and (2) prove or disprove the property that output $f$ can remain at value 1 for two cycles.

First, we will simulate using the traditional simulation methodology. At cycle 1, the values at nodes $h, g, k, n, j$, and $f$ are 1, 0, 0, 0, 1, and 1 respectively. At cycle 2, the values at nodes $h, g, k, n, j$, and $f$ are 0, 0, 1, 1, 1, and 0 respectively.

**Example 8.22    (Continued)**

With symbolic simulation, symbolic values are used at nodes and each cycle has its own free variables. Let's use subscript labeling to denote time for a symbol. For instance, for input $a$, $a_1$ denotes the symbol at input $a$ at cycle 1; and $a_2$, the symbol at cycle 2. Let's assume that the initial value of the FF is $k_1$.

At cycle 1 the symbolic value for node $h$ is computed by NORing the inputs, giving $\overline{a_1 + b_1}$; at node $g$, the value is $b_1 \cdot c_1$; at node $k$, the value is $k_1$. Thus, the value at node $f$ is $\overline{k_1}(\overline{a_1 + b_1}) + k_1(b_1 \cdot c_1)$. The values of nodes $n$ and $j$ are $\overline{c_1 \cdot d_1}$ and $(\overline{c_1 \cdot d_1}) \oplus f_1 = (\overline{c_1 \cdot d_1}) \oplus (\overline{k_1}(\overline{a_1 + b_1}) + k_1(b_1 \cdot c_1))$ respectively. These values are listed in Table 8.6. Variables $a_1$, $b_1$, $c_1$, $d_1$, and $k_1$ are independent of each other (in other words, they are free variables).

Similar computations are done for cycle 2. $k_2 = j_1$ and $f_2$ is $\overline{k_2} \cdot h_2 + k_2 \cdot g_2$. To express this result in terms of the inputs, we simply substitute the expressions for $k_2$, $h_2$, and $g_2$. The results are tabulated in Table 8.7. Again, $a_1$, $b_1$, $c_1$, $d_1$, $k_1$, $a_2$, $b_2$, $c_2$, and $d_2$ are free variables.

To compare this with traditional simulation for the given input vectors, we need to substitute the input values and compute the output at cycles 1 and 2. For the first input vector (0011), we have $a_1 = 0$, $b_1 = 0$, $c_1 = 1$, and $d_1 = 1$. For the second input vector (1101), $a_2 = 1$, $b_2 = 1$, $c_2 = 0$, and $d_2 = 1$. The initial value of the FF is 0, $k_1 = 0$. Evaluating the expressions for $f_1$ and $f_2$, we have $f_1 = 1$ and $f_2 = 0$, as expected.

For part 2, because $f = 1$ at cycle 1, if we can show $f = 1$ at cycle 2, the property is affirmed. With traditional simulation this is difficult to determine without simulating other input vectors, and it is not obvious which input vectors can produce $f$ to be 1 at cycle 2. With symbolic simulation, we just need to check whether $f_2$ is satisfiable under the constraint that $a_1 = 0$, $b_1 = 0$, $c_1 = 1$, $d_1 = 1$, and $k_1 = 0$. To do this, we substitute in the constraint $a_1 = 0$, $b_1 = 0$, into $f_2$ and it simplifies to $b_2 \cdot c_2$, which becomes 1 when $b_2 = 1$ and $c_2 = 1$. Therefore, it is possible for output $f$ to remain at 1 for two cycles.

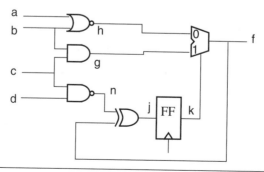

**Figure 8.38** Simulate using traditional and symbolic simulation

**Table 8.6**    Symbolic Simulation Result at Cycle 1

| Node | Cycle 1 |
|------|---------|
| $h$ | $\overline{a_1 + b_1}$ |
| $g$ | $b_1 \cdot c_1$ |
| $k$ | $k_1$ |
| $f$ | $\bar{k}_1(\overline{a_1 + b_1}) + k_1(b_1 \cdot c_1)$ |
| $n$ | $\overline{c_1 \cdot d_1}$ |
| $j$ | $(\overline{c_1 \cdot d_1}) \oplus (\bar{k}_1(\overline{a_1 + b_1}) + k_1(b_1 \cdot c_1))$ |

**Table 8.7**    Symbolic Simulation Result at Cycle 2

| Node | Cycle 2 |
|------|---------|
| $h$ | $\overline{a_2 + b_2}$ |
| $g$ | $b_2 \cdot c_2$ |
| $k$ | $(\overline{c_1 \cdot d_1}) \oplus (\bar{k}_1(\overline{a_1 + b_1}) + k_1(b_1 \cdot c_1))$ |
| $f$ | $\overline{(\overline{c_1 \cdot d_1}) \oplus (\bar{k}_1(\overline{a_1 + b_1}) + k_1(b_1 \cdot c_1))}(\overline{a_2 + b_2}) +$ <br> $((\overline{c_1 \cdot d_1}) \oplus (\bar{k}_1(\overline{a_1 + b_1}) + k_1(b_1 \cdot c_1)))(b_2 \cdot c_2)$ |
| $n$ | $\overline{c_2 \cdot d_2}$ |
| $j$ | $(\overline{(\overline{c_1 \cdot d_1}) \oplus (\bar{k}_1(\overline{a_1 + b_1}) + k_1(b_1 \cdot c_1))}(\overline{a_2 + b_2}) +$ <br> $((\overline{c_1 \cdot d_1}) \oplus (\bar{k}_1(\overline{a_1 + b_1}) + k_1(b_1 \cdot c_1)))(b_2 \cdot c_2)) \oplus (\overline{c_2 \cdot d_2})$ |

From the previous example, we see that verifying properties can be achieved easier using symbolic simulation than traditional simulation. This advantage comes directly from the fact that many vectors are simulated at once. However, on the negative side, the complexity of symbolic values grows rapidly as the number of cycles increases, resulting in massive memory use and computational time, thus limiting simulation capacity. Furthermore, note that the nodes with a high-complexity growth rate are the ones that keep track of the history of simulation (in other words, in feedback loops).

Because node values are in symbolic form, the concept of transition and event is no longer as precisely defined as in traditional simulation; therefore, symbolic simulation is mainly used in functional verification as opposed to timing verification. However, an event-driven simulator and a symbolic simulator can couple together to verify both functional and timing properties. In addition, even in functional verification, a traditional cycle-based simulator and a symbolic simulator can team up to verify long tests. The cycle-based simulator runs the initial portions of the tests, leading to the crucial intervals for the properties under verification, and lets the symbolic simulator take over verification during the intervals. Using this methodology, the high complexity from symbolic simulation over the long tests can be avoided.

Symbolical simulation of a circuit for $n$ cycles can be regarded as unrolling the circuit $n$ times. The unrolled circuit is a combinational circuit, and the $i$th copy of the circuit represents the circuit at cycle $i$. Thus, the unrolled circuit contains all the symbolic results from the $n$ cycles. Figure 8.39B, which illustrates this unrolling representation, is a three-cycle unrolled combinational circuit of the sequential circuit in Figure 8.39A, equivalent to three cycles of symbolic simulation. Each signal generates a copy for each cycle. For example, $x$ becomes $x_1$, $x_2$, and $x_3$ for the three cycles. In the unrolled circuit, the FF model is a

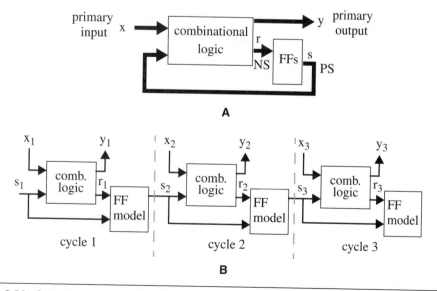

**Figure 8.39** Conceptual representation of symbolic simulation. (A) Original circuit (B) Unrolled circuit representing three cycles of symbolic simulation

combinational circuit representing the current state $s$ as a function of its previous state and the input $r$. For DFF, the output is a copy of the input, and is modeled by a wire connecting the input $r$ to the current state $s$. An unrolled circuit for two cycles of symbolic simulation of the circuit in Figure 8.38 is shown in Figure 8.40. The inputs to this unrolled circuit are the primary inputs at each cycle and the initial state. Figure 8.40 reveals that symbolically simulating one cycle is tantamount to replicating the original circuit once. This is the cause of the rapid growth in complexity.

## 8.5.1  Symbolic Verification

The ability to simulate multiple inputs simultaneously allows one to verify conclusively and efficiently some properties or assertions that a traditional simulation methodology would have to enumerate explicitly for all possibilities. The properties, sometimes called *assertions*, are in the form of Boolean functions of states, internal variables, and external inputs. An example property is that if signal $A$ is high then $B$ must be low after, at most, two cycles.

A property or assertion can be categorized as *bound* or *unbound*. A bound property is defined over a finite time interval or a bound number of cycles; otherwise, it is an unbound property. An example of a bound property is that from cycle $n$ to cycle $m$, at least one packet must arrive. The interval in which this property is defined is from cycle $n$ to $m$. An example of an unbound property is that after a request has become active, the grant will become active eventually.

To verify a bound property, the circuit can be simulated symbolically for the interval of the property. If the property holds within the interval, the property is affirmed. Suppose that a property is $f(x, y, z) = 1$ from cycle 5 to cycle 6, where $x$, $y$, and $z$ are nodes of a circuit. If $f(x_5, y_5, z_5) \cdot f(x_6, y_6, z_6) = 1$, where $x_5$, $y_5$, $z_5$, $x_6$, $y_6$, and $z_6$ are results at cycles 5 and 6 from a symbolic simulation, then the property is affirmed.

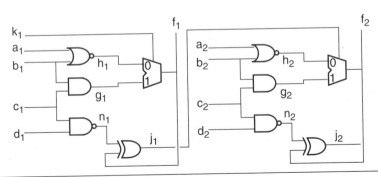

**Figure 8.40** Equivalent circuit for two cycles of symbolic simulation

An unbound property, sometimes called an *invariant*, is verified if it can be shown that the property holds for all possible inputs and states. A standard verification strategy is to apply mathematical induction on the cycle number: If the invariant holds at the initial cycle and at the $n^{th}$ cycle, given that it holds at the $n-1$ cycle, then it holds for all cycles. However, not all invariants are easy to recast into a form amicable for mathematical induction.

## 8.5.2 Input Constraining

Variable constraining restricts some free variables, which can take on any value when unconstrained, to certain values. Two situations in which variable constraining is useful are case splitting and environment modeling.

As seen in the previous example, the complexity of node expressions grows rapidly if all inputs are allowed to be free variables. To reduce the complexity, we can restrict some free variables to take on specific values and, by doing so, the unconstrained become constants and the free variables are eliminated from the symbolic expressions. By constraining variables, we sacrifice the number of input points simulated simultaneously, and thus reduce the coverage of the symbolic run. To cover the entire input space, the other cases of the constrained variables have to be simulated on a separate run. For example, suppose that input $c$ has been constrained to be 0 for the first two cycles in a symbolic run. Then the other cases of variable $c$, 10, 01, and 11 in the first two cycles, must be simulated in other runs. Furthermore, input variables can be constrained to an expression other than constants. For example, input $a$ can be constrained to be equal to input $c$.

In a general setting of case splitting, input space is partitioned so that instead of verifying a circuit under the most general input conditions, one can verify the circuit under inputs in each of the subspaces. Within a subspace, the circuit simplifies, and thus verification is accelerated. An example of case splitting is to divide inputs to an ALU into four subspaces: logical operations, arithmetic operations, shift and rotation, and all other operations. For logical operations, the ALU is reduced only to the units responsible for logical operations. Therefore, in case-splitting verification, the input must be constrained to the subspace for the case. Case splitting is complementary to structural partitioning. Structural partitioning breaks down a system into components and verifies each component individually, and then verifies at the system level the components' interoperation, whereas case splitting keeps the structure and partitions the data space.

Therefore, case splitting is a compromise between complexity and coverage. The more input space partitioned, the simpler the simulation complexity and the less the coverage of each run. Case splitting is the transition mechanism bridging traditional and symbolic simulation. If all free variables are constrained, symbolic simulation degenerates to traditional simulation.

**Example 8.23**

Consider the circuit in Figure 8.38. Let us constrain inputs $a$ to 1 at cycle 1, and $c$ to 0 at cycle 2. In other words, $a_1 = 1$ and $c_2 = 0$. In the first cycle, node $h$ becomes 0. All node results are summarized in Table 8.8. In the second cycle, input $c$ is constrained to 0, forcing node $n$ to 1 and node $g$ to 0. All node results are shown in Table 8.9. Note how the expressions have simplified compared with the ones in Example 8.22. Of course, we are now simulating only one fourth of the previous space.

**Table 8.8**    Symbolic Simulation Result at Cycle 1 with Input Constraining

| Node | Cycle 1 |
|------|---------|
| $h$ | 0 |
| $g$ | $b_1 \cdot c_1$ |
| $k$ | $k_1$ |
| $f$ | $k_1(b_1 \cdot c_1)$ |
| $n$ | $\overline{c_1 \cdot d_1}$ |
| $j$ | $(\overline{c_1 \cdot d_1}) \oplus (k_1(b_1 \cdot c_1))$ |

**Table 8.9**    Symbolic Simulation Result at Cycle 2 with Input Constraining

| Node | Cycle 2 |
|------|---------|
| $h$ | $\overline{a_2 + b_2}$ |
| $g$ | 0 |
| $k$ | $\overline{c_1 \cdot d_1} \oplus k_1 b_1 c_1$ |

**Table 8.9**   Symbolic Simulation Result at Cycle 2 with Input Constraining (Continued)

| Node | Cycle 2 |
|------|---------|
| $f$ | $\overline{\overline{c_1 \cdot d_1} \oplus k_1 b_1 c_1 \overline{a_2} + b_2}$ |
| $n$ | 1 |
| $j$ | $\overline{c_1 \cdot d_1} \oplus k_1 b_1 c_1 + a_2 + b_2$ |

The second situation in which constraining is applicable is modeling the input environment of the circuit under consideration. For instance, if the input bits to an encoder are always one-hot, then, in verifying the encoder, its inputs cannot be totally free variables, but must be constrained to the one-hot condition. To constrain, we create a set of new input variables (which are free variables) and a mapping from the new inputs to the original inputs. With the mapping, the circuit sees only the permissible inputs while the new inputs are allowed to take on any value. This is achieved by crafting the mapping such that every point in the new input space produces a point in the restricted space, and every point in the restricted space is produced by at least one point in the new input space. The mapping is then composed with the circuit. Assume that a circuit has $n$ inputs, and $R$ is the constrained input space, where $R \subseteq B^n$. That is, only inputs in $R$ are permissible. The mapping, from $B^m$ to $B^n$, maps $B^m$ to $R$. This constraining process is conceptually illustrated in Figure 8.41. This process is also called *parameterization*.

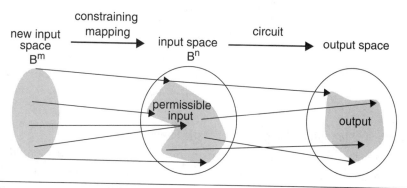

**Figure 8.41** Constraining input space with a onto mapping

**Example 8.24**

The environment feeding the inputs of an embedded circuit of three inputs $x, y, z$, and outputs $p$ and $q$, where $p = f(x, y, z)$ and $q = h(x, y, z)$, can only produce 100, 111, 010, and 011. These values form the permissible input space. If we take out the embedded circuit and simulate it as a stand-alone, we must ensure that the inputs to the circuit are permissible. To constrain the inputs, we need to introduce two new inputs, $v$ and $w$, and map the four permissible inputs to the inputs of the circuit. The mapping is shown in Table 8.10.

Translating the mapping into Boolean functions, we get $x = \bar{v}, \ y = v + w, \ z = w$. Now, substitute these functions into the circuit equations. We obtain

$$p = f(x, y, z) = f(\bar{v}, v + w, w)$$

$$q = h(x, y, z) = h(\bar{v}, v + w, w)$$

The new inputs $v$ and $w$ are free variables that always produce permissible inputs to the circuit.

**Table 8.10**   Constraining Mapping

| (v, w) | (x, y, z) |
|--------|-----------|
| 00     | 100       |
| 01     | 111       |
| 10     | 010       |
| 11     | 011       |

## 8.5.3   Symbolic Simulation Using Characteristic Functions

Thus far in our discussion, circuits and constraining mappings are represented using I/O equations. An alternative is representation using characteristic functions. An I/O equation, say $y = f(x, q), z = h(x, q)$, where $x$ is input, $y$ is output, $q$ is state, and $z$ is next state, can be represented by the following characteristic function:

$$\lambda(y, z, x, q) = \begin{cases} 1, \text{ if } y = f(x,q) \text{ and } z = h(x,q) \\ 0, \text{ otherwise} \end{cases}$$

If we use $(a = b)$ to denote the characteristic function that produces 1 if $a = b$ and 0 otherwise (XNOR operator), then the previous characteristic function can be written as $(y = f(x, q))(z = h(x, q))$. Using this representation, the values of the output $y$ and the next state $z$ after symbolically simulating one cycle are all the values $y_1$ and $z_1$ satisfying the following expression:

$$(y_1 = f(x_1, q_1))(z_1 = h(x_1, q_1))$$

If state devices are DFFs, then the state value is equal to the next-state value computed during the previous cycle. Therefore, the result of symbolically simulating two cycles is

$$(y_2 = f(x_2, q_2))(z_2 = h(x_2, q_2))(z_1 = h(x_1, q_1))(z_1 = q_2)$$

The first two terms are governed by the circuit's output and next-state function. The third term computes the next-state function at cycle 1, which becomes the state value at cycle 2. The last term simply forces the state at cycle 2 to be equal to the next-state value computed at cycle 1.

---

### Example 8.25

Let's simulate the circuit in Figure 8.42 using characteristic functions. The circuit equations are $y = xq$ and $z = x \oplus q$.

Using a characteristic function, we obtain $\lambda(y, z, x, q) = (\overline{y \oplus xq})(\overline{z \oplus (x \oplus q)})$. The result of one cycle of the simulation is given by $\lambda(y_1, z_1, x_1, q_1) = (\overline{y_1 \oplus x_1 q_1})(\overline{z_1 \oplus (x_1 \oplus q_1)})$. Comparing this result with that from circuit equations, we assert that the characteristic function is 1 if and only if $y_1 = x_1 q_1$ and $z_1 = x_1 \oplus q_1$. So values produced by the characteristic function are identical to those derived using circuit equations.

The simulation result based on circuit equations at cycle 2 is $y_2 = x_2(x_1 \oplus q_1)$ and $z_2 = x_2 \oplus (x_1 \oplus q_1)$. The result of the simulation at cycle 2 using characteristic functions is $(\overline{y_2 \oplus x_2 q_2})(\overline{z_2 \oplus (x_2 \oplus q_2)})(\overline{z_1 \oplus (x_1 \oplus q_1)})(\overline{z_1 \oplus q_2})$.

*continues*

---

**Example 8.25    (Continued)**

If the characteristic function is 1, then each of the four terms is equal to 1. That is,

$\overline{y_2 \oplus x_2 q_2}$, $\overline{(z_2 \oplus (x_2 \oplus q_2))}$, $\overline{z_1 \oplus (x_1 \oplus q_1)}$, and $\overline{z_1 \oplus q_2}$ are all 1, which hold if and only if $y_2 = x_2 q_2$, $z_2 = x_2 \oplus q_2$, $z_1 = x_1 \oplus q_1$, and $q_2 = z_1$. Eliminating intermediate variables $q_2$ and $z_1$, we have

$$y_2 = x_2 q_2 = x_2 z_1 = x_2 (x_1 \oplus q_1)$$

and

$$z_2 = x_2 \oplus q_2 = x_2 \oplus z_1 = x_2 \oplus (x_1 \oplus q_1)$$

which are identical to the result derived based on circuit equations. Therefore, we conclude that the simulation using characteristic functions yields the same result.

---

Constraining inputs is readily done when characteristic functions are used in symbolic simulation. Let $\mu(x)$ be a characteristic function representing the permissible input space. In other words, $\mu(x) = 1$ if and only if $x \in R$. And let $\lambda(y, z, x, q)$ be the characteristic function of the circuit. Then, the circuit with its inputs constrained to the permissible space $R$ is $\mu(x) \cdot \lambda(y, z, x, q)$. This is because the product is 1 if and only if the input lies in the permissible space and it satisfies the circuit. Notice how easily constraining is done using characteristic functions, compared with the method using circuit equation representation discussed earlier. This illustrates an advantage of using characteristic functions in symbolic manipulations.

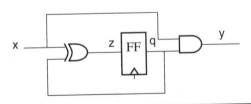

**Figure 8.42** Symbolic simulation using characteristic functions

**Example 8.26**

Let's repeat Example 8.24 using characteristic functions. The restricted inputs are represented by

$$\mu(x, y, z) = \begin{cases} 1, \text{ if } (x,y,z) = 100, 111, 010, 011 \\ 0, \text{ otherwise} \end{cases} = x\bar{y}\bar{z} + xyz + \bar{x}y\bar{z} + \bar{x}yz$$

The characteristic function for the unconstrained circuit is

$$\lambda(x, y, z, p, q) = (p = f(x, y, z))(q = h(x, y, z))$$

Therefore, the characteristic function of the constrained circuit is

$$\Lambda(x, y, z, p, q) = (x\bar{y}\bar{z} + xyz + \bar{x}y\bar{z} + \bar{x}yz)(p = f(x, y, z))(q = h(x, y, z))$$

As a check, if input $(x,y,z) = 101$, we want to know what output value the previous characteristic function would produce. Substituting the input vector, we have

$$\Lambda(1, 0, 1, p, q) = (1\bar{0}\bar{1} + 101 + \bar{1}0\bar{1} + \bar{1}01)(p = f(x, y, z))(q = h(x, y, z)) = 0$$

which means no values of $p$ and $q$ are valid outputs. This is because 101 is not a permissible input. For any permissible input, $\Lambda(x, y, z, p, q)$ reduces to $\lambda(x, y, z, p, q)$.

## 8.5.4  Parameterization

In addition to being used in constraining input space, parameterization is also used to reduce representation complexity by taking advantage of functional dependency. A case in point is the functional dependency among the components of the state vector in a finite-state machine. For a vector Boolean function

$$F(x_1, ..., x_n) = (f_1(x_1, ..., x_n), ..., f_k(x_1, ..., x_n))$$

parameterization of $F(x_1, ..., x_n)$ creates another Boolean function

$$G(t_1, ..., t_m) = (g_1(t_1, ..., t_m), ..., g_k(t_1, ..., t_m))$$

where variables $t_1, ..., t_m$ are called *parameters*, and $g_i: B^m -> B, i = 1, ..., k$.

Furthermore, $F$ and $G$ satisfy the following condition. For every $(x_1, ..., x_n)$ there is a $(t_1, ..., t_m)$ such that $F(x_1, ..., x_n) = G(t_1, ..., t_m)$, and vice versa. In other words, $G(t_1, ..., t_m)$ preserves the range of $F(x_1, ..., x_n)$. Note that $f_i$ and $g_i$ may have different numbers of arguments. For practical purpose, $G(t_1, ..., t_m)$ should be simpler than $F(x_1, ..., x_n)$.

---

**Example 8.27**

Consider $F(a, b, c) = (a + b, \bar{c}(a + b), \bar{a}\bar{b})$. We can create parametric variables $r$ and $s$ such that $r = a + b$ and $s = c$. Then $G(t, s) = (r, \bar{s}r, \bar{r})$. To see whether the range of $F(a, b, c)$ is preserved, we note that for each value $(a, b, c)$, the corresponding $(r, s)$ is computed using the parametric equation. In other words, $r = a + b$ and $s = c$. For each value $(r, s)$, we set variable $a$ to 0 and $b = r$, and $c = s$. The values of $F(a, b, c)$ and $G(t, s)$ are equal when the values of the variables and the parametric variables are determined as noted earlier. $G(t, s)$ is simpler than $F(a, b, c)$.

Let us consider function $F(x, y, z) = (x + y + z, x + y, \bar{x})$. We cannot parameterize it into $G(a, b, c) = (a + b, a, \bar{c})$, where $a = x + y$, $b = z$, $c = x$, because the parameterized function has in its range the point (0,0,0), when $a = 0$, $b = 0$, and $c = 1$, which is not in the range of the original function. That is, there is no $(x, y, z)$ such that $F(x, y, z) = (0, 0, 0)$. The reason for this problem is that $a$ and $c$ are not independent of each other; they are related through variable $x$.

---

Before delving into parameterization on functional dependency, let's first look at the simplifying effect of using parameterization with Example 8.28.

---

**Example 8.28**

Consider the circuit in Example 8.22. Because the state is only 1 bit, it can be replaced with a free variable at each cycle. (Here we assume the state is not a constant; thus, it can take on 1 or 0, just as a free variable.) The new variable is a parameter. For example, in cycle 2, the expression for node $k$ is simply $k_2$, a free variable, as opposed to the computed value $(\overline{c_1 \cdot d_1}) \oplus (\bar{k}_1(a_1 + b_1) + k_1(b_1 \cdot c_1))$. Using this method of parameterization, the symbolic node values at cycle 2 are as shown in Table 8.11. The second column lists the values with state parameterization, and the last column lists the results without state parameterization. Note the reduction in complexity.

---

**Example 8.28    (Continued)**

---

When simulation detects an error, a counterexample, which is just an input sequence that would produce the same error when run on a logic simulator, has to be generated. To generate a counterexample in symbolic simulation with parameterization, we may have to use the equations that replaced the parameters. A counterexample input sequence is formed by back tracing cycle by cycle from the time of error back to the first cycle. At the $n^{th}$ cycle, a binary constant, say $V_n$, is picked that is consistent with the symbolic result at the current cycle such that $V_n$ would cause $V_{n+1}$ in the $n + 1^{th}$ cycle. Then time is moved back one cycle, and the same procedure is repeated. After the initial cycle is reached, the input sequence made of the $V$s in reverse order is a counterexample.

As an example, suppose at cycle 2, the output $f$ should not be 1. Based on our simulation, the value of $f$ at cycle 2 is $\bar{k}_2(\overline{a_2 + b_2}) + k_2(b_2 \cdot c_2)$, which can be 1. Therefore, an error has occurred. To generate a counterexample, let's search a value at cycle 2 at which node $f$ is 1. Such a value is $k_2 = b_2 = c_2 = 1$. Because $k_2$ is a parameter, we have to use its original equation to back trace further. The equation that $k_2$ was replaced with is

$$(\overline{c_1 \cdot d_1}) \oplus (\bar{k}_1(\overline{a_1 + b_1}) + k_1(b_1 \cdot c_1))$$

From which we determine that if $a_1 = 1$, $c_1 = 0$, then $k_2 = 1$. Therefore, a counterexample is as follows:

At cycle 1, $a = 1, c = 0$.

At cycle 2, $b = 1, c = 1$.

---

**Table 8.11**  Symbolic Simulation Value at Cycle 2 with and without State Parameterization

| Node | With Parameterization | Without Parameterization |
|------|----------------------|--------------------------|
| $h$ | $\overline{a_2 + b_2}$ | $\overline{a_2 + b_2}$ |
| $g$ | $b_2 \cdot c_2$ | $b_2 \cdot c_2$ |
| $k$ | $k_2$ | $(\overline{c_1 \cdot d_1}) \oplus (\bar{k}_1(\overline{a_1 + b_1}) + k_1(b_1 \cdot c_1))$ |

*continues*

**Table 8.11**   Symbolic Simulation Value at Cycle 2 with and without State Parameterization (Continued)

| Node | With Parameterization | Without Parameterization |
|------|----------------------|--------------------------|
| $f$ | $\bar{k}_2(\overline{a_2 + b_2}) + k_2(b_2 \cdot c_2)$ | $\overline{(\overline{c_1 \cdot d_1}) \oplus (\bar{k}_1(\overline{a_1 + b_1}) + k_1(b_1 \cdot c_1))}$ $(a_2 + b_2) + ((\overline{c_1 \cdot dw_1}) \oplus (\bar{k}_1(\overline{a_1 + b_1})$ $+ k_1(b_1 \cdot c_1)))(b_2 \cdot c_2)$ |
| $n$ | $\overline{c_2 \cdot d_2}$ | $\overline{c_2 \cdot d_2}$ |
| $j$ | $\overline{c_2 \cdot d_2} \oplus (\bar{k}_2(\overline{a_2 + b_2}) + k_2(b_2 \cdot c_2))$ | $((\overline{c_1 \cdot d_1}) \oplus (\bar{k}_1(\overline{a_1 + b_1}) + k_1(b_1 \cdot c_1)))$ $(a_2 + b_2) + ((\overline{c_1 \cdot d_1}) \oplus (\bar{k}_1(\overline{a_1 + b_1}) +$ $k_1(b_1 \cdot c_1)))(b_2 \cdot c_2)) \oplus (\overline{c_2 \cdot d_2})$ |

Parameterization when the state vector has more than 1 bit is more complicated. First, functional dependency among the state bits must be detected, and then parameters are created to minimize the state representation. It is important that the range of the original function be preserved. Example 8.29 illustrates this concept and process by parameterizing the next-state function.

---

**Example 8.29**

Let's parameterize the state expressions for the circuit in Figure 8.43. The next-state function has 2 bits, $S(x, y, p, q) = (u, v)$, and is $(x\overline{(p + q + y)}, \bar{x} + q(p + y))$. The functional dependency between the 2 bits is $(p + y)$. We want to introduce a vector parametric function $R(a, b, c)$ for $S(x, y, p, q)$ so that the resulting representation is simpler, while preserving the range of $S(x, y, p, q)$. In other words, for each state $s$ that is in the range of $S(x, y, p, q)$ there are $a, b, c$ such that $s = R(a, b, c)$ and vice versa. In this example, we let $a = x$, $b = p + y$, and $c = q$. The resulting parameterized function becomes $(a\overline{(b + c)}, \bar{a} + cb)$. Note that the parametric function has one fewer variables than the original.

---

As seen in the previous examples, there are many ways to parameterize a function, and parameterized functions of various complexity result. For a more complete study of parameterization, consult the bibliography.

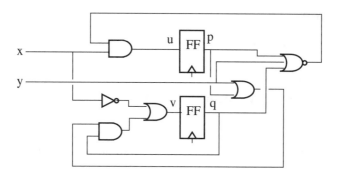

**Figure 8.43** Parameterization based on functional dependency

## 8.6  Summary

We started the chapter with decision diagrams. First, the BDD was studied. We showed how to construct a BDD and reduce it to canonical form. The reduction procedure consists of two steps: removing nodes with both edges that point to the same node and merging isomorphic subgraphs. Then we discussed operations on BDDs and how various Boolean manipulations can be cast into *ITE* operations, and provided algorithms for computing the Boolean operations. Next we examined the effect of variable ordering on BDD size, and introduced static and dynamic variable ordering algorithms.

We then discussed other variant forms of decision diagrams, which include SBDDs, edge-attributed BDDs, ZBDDs, OFDDs, and pseudo Boolean functions. Of the pseudo Boolean functions, we studied ADDs and BMDs.

We next examined how decision diagrams can be used to check the functional equivalence of circuits. In particular, two practical techniques to reduce complexity were mentioned: node mapping and constraining.

As an alternative to decision diagrams, we introduced the Boolean satisfiability problem and presented two methods. The first method, the resolvent method, relies on generating clause resolvents to eliminate variables until the formula or clauses can be easily inferred using the pure literal rule or the unit clause rule. The second method, a search-based method, has the main components of selecting branches, inferring results, and backtracking.

Symbolic simulation, contrary to traditional simulation, is capable of simulating multiple traces simultaneously. We introduced the basic idea of symbolic simulation and common techniques to reduce complexity by constraining variables and parameterizing states.

## 8.7 Problems

**1.** For the BDD in Figure 8.44, do the following:

    a. Is the BDD reduced?
    b. Derive the function represented by the BDD in sum-of-products form.
    c. Repeat (b) for product of sums.

**2.** Reduce the BDD in Figure 8.45. Which rules of reduction did you use?

**3.** Synthesize from BDDs.

    a. Synthesize the function represented by the BDD in Figure 8.45 using only 2 – 1 multiplexors. You may want to reduce the BDD first. State your synthesis algorithm.
    b. Suppose that you have found the best variable ordering for a function that results in an ROBDD with the minimum number of nodes. Can you claim that your synthesis algorithm based on the optimal ROBDD produces the minimum of 2 – 1 multiplexors?

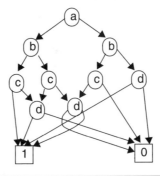

**Figure 8.44** BDD for obtaining sum-of-products and product-of-sums

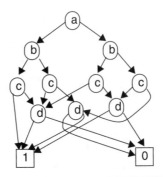

**Figure 8.45** BDD for reduction

**4.** Perform the following operations on the BDDs in Figure 8.46:

   a. Construct an ROBDD for the complement of the function represented by the BDD
   in Figure 8.46A.
   b. Restrict the BDD in Figure 8.46A to $c = 1$.
   c. Suppose variable $c$ of the BDD in Figure 8.46B is replaced by its complement and a
   new function results. Find an ROBDD for this new function.
   d. Construct the BDD that is the AND of the BDDs in Figures 8.46A and 8.46B.

**5.** Find a good variable ordering for the circuit in Figure 8.47 and build an ROBDD using
this variable ordering.

**6.** Apply sifting operations to the BDD constructed in the previous problem. Did you find
a better ordering?

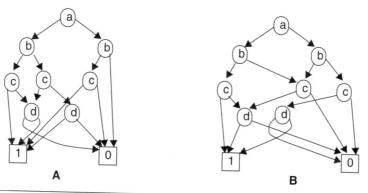

**Figure 8.46**  (A, B) BDDs for Boolean operations

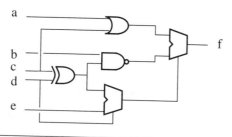

**Figure 8.47**  Circuit for variable ordering

**7.** Find an optimal variable ordering for the following function. Can you prove that it is an optimal variable ordering?

$$f(a, b, c, d) = \bar{a}\bar{b}\bar{c}d + a\bar{b}\bar{c}d + a\bar{b}cd + a\bar{b}\bar{c}\bar{d} + a\bar{b}c\bar{d}$$

**8.** Make the BDDs in Figure 8.48 into an SBDD. What is the ratio of the number of nodes before and after the merge?

**9.** Convert the BDD in Figure 8.44 into a BDD using edge attributes. What is the savings in using an edge-attributed BDD?

**10.** Transform the edge-attributed BDD in Figure 8.49 to the standard form.

**11.** Convert the function in Figure 8.50 to a ZBDD.

**12.** Derive the function of the ZBDD in Figure 8.51, assuming the variables of the functions are $a$, $b$, $c$, and $d$.

**13.** Five bits are used to encode 32 objects: bits $a$, $b$, $c$, $d$, and $e$. For the following, use variable ordering $a$, $b$, $c$, $d$, and $e$.

    a. Derive the characteristic function that represents the set consisting of objects 00010, 11000, and 10010.

    b. Construct an ROBDD for the characteristic function.

    c. Construct a ZBDD for the characteristic function. What is the ratio of the sizes?

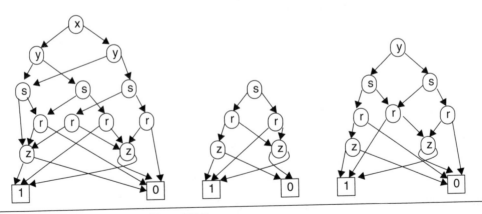

**Figure 8.48** BDDs to be converted to an SBDD

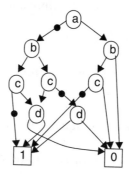

**Figure 8.49** Nonstandard edge-attributed BDD

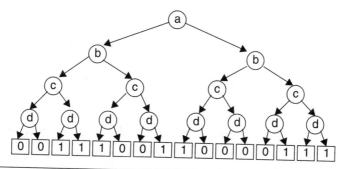

**Figure 8.50** A complete decision diagram

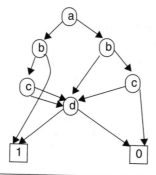

**Figure 8.51** A ZBDD

**14.** Derive a Reed–Muller form for the following function by first constructing an OFDD:

$$f(a, b, c, d) = (\bar{a}b + a\bar{d})(b\bar{d} + \bar{a}\bar{c}) + b\bar{c}(a + \bar{d})$$

**15.** Reduce the OFDD in Figure 8.52 and derive its function.

**16.** Consider the modulo operator $x \% y$ where $x$ and $y$ are integers that can take on any value from 1 to 4. Use an MDD to describe the operation of this modulo function.

**17.** The variables in the following polynomial can have value of either 0 or 1. Represent the polynomial using a decision diagram.

$$f(x, y, z, r) = x^2y - z(r - x) + 3(xy^2z - r^2)$$

**18.** Derive a BMD for the following function:

$$f(a, b, c, d) = \bar{a}b\bar{d} + a\bar{c}\bar{d} + \bar{a}b\bar{c} + b\bar{c}\bar{d}$$

In the context of a BMD, does every polynomial have a corresponding Boolean function? Why?

**19.** Determine whether the two circuits, $F$ and $G$, in Figure 8.53 are equivalent.

   a. Find a set of cut points to break each circuit into three pieces.
   b. Using the cut points in (a), build BDDs for the subcircuits. Are the subcircuits equivalent to their corresponding subcircuits?
   c. If the answer to (b) is no, how should you proceed? Are $F$ and $G$ functionally equivalent?

**20.** Use the resolvent algorithm to determine whether the following function is satisfiable:

$$(\bar{a} + \bar{b} + c)(a + b + \bar{c})(a + \bar{b})(a + \bar{c})(\bar{a} + b)(\bar{b} + \bar{c})$$

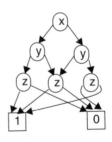

**Figure 8.52** An OFDD for reduction

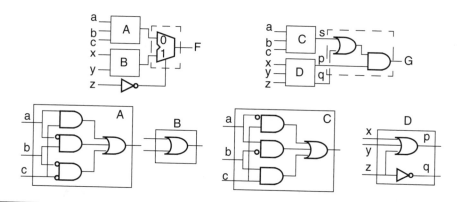

**Figure 8.53** Circuits for equivalence checking

21. Use a search-based algorithm to determine satisfiability of the following function. Repeat using a BDD.

$$(\bar{a} + \bar{b} + c)(b + c)(\bar{a} + \bar{c})(a + \bar{b} + \bar{c})(a + \bar{b})(a + \bar{c})(\bar{a} + b)(\bar{b} + \bar{c})(a + b + c)$$

22. Prove, using symbolic simulation, that for the output $f$ in Figure 8.54, zeros at output $f$ occur in pairs. The initial state is 11.

  a. Show that the property that zeros occur in pairs can be expressed as the following two Boolean assertions: $\overline{f_n}f_{n-1} \to \overline{f_{n+1}}$ and $\overline{f_n}\overline{f_{n-1}} \to f_{n+1}$, where $f_n$ is the value of node $f$ at cycle $n$.
  b. Show that the assertions are true based on the symbolic simulation result.

23. If the inputs to the circuit in Figure 8.55 are of even parity, parameterize the circuit.

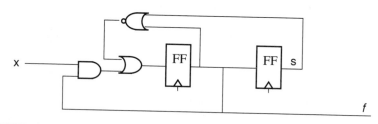

**Figure 8.54** Circuit for symbolic simulation

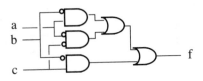

**Figure 8.55** Input to the circuit is of even parity

24. Suppose the next-state vector of a finite-state machine at a cycle is given by $(x, y, z) = (a + xyz + c, bz + xy, z + \overline{(a + c)})$, where the state bits are $x$, $y$, and $z$. Variables $a$, $b$, and $c$ are inputs.

   a. Derive the characteristic function of the states reachable at cycle 3, assuming the initial state is $(r, s, t)$.

   b. Repeat (a) when the inputs are constrained: $abc = \{001, 101, 010\}$.

   c. Parameterize the state vector.

## CHAPTER 9

# Model Checking and Symbolic Computation

## Chapter Highlights

- Properties, specifications, and logic

- Property checking

- Symbolic computation and model checking

- Symbolic CTL model checking

- Computational improvements

- Using model-checking tools

C hecking equivalence between two circuits, proving or disproving that the two circuits are functionally equivalent, is one aspect of formal verification. If two circuits are sequential and the mapping between their states is not available, the techniques we discussed earlier fail, and model-checking techniques are required. In general, model checking is required when verifying partial and abstract specifications. Modeling checking, the main subject of formal verification, deals with more general problems than equivalence checking and its goal is to prove or disprove that a circuit possesses a property that is a part of a specification. Model checking exhaustively searches the entire state space, constrained or unconstrained, and determines the property's validity in the space. The basic components of model checking consist of a circuit model, property representations, and checking the properties' validity. Circuits are modeled as finite state automata and Kripke structures, which, unlike automata, do not have input symbols labeled on transitions and have states that are atomic propositions. In a later section, we will discuss Kripke structures, as well as algorithms for transforming finite state automata to Kripke structures. A previous chapter reviews finite state automata; so we will not elaborate on circuit modeling. In this chapter, we introduce mechanisms for capturing properties, then study algorithms for verifying properties, and finally examine efficient implementations of the algorithms using BDDs, namely, symbolic model checking.

## 9.1   Properties, Specifications, and Logic

RTL code of a circuit is an implementation of a specification. To verify that the implementation meets the specification, subsets of the specification are checked against the implementation. If the subsets that constitute the specification pass, the implementation meets the specification. In some cases, the entire specification, as opposed to subsets, is checked against the implementation. A case in point is equivalence checking, where one circuit serves as a specification of the other. In model checking, due to practical computational limitations, the specification is partitioned into subsets and each subset is verified individually. A specification or a subset of it is called a *property*. Ideally, the way a property is written should be as different from the circuit implementation as possible, because the more different they are the less likely it is for them to hide the same bugs. Therefore, properties are often expressed at a higher level of abstraction or using a more abstract language. Abstraction is a measure of freedom or nondeterminism in an expression. On the spectrum of abstraction, an example of a less abstract expression is an implementation of a circuit in which all gate types and connections are completely determined to perform a certain functionality. A more abstract expression is an architectural specification that dictates only what functionality needs to be implemented, but not how it is to be accomplished. In other words, a more abstract description has fewer details and thus more nondeterminism about implementation. As an example, a gate-level description of a carry bypass 64-bit adder is less abstract than the Verilog statement $x = y + z$, where $y$ and $z$ are 64-bit "regs" and $x$ is a 65-bit reg.

There are three common types of property in practice:

1. Safety property. This type of property mandates that certain undesirable conditions should not happen. The property that a cache being written back cannot be updated at the same time is a safety property.
2. Liveness property. This type of property ensures that some essential conditions must be reached. An example is that a bus arbiter will be in a grant state.

**3.** Fairness property. This type of property requires that certain states be reached or certain conditions happen repeatedly. A design in operation can be viewed, from the perspective of a state diagram, as making an infinite sequence of state transitions in time. A fairness condition consisting of a set of states imposes that only the transitions that include that set of states be checked against a property. A fairness constraint on a bus arbiter is that the request state and the grant state for each client must be visited infinitely often.

## 9.1.1 Sequential Specification Using Automata

The simplest kind of property involves node values at the current cycle and consists of compositions of propositional operators, called *Boolean operators*, with operands being the node values. This kind of property is combinational in nature. An example property is that, at most, one bus enable bit is high at any time. Most interesting properties are sequential and involve values spanning over a period of time. A propositional property cannot describe variables over time. For example, it cannot express the value of a node from the last clock cycle.

A sequential property can be described using a finite-state automaton, called a *property automaton*, which monitors the behavior of the circuit by taking as inputs the inputs and states of the circuit. If the property under verification succeeds, then certain designated states in the property automaton are reached. The property automaton can also be constructed to represent failures when certain states are reached if the property fails. Example 9.1 illustrates property specification with an automaton.

---

**Example 9.1**

---

The design under verification is a two-way traffic controller. A property for this design is that, at any time, the light for south-bound or north-bound traffic cannot be the same color as the west-bound or east-bound light. (It is arguable that a red light in both directions may be acceptable, even though access to the intersection is not best used.) This property is combinational because the values—light colors of both directions—are available at the current cycle. Therefore, a Boolean function will suffice. We further assume, for simplicity, that the south-bound and north-bound lights are the same all the time. The assumption also applies to west-bound and east-bound lights. Let $S_1 S_2$ and $W_1 W_2$ be the bits representing the light colors for the south-bound traffic and the west-bound traffic respectively. Then, this property can be written as $P_1 = (S_1 \oplus W_1) + (S_2 \oplus W_2)$. $P_1$ is 1 if the property succeeds and is 0 otherwise.

*continues*

---

**Example 9.1    (Continued)**

A second property about the traffic light controller is that the light sequence must follow the order R, G, Y, R, G, Y, and so on. Because past values are involved, this property is a sequential property. For simplicity, let us assume that the traffic light controller changes colors every clock cycle. We need to construct a property automaton that takes in as input the output of the traffic light controller. This property automaton monitors the north–south direction; a similar property automaton can be constructed for the west–east direction. The automaton has three states—$S_R$, $S_G$, and $S_Y$—each of which represents the current color of the traffic light. If the light sequence deviates, the automaton moves to an error state, $S_e$, and stays there forever. The property automaton and its interaction with the traffic light controller is shown in Figure 9.1. This property is verified if it can be proved that the property automaton never reaches state $S_e$ for all operations of the traffic light controller.

---

Describing temporal behavior using finite-state automata can be cumbersome for many properties and thus error prone. Therefore, a more concise and precise language is needed for specifying temporal properties.

## 9.1.2    Temporal Structure and Computation Tree

The temporal logic to be introduced in this section is well developed and used widely in verification literature. Although rarely used as a property language in practice, it serves as

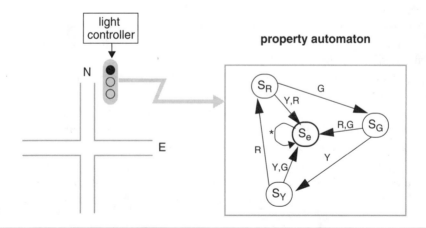

**Figure 9.1**    Property automaton and its interaction with the traffic light controller

a means for explaining the underlying theory in model checking. Knowledge of the logic is instrumental in understanding the essential concepts of temporal properties.

Time can be modeled as continuous or discrete. In a discrete model, time is represented by an integer; in a continuous model, by a real number. In this section we are concerned only with discrete time. Furthermore, let's assume the discrete time points are evenly spaced, and each state transition in an automaton occurs in exactly one time unit. So instead of explicitly stating time, a sequence of state transitions represents the progress of time. Example 9.2 demonstrates the use of state transitions to represent progress of time.

---

### Example 9.2

The finite-state automaton in Figure 9.2A accepts input symbols *a*, *b*, or *c*. Consider a run on input sequence *bcabbccaaa...* Each time an input symbol is taken, the automaton makes a transition. The sequence of state transitions activated by the input sequence, shown in Figure 9.2B, represents the progress of time. Each state denotes a time instance. In this run, every state has only one successor state. Translating to time instance, we say that every time instance has only one successor time instance.

Now, consider the input sequence *c{a,b}*{b,c}* ..., where the symbols enclosed within {} are possible inputs at that time instance and * denotes any input symbol. Thus, the input symbol at time 1 is *c*; at time 2, either *a* or *b*; at time 3, any input symbol that can be either *a* or *b* or *c*; at time 4, either *b* or *c*; and so on. This phenomenon of having multiple possible input symbols is a form of nondeterminism. At time 1, on input *c*, the transition is from state $S_1$ to $S_4$. At time 2, because the input can be either *a* or *b*, $S_4$ can transit to either $S_2$ on *a* or $S_3$ on *b*. At time 3, because there are two present states, $S_2$ and $S_3$, we need to consider them separately. At each of the present states there are three possible inputs. Therefore, there are a total of six next states, although they are not necessarily distinct. At time 4, six present states on two possible inputs produce 12 next states, and so on. This computation of next states represents all computational possibilities of the automaton on the input, and these possibilities are shown in Figure 9.3. Note that every time instance can have more than one successor time instance, and this branching phenomenon is caused by nondeterminism in the input sequence. Other forms of nondeterminism, such as multiple initial states and multiple transitions from a state on the same symbol, also cause branching.

To enumerate all possible computational behavior of a finite-state automaton, we use the input sequence that consists of all possible symbols at each and every time step: *** .... The resulting state traversal is all possible computations by the finite-state automaton and is an infinite tree of state transitions.

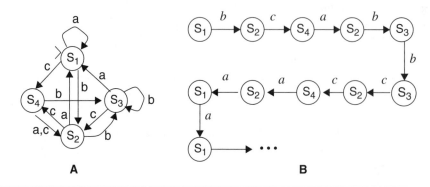

**Figure 9.2**  This sequence of state transitions represents a progress of time. (A) A state transition diagram (B) Sequence of state transitions on input sequence bcabbccaaa...

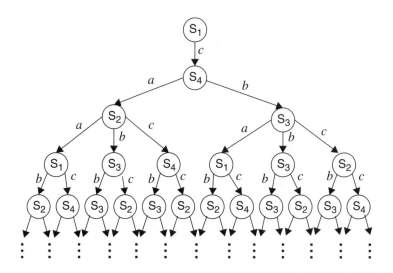

**Figure 9.3**  A branching state transition tree

Although engineers are more familiar with finite-state machines, an equivalent temporal structure, called a *Kripke structure*, is widely used in formal verification literature. A Kripke structure is intuitively a state diagram without inputs and outputs, and its transition function depends on the present state only. Associated with each state is a set of variables that are true at the state. More precisely, a Kripke structure $K$ over a set of propositional variables is a 4-tuple $K = (S, S_0, T, L)$, where $S$ is a set of states, $S_0$ is a set of initial states, $T$ is a total transition relation that maps the present state to a set of next state (a transition rela-tion is *total* if for every state there exists a next state), and $L$ is a labeling function that asso-ciates each state with a set of propositional variables that are true at that state.

An example Kripke structure over propositional (Boolean) variables $a$, $b$, and $c$ is shown in Figure 9.4. We use the convention that a variable is true at a state if it is labeled at the state; otherwise, it is false. In the Kripke structure, for example, variables $b$ and $c$ are true and variable $a$ is false at state $S_1$ because $b$ and $c$ are labeled but not $a$.

Although without inputs, a Kripke structure captures the operations of a finite-state machine by encoding the inputs of the finite-state machine into states in the Kripke struc-ture. Conversely, a Kripke structure can be transformed into a finite-state machine. To transform a finite-state machine into a Kripke structure, each state in the automaton is processed one by one. At a given state, all incoming transitions are partitioned according to their input symbols. Transitions on the same input symbol are grouped together. Then, the state is split into a number of states equal to the number of groups. Each newly formed state then includes the input symbol in its set of labels, and the symbol on the transitions is removed.

If an input has more than one selection (in other words, is nondeterministic, such as {a,b}), all possible values are considered. This can be achieved by transforming the transi-tion with the nondeterministic input into multiple transitions, each taking on one of the possible values. Example 9.3 illustrates this transforming process.

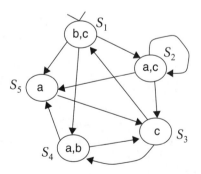

**Figure 9.4** An example Kripke structure

**Example 9.3**

Transform the finite-state machine in Figure 9.5A into a Kripke structure. The state machine has two binary inputs, $a$ and $b$. First, because state $S_1$ has only one deterministic incoming transition—namely, $a = 1$, $b = 0$—it does not need to be split. State $S_2$ has two incoming transitions—namely, $a = 0$, $b = 1$ and $a = 1$, $b = 1$. Therefore, $S_2$ splits into two states, each corresponding to an input value, $S_{20}$ and $S_{21}$, where $S_{20}$ is the state when $a = 0$, $b = 1$ and $S_{21}$ is the state when $a = 1$, $b = 1$. This state splitting is illustrated in Figure 9.5B. For state $S_3$, there are four incoming transitions and some inputs are nondeterministic. Therefore, we partition the transitions according to the input values. Because there are three possible input values—$a = 0$, $b = 0$; $a = 1$, $b = 1$; and $a = 0$, $b = 1$—$S_3$ splits into three states—$S_{30}$, $S_{31}$, and $S_{32}$—as shown in Figure 9.6A. At this stage, all inputs to a state are identical and can be removed from the edges to become a part of the state label. Furthermore, we use 3 bits $(r, s,$ and $t)$ to encode the six states as shown in Table 9.1. The state bits that are true are then labeled on the states with the input bits.

For example, at state $S_{20}$, the input value is $\bar{a}b$ and the state bits are 001. The bits that are true are $b$ and $t$. Hence, the Kripke state for $S_{20}$ has labels $b$ and $t$. The resulting Kripke structure is shown in Figure 9.6B.

**Table 9.1**    State Encoding

| rst | State |
|-----|-------|
| 000 | $S_1$ |
| 001 | $S_{20}$ |
| 010 | $S_{21}$ |
| 011 | $S_{30}$ |
| 100 | $S_{31}$ |
| 101 | $S_{32}$ |

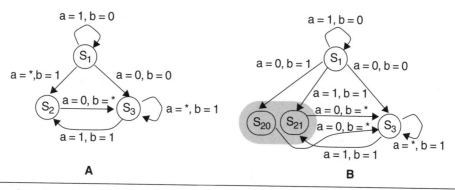

**Figure 9.5** Transforming a finite-state machine to a Kripke structure. (A) A state transition diagram. (B) State splitting of $S_2$ in the process of transforming to a Kripke structure.

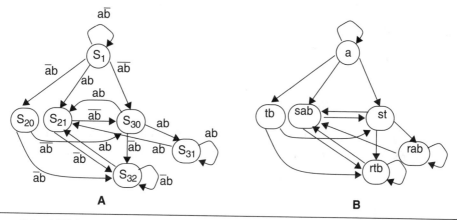

**Figure 9.6** A Kripke structure evolved from a finite-state machine. (A) An intermediate state diagram in the transforming to a Kripke structure. (B) A Kripke structure derived from the state transition diagram.

The procedure of transforming a finite-state machine to a Kripke structure is summarized here. Conversely, a Kripke structure is a finite-state machine with $\varepsilon$ transitions. An $\varepsilon$ transition is activated without an input symbol and is a form of nondeterminism.

---

**Transform a Finite-State Automaton to a Kripke Structure**

For each state in the automaton, do the following:

1. Partition the incoming transitions according to input symbols.
2. Create a state for each group of transitions.
3. Label the state with the state bits and input bits that are true.
4. Remove input symbols from transitions.

Continuously unrolling a Kripke structure traces out all paths from the initial states. If the Kripke structure has loops, paths will be infinite and will never reconverge, producing a tree. This tree is called *computation tree*. Unrolling the Kripke structure in Figure 9.4 produces the computation tree in Figure 9.7. The initial state is the state labeled *bc*. From the initial state, there are three next states, labeled *a, ab*, and *ac*. From each of the next states, come more states. The process continues indefinitely. Note that states that are encountered along different paths are not merged together. Therefore, the resulting structure is a tree structure. Along a path in a computation tree, crossing a state means the passage of one time unit and the variables labeled at the state at that time evaluate to true. Computation trees are an aid to illustrate concepts and algorithms; they are never used in implementing verification algorithms.

The time model can be further classified as linear time and branching time. In a *linear time* model, a time instance has exactly one successor time instance, an example being that of Figure 9.2B; whereas in a *branching time* model, a time instance can have more than one successor time instance, as in Figures 9.3 and 9.7. In the following discussion on temporal logic, a *path* is defined as a path in a computation tree and it consists of a sequence of state transitions. A computation tree in a linear time model has only one path, but a computation tree in a branching time model has many paths.

With the temporal structure defined here, the concept of time is captured in a computation tree. To reason about temporal behavior, we define temporal operators and logic based on computation trees, as presented in the following section.

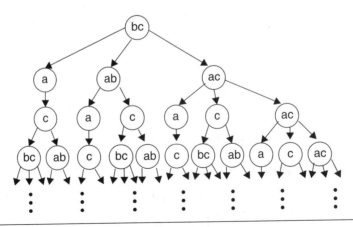

**Figure 9.7**   A computation tree for a finite-state automaton

### 9.1.3 Propositional Temporal Logic: LTL, CTL*, and CTL

In this section we will study three types of temporal logic: linear temporal logic (LTL), computation tree logic (CTL), and CTL*. In temporal logic, there are two types of formulas: state and path formulas. A state formula reasons about a state (that is, the validity of a formula at a state). A path formula reasons about a property of a path in a computation tree (that is, whether a formula holds along a path). An example of a state formula is $b + c$ at the present state. An example path formula is as follows: If variable $a$ is 1 at the present state, then $b + c$ will be 1 in the future.

LTL assumes a linear time structure and consists of temporal operators to reason about behavior along the path. An LTL formula is either a state formula that is an atomic formula or a path formula.

The two basic temporal operators in LTL are X (neXt) and U (Until). Let the underlying path in the linear time model be $\pi = s_1 \to s_2 \to \ldots \to s_i \to \ldots$, where $s_i$ is a state and the arrows are transitions. Suppose $s_1$ is the present state and $f$ and $g$ are state formulas. Operation X$f$ on $\pi$ succeeds if formula $f$ holds at the next state, $s_2$. Operation $f$U$g$ succeeds if $f$ holds until $g$ holds. More precisely, there is $k \geq 0$ such that $f$ holds at $s_j$ $1 \leq j < k$ and $g$ holds at $s_k$. Suppose $f$ and $g$ are path formulas. Let us define $\pi_i$ as the subpath starting at state $s_i$. X$f$ on $\pi$ succeeds if formula $f$ holds at $\pi_2$. $f$U$g$ succeeds if $f$ holds until $g$ holds. More precisely, there is $k \geq 0$ such that $f$ holds at $\pi_j$ $1 \leq j < k$ and $g$ holds at $\pi_k$.

LTL formulas are constructed using Boolean operators and temporal operators X and U, following these rules:

   **1.** Every propositional variable (Boolean variable) is an LTL formula.
   **2.** If $f$ and $g$ are LTL formulas, then ~$f$ and $f + g$ are LTL formulas.
   **3.** If $f$ and $g$ are LTL formulas, $f$U$g$ and X$g$ are LTL formulas.

Rule 2 says that a formula built by applying Boolean operators to existing LTL formulas is also an LTL formula. Rule 3 states the use of the X and U operators. Although X and U are the minimum LTL operators to build LTL formulas, for convenience other operators are used, despite the fact they can be expressed in terms of X and U. Some such operators are

   **1.** F$g$. This operation succeeds if $g$ holds at some Future state, F. It can be expressed in terms of X and U as ($TRUE$ U $g$).
   **2.** G$f$. This operation succeeds if $f$ holds at every state along the path. In other words, $f$ is Globally true. It can be expressed as (~(F~$f$)).

**3.** *f*R*g*. This is the Release operator and it is equivalent to ~(~*f* U ~*g*). To understand its meaning, consider its inverse, (~*f* U ~*g*), which has the interpretation of "*f* held by *g*." If *f* and *g* are mutually exclusive (at most, one can be high at any time), then (~*f* U ~*g*) means that *f* must be low until *g* becomes low. In other words, *f* must be held low when *g* is high. The inverse of "*f* held by *g*" is *f* released by *g* or (*f* R *g*).

The meaning, or semantics, of LTL operators is shown in Figure 9.8. The formula above a state denotes the validity of the formula at the state. For instance, in X*f*, *f* is false at $s_1$ but becomes true at $s_2$. This temporal behavior of *f* makes X*f* succeed. The trace shown for (*f* R *g*) is one of several situations in which (*f* R *g*) succeed.

With a branching time model, in which there are multiple paths, CTL* is applicable. CTL* uses the same temporal operators as LTL, but it adds two path quantifiers to cope with multiple paths: A (for all paths) and E (for some path). Quantifiers A and E mirror the universal and existential quantifiers, $\forall x$ and $\exists x$, in first-order logic, and apply over paths instead of variables. The two path quantifiers are prefixed to formulas, state or path, to create CTL* formulas. The rules for constructing CTL* formulas are divided into two parts: one for state formulas and the other for path formulas.

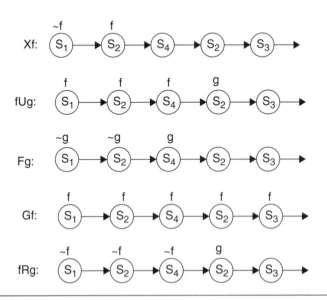

**Figure 9.8**   Semantics of LTL operators

For state formulas

> **1.** If $f$ is a propositional variable, then $f$ is a state formula
> **2.** If $f$ and $g$ are state formulas, then $\sim f$ and $f + g$ are state formulas
> **3.** If $f$ is a path formula, then $Ef$ and $Af$ are state formulas

For path formulas

> **1.** If $f$ is a state formula, then $f$ is also a path formula
> **2.** If $f$ and $g$ are path formulas, then $\sim f$, $f + g$, $Xf$, and $fUg$ are path formulas

Note that path quantifier $A(f)$ is not essential and can be expressed as $\sim E(\sim f)$; that is, "$f$ is true for all paths" is equivalent to "there is no path such that $f$ is false." Therefore, only three operators are basic in CTL*: X, U, and E. Rule 3 of the state formula says that $Ef$ is a state formula and it has the interpretation that if $Ef$ holds at a state, then there is a path starting from the state such that $f$ holds along the path. Similarly, $Af$ holds at a state if all paths starting from the state satisfy $f$.

---

**Example 9.4**

Examples of CTL* formulas are $A(fUg)$, $E(FG\,g)$, and $AG(EF\,f)$. If state $s$ has formula $A(fUg)$, then the formula succeeds if for all paths starting at state $s$, $fUg$ succeeds. $E(FG\,g)$ succeeds at state $s$ if there is a path starting at $s$ such that it eventually reaches a state $r$, at which $Gg$ succeeds. $AG(EF\,f)$ states that at all nodes along all paths starting at state $s$ $(EF\,f)$ succeed.

Consider the computation tree shown in Figure 9.9. A formula next to a node means that the formula succeeds at the node. Assume that formula $f$ succeeds at all descendent nodes of nodes 1, 2, 5, and 6; and formula $g$ succeeds at all descendent nodes of nodes 3 and 4. Thus, $A(fUg)$ succeeds at node $s$ because all paths starting at $s$ satisfy $f$ along the paths until $g$ succeeds. $E(FG\,g)$ succeeds at nodes $s$, $v$, and $r$, because there is a path—the path that ends at node $u$—that leads to the subtree where $g$ succeeds globally. However, $E(FG\,g)$ fails at node $t$, because all paths from node $t$ eventually lead to nodes where only $f$ holds. Formula $AG(EF\,f)$ succeeds at node $t$, because all path nodes eventually lead to success of $f$, but fail at node $v$, because the paths passing through node $u$ have nodes that never see $f$ succeed—namely, all the nodes below, including node $u$.

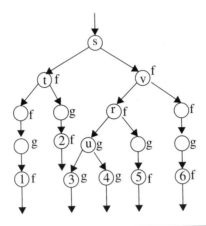

**Figure 9.9**    Examples of successful CTL* formulas

CTL derives from CTL* by allowing only certain combinations of path quantifiers and temporal operators, and hence is less expressive but easier to verify. The syntax of CTL is defined as follows.

State formulas:

     **1.** If $f$ is a propositional variable, then $f$ is a state formula.
     **2.** If $f$ and $g$ are state formulas, then $\sim f$ and $f + g$ are state formulas.
     **3.** If $f$ is a path formula, then $Ef$ and $Af$ are state formulas.

Path formulas:

     **1.** If $f$ and $g$ are state formulas, then $Xf$, and $fUg$ are path formulas.

Comparing these rules with those of CTL*, we notice that in CTL if $f$ is a path formula, then the only way to expand the formula is through path quantifiers E and A (rule 3). Furthermore, the only way to obtain a path formula is through temporal operators X and U.

If we include derived temporal operators F and G, then CTL has only eight base operators—namely, AX, EX, AG, EG, AF, EF, AU, and EU. Furthermore, five of these base operators can be expressed in terms of EX, EG, and EU. That is,

$AXf = \sim EX(\sim f)$

$AF(f) = \sim EG(\sim f)$

$AG(f) = \sim EF(\sim f)$

$$EF(f) = E(\text{TRUE U f})$$

$$A(fUg) = (\sim E(\sim g \text{ U } (\sim f)(\sim g))) (\sim EG(\sim g))$$

---

**Example 9.5**

Now let us determine which of the CTL* formulas in Example 9.14 are CTL formulas. Because $fUg$ is a path formula, and by prefixing path quantifier A, we obtain CTL formula $A(fUg)$. In E(FG $g$), (G $g$) is a path formula and the only way to expand is to add path quantifier. Thus, the CTL rules do not allow (FG $g$). Therefore, E(FG $g$) is not a CTL formula. Because (EF $f$) is a CTL state formula, and AG(EF $f$) can be generated by applying to (EF $f$) rule 1 of the CTL path formula followed by rule 3 of the CTL state formula, AG(EF $f$) is a CTL formula.

---

The semantics of these eight base operators and computation trees that succeed them are enumerated in Table 9.2.

**Table 9.2**  CTL Base Operators and Semantics

| CTL Operator | Semantics | Computation Tree Satisfying the Operator |
|---|---|---|
| AX($f$) | For all paths, $f$ holds at the next state. | 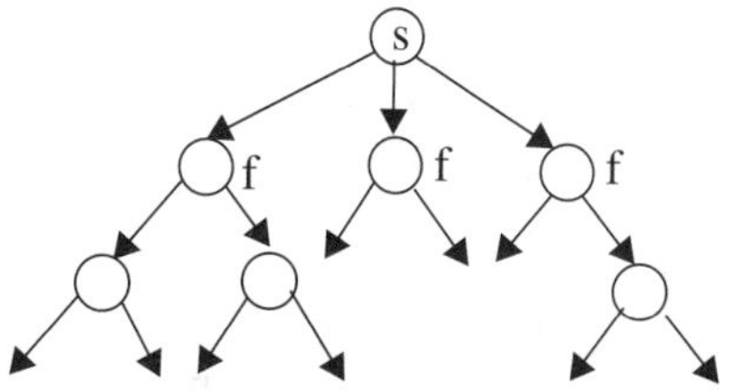 |
| EX($f$) | There is a path such that $f$ holds at the next state. | 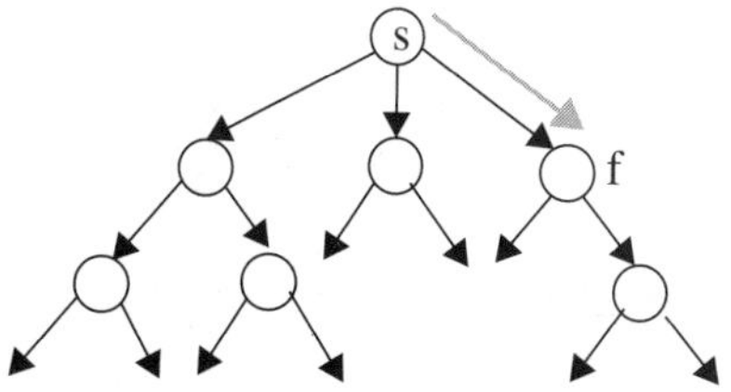 |

*continues*

**Table 9.2**    CTL Base Operators and Semantics (Continued)

| CTL Operator | Semantics | Computation Tree Satisfying the Operator |
|---|---|---|
| AG(*f*) | For all paths, *f* holds at every node of the path. | |
| EG(*f*) | There is a path along which *f* holds at every state. | |
| AF(*f*) | For all paths, *f* holds eventually. | |
| EF(*f*) | There is a path along which *f* holds eventually. | |

**Table 9.2**   CTL Base Operators and Semantics (Continued)

| CTL Operator | Semantics | Computation Tree Satisfying the Operator |
|---|---|---|
| A(fUg) | For all paths, f holds until g holds. | |
| E(fUg) | There is a path along which f holds until g holds. | |

---

**Example 9.6**

In a multiple-core CPU design, a multiplication unit is shared among four computation cores. To use the multiplication unit, a core must make a reservation first. Granting use of the unit is based on availability and priority, which select nonspeculative instructions over speculative instructions. Let's formulate three properties about the arbiter using CTL operators.

1. The property is as follows: A request, no matter how low its priority, will be granted eventually. Let $r$ and $g$ denote the request and grant signal for the $i$th core respectively. The statement "it will be eventually granted" translates to a statement about the underlying Kripke structure: "along all paths (from the present state) it will be true in future that request implies grant, symbolically, $r \to AFg$. This must be true at any state. Thus, we add AG to the previous formula $r \to AFg$. Iterating over four cores, the property in CTL is

$$\prod_{i=1}^{4} AG(r_i \to AFg_i)$$

*continues*

---

**Example 9.6    (Continued)**

**2.** Only one core has access to the multiplication unit at any time. If more than one core has access, then $g_i \cdot g_j$ is true where $i \neq j$. To prevent this at all states, we invert the expression and add the qualifier AG. The property is expressed as

$$AG\left(\prod_{i \neq j} \overline{(g_i \cdot g_j)}\right)$$

**3.** Furthermore, let $r_i^s$ and $r_i^n$ denote a request of speculative and nonspeculative instructions respectively. If both speculative and nonspeculative requests occur at the same time, then the nonspeculative request is granted no later than the speculative request. Symbolically, it is $r_i^s \cdot r_j^n \to (\neg g_i^s) U g_j^n$. This formula must be true at all states. By adding CTL operator AG, the property that nonspeculative instructions have a higher priority over speculative ones translates to

$$\prod_{i \neq j} AG(r_i^s \cdot r_j^n \to (\neg g_i^s) U g_j^n).$$

---

## 9.1.4   Fairness Constraint

Thus far we do not distinguish paths to which temporal formulas apply, but there are situations when we are only interested in verifying formulas along certain paths. One common such path constraint is a fairness constraint. The term *fairness* originates from protocol verification, in which a protocol is fair if a persistent request is eventually granted so that an infinite number of requests would warrant an infinite number of grants. Now the usage scope of "fairness" has been extended beyond protocol verification, and it is often associated with situations in which some states or transitions are visited infinitely often. As an example of checking a formula only along fair paths, consider verifying that a bus arbitration protocol handles prioritized requests correctly. Because this priority property is built on top of a fair bus arbitration protocol—fair in the sense that every request will be eventually granted—we should assume that the protocol is fair and will be interested in checking the priority property along fair paths only.

A fairness constraint may be expressed in several ways. In a Buchi automaton, a fairness constraint $F$ is defined as a subset of states. Thus, a path is fair if the path visits some state in $F$ infinitely often. At first it may be puzzling to see the term *infinitely often* used in a finite-state automaton. The term *infinitely often* makes more sense if we recall that a path in a computation tree is always an infinite path, even though the set of states is finite. The infinite paths are a result of unrolling loops an infinite number of times. Furthermore, an infinite path must visit some states infinitely often. If we denote the set of states visited infinitely often along path $\pi$ by $inf(\pi)$, then $\pi$ is fair if $inf(\pi) \cap F \neq \varnothing$. Another way of expressing fairness is the definition used in a Muller automaton in which a fairness constraint $F$ is a set of edge subsets. A path is fair if the edges traversed infinitely often by the path are an edge subset of $F$.

Furthermore, a fairness constraint can be in the form of a CTL formula. In this case, a fair path makes the formula true infinitely often along the path. Let $\phi$ be a CTL* formula for a fairness constraint and $f$ be a CTL* formula to be verified for all fair paths with respect to $\phi$. The expression for $\phi$ to be true infinitely often is $GF(\phi)$, which means that, along a fair path, $\phi$ will eventually hold at all nodes along the path. Thus, $\phi$ holds infinitely often. A weaker form of this concept of "infinitely often" is $FG(\phi)$, which means "almost always." Given this interpretation, the statement that $f$ succeeds with respect to fairness constraint $\phi$ is $A(GF\phi \rightarrow f)$, which says that along all paths for which $GF(\phi)$ succeeds (in other words, all fair paths), $f$ also succeeds. Similarly, a weaker form is $A(FG\phi \rightarrow f)$.

Note that the previous fairness expressions are not CTL formulas. Thus, fairness constraints can be expressed in CTL* but cannot be directly expressed in CTL. Thus, when checking CTL formulas with fairness constraints, the checking algorithms must ensure that only fair paths are examined. We will look at checking CTL formulas with fairness constraints later.

## 9.1.5   Relative Expressiveness of CTL*, CTL, and LTL

Expressiveness of logic is the scope of properties the logic can express. Before we can compare, we should address the differences in temporal models underlying CTL*/CTL and LTL. The former assumes a branch time model whereas the latter assumes a linear time model. To have a common ground, we interpret an LTL formula $f$ as A$f$ when it is used in a branching time model. An LTL formula must hold for all paths in a branching time model.

Clearly, CTL and LTL are proper subsets of CTL*, because CTL follows a restricted set of formula construction rules and LTL cannot have path quantifier E. Hence, the real question is the relative expressiveness between CTL and LTL. Because in a CTL formula every temporal operator must be preceded by a path formula, it can be shown that the LTL formula A(FG$g$) cannot be expressed as a CTL formula, but it is an LTL formula. Conversely, because an LTL formula can only have path quantifier A, it can be proved that there is no LTL equivalent formula for CTL formula AG(EF$g$). Obviously, there are formulas that are both LTL and CTL formulas. Finally, by ORing formulas A(FG$g$) and AG(EF$g$), A(FG$g$) + AG(EF$g$), we have a CTL* formula that is expressible in neither LTL nor CTL. Therefore, we conclude that (1) CTL and LTL are strictly less expressive than CTL* and (2) neither LTL nor CTL subsumes the other. Figure 9.10 is a visualization of this conclusion.

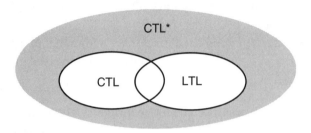

**Figure 9.10** Relative expressiveness of CTL*, CTL, and LTL

### 9.1.6    SystemVerilog Assertion Language

SystemVerilog extends the Verilog language into the temporal domain by including sequences and operators for sequences. A sequence specifies the temporal behavior of signals. The key constructor of sequence is the cycle delay operator ##. Operand of ## can be a single integer or a range of integers. For instance, sequence *x ##3 y* means that *y* follows *x* after three cycles, and sequence *x ##[1,3] y* means that *y* follows *x* after a number of cycles ranging from one to three. For instance, the next operator X in CTL* is simply *##1* and the future operator F is *##[1,$]*, where *$* stands for infinity. Sequence operators create complex sequences from simple ones. Example sequence operators are AND, OR, intersect, implication, throughout, and within. For more details on SystemVerilog temporal language, refer back to Section 5.5, SystemVerilog Assertion.

## 9.2    Property Checking

A property about a design is a set of functionality that a correct design must have or avoid. A property can be described with an automaton or a temporal logic formula. To check a property is to verify that the design satisfies the property. In this section we first study the basic principles behind property checking by describing the graph-based algorithms. In this case, the state diagram or Kripke structure is *explicitly* represented by a graph, and property checking is cast as graph algorithms. The specifics of property verification algorithms depend on the language describing the property. Checking a property described by a CTL formula uses one algorithm, whereas checking a property automaton uses another. We will consider verification algorithms for property described using automata and CTL formulas. In both cases, the design is described as a Moore machine. Then we will study BDD-based property-checking algorithms, which are far more efficient than their graph-based counterparts, because the state diagram or Kripke structure is *implicitly* represented.

## 9.2.1 Property Specification Using an Automaton

Let us assume that a property automaton has certain states marked as error states that indicate, when reached, the design has an error. The property automaton accepts as input the design's output and state, and transits accordingly. To verify that an error state can or can never be reached, we need to form a product machine by composing the design machine with the property machine. A state in this product machine consists of two components: one from the design machine and the other from the property machine. Therefore, a state of the product machine is an error state if it contains an error state from the property machine. So, the design has an error if and only if an error state in the product machine is reachable from an initial state. Therefore, the problem of verifying whether a design has an error becomes the problem of deciding whether the set of reachable states of the product machine contains an error state. This procedure is summarized here:

---

**Checking Property in Automata**

---

**1.** Describe the property using an automaton. Assume that the property is such that some states represent success of the property and some represent failure of it.
**2.** Compose the design automaton with the property automaton.
**3.** The property succeeds if and only if no failure composite state is reachable.

---

**Example 9.7**

---

In a game of two players, each player tosses a die and compares the face values. If player A has a higher value, player A's score increases by 1, unless it is 2, in which case it becomes 0. Furthermore, B's score is decreased by 1, unless it is 0, in which case it remains the same. On the other hand, if player B has a higher value, these scoring rules apply except the roles of A and B are reversed. If both dice values are equal, their scores stay unchanged. The properties to verify are

    a. Is it possible for both players to tie at the score of 1?
    b. How about for both to tie at 2 apiece?

First, we need to construct two automata describing the scores of players A and B. The input to the automata is either 0 or 1. Input 0 means player A has won the toss. Input 1 means player B has won the toss. The states represent the score of the player. Based on the rules, the automata are as shown in Figure 9.11A. Next we construct an automaton to describe both properties. The automaton has three states: State $a$ is that both players' scores are equal to 1; state $b$, both players' scores are equal to 2; and state $c$, all other cases. The property automaton takes as its input the player automata's present states.

*continues*

**Example 9.7     (Continued)**

To determine the properties we need to compose the three automata and check reachability. The composite automaton is shown in Figure 9.12. A state in the composite automaton consists of three components, one from each of the three automata. Therefore, property *a* is affirmed if and only if there is a composite state that contains component *a*. Similarly, property *b* is verified if a composite state containing component *b* is reachable. Because state *(1,1,a)* is reachable from initial state *(0,0,c)*, property *a* is true. Indeed, input sequence *1,1,0* will make both players' scores equal to 1, and there are infinitely many such sequences. On the other hand, no state with component *b* is reachable; thus, property *b* is false. It is impossible for both players' scores to be tied at 2.

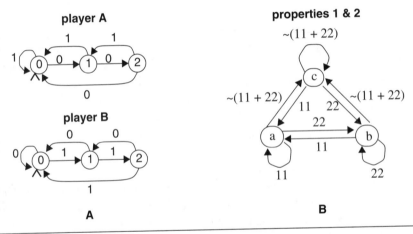

**Figure 9.11** (A, B) Automata for players' score (A) and property automata (B)

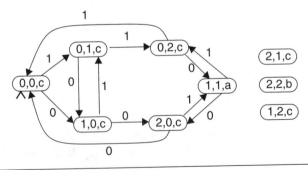

**Figure 9.12** Composite automaton for checking properties

## 9.2.2 Language Containment

The design process can be viewed conceptually as a transformation from specifications to final implementation by gradually refining each part of the specification until all parts have been implemented. Refinement turns a more abstract form of design specification into a more specific form. A more abstract version of a design has more nondeterminism, whereas a more specific version contains more concrete details, and hence less freedom. For instance, a 32-bit adder specification is just $x = a + b$, where $a$ and $b$ are 32-bit variables and $x$ is a 33-bit variable. This specification leaves it open as to how it should be implemented, be it a carry bypass or a ripple adder. This "unspecified" part is freedom, or nondeterminism, in specification. An example of refinement is turning a 32-bit adder specification into a concatenation of two 16-bit adders. Even though the 16-bit version still has freedom for further refinement, it is less abstract than the 32-bit version of the specification. Therefore, as an initial set of specifications goes through stages of refinement, the design becomes more and more specific, and eventually all nondeterminism is removed and a final implementation takes form. At each stage of the transformation, it is necessary that a refined version preserve all the properties in a more abstract version. Using the adder example, the adder at each refinement stage must add correctly. This method of verifying properties preserved at each refinement stage is called *refinement verification*, and it can be done using language containment.

A design is characterized by the sequences of state transitions it can produce. Because a design can be regarded as a forever-running automaton, all sequences of state transitions in the design are infinite. An automaton accepting infinite input sequences or producing infinite state transition sequences is called an ω *automaton*. Because the sequences are infinite, the usual acceptance condition that a final state is reachable has to be expanded to cope with sequences of infinite length. One such enhancement is to designate some states or edges as acceptance states or edges. An input string is accepted if the input drives the automaton to visit an acceptance state or edge infinitely often.

---

**Example 9.8**

The ω *automaton* in Figure 9.13 accepts infinite sequences, and edge E is an acceptance condition so that those input sequences traversing edge E infinitely often are accepted. Input sequence 0 101 101 101 101 101 ... traverses E infinitely often and thus is accepted, but neither sequence 0011111111 ... nor 0 1000 1000 1000 ... is accepted.

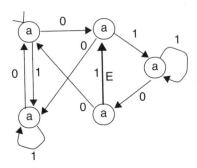

**Figure 9.13** A ω *automaton* and its acceptance condition

In the following discussion we will refer to an ω *automaton* simply as an automaton. The set of all accepted sequences is the language of the automaton. Furthermore, we will refer to a more abstract version of a design as the property, and a less abstract version as the design. This is consistent with the fact that a property, being part of a specification, is usually more abstract than the design.

Therefore, a refined design (design) preserves the specifications of a more abstract design (property) if every string accepted by the design is also accepted by the property. If not, there is a design behavior that deviates from the property behavior, and the design fails the property. In other words, a property is satisfied if the language of the design is contained in the language of the property. Let $L(D)$ and $L(P)$ be the languages of the design and property respectively. To check language containment $L(D) \subseteq L(P)$, we need to make use of the fact that $L(D) \subseteq L(P)$ if and only if $L(D) \cap \overline{L(P)} = \varnothing$. $\overline{L(P)}$ is equivalent to the language of the complemented automaton, $L(\bar{P})$, if it exists. For Buchi automata, complemented Buchi automata are also Buchi automata. $L(D) \cap \overline{L(P)}$ is equivalent to the language of product machine $D \times \bar{P}$, or $L(D \times \bar{P})$. Consequently, $L(D) \subseteq L(P)$ if and only if $L(D \times \bar{P}) = \varnothing$. Therefore, checking language containment reduces to checking language emptiness of the product automaton. This identity is visualized in Vann diagrams, as shown in Figure 9.14.

A machine has empty language if no string is accepted by the machine—namely, no acceptance state or edge is traversed infinitely often. To test whether a state or an edge is traversed infinitely often, we need to determine whether there is a loop in the state diagram that contains the acceptance state or edge, and to determine whether the loop is reachable from an initial state. If such a loop exists, a run that starts from the initial state, enters the loop, and traverses the loop infinitely often produces an acceptable infinite

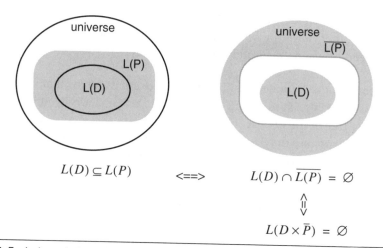

**Figure 9.14** Equivalence between containment and empty intersection

sequence. Hence, the language is not empty. On the other hand, if no such loops exist, the language is empty. Therefore, language containment becomes a reachability problem. This verification algorithm using language containment is as follows:

---

**Verification Using Language Containment**

Given a design automaton $D$ and a property automaton $P$, verify that $L(D) \subseteq L(P)$.

1. Construct an automaton $\bar{P}$ complementing the property automaton $P$.
2. Compose automaton $D$ with $\bar{P}$, giving product automaton $D \times \bar{P}$.
3. $L(D) \subseteq L(P)$ if $L(D \times \bar{P}) = \emptyset$.

---

## 9.2.3  Checking CTL Formulas

When verifying properties expressed in CTL formulas, we assume a Kripke structure for the design and present the algorithms based on the structure. Hence, verifying a CTL formula means determining all nodes in the Kripke structure at which the CTL formula holds, and if there are initial states, all initial states should be among those nodes. Therefore, checking a CTL formula is done in two steps. The first step finds all nodes at which the formula holds and the second step determines whether all initial states are contained in the set of nodes. Because the second step is trivial in comparison, we will only discuss the first step in this section. Verifying a CTL formula is done by induction on the level of the formula—in other words, by verifying its subformulas level by level, starting from the variables and ending when the formula itself is encountered.

Recall that labeled at each node of a Kripke structure are variables that are true, and the missing variables are false. Let's extend this labeling rule to include formulas or subformulas that evaluate true at the node. To begin, the subformulas at the bottom level of a CTL formula (made of variables only) are simply the labeled variables at the nodes. At the next level we need to consider two separate cases. In the first case, the operators are Boolean connectives, such as AND and NOT. In this case we will only consider AND and NOT operators because all other Boolean operators can be expressed in these two. At a node, if both operand formulas are true at the node, the resulting AND formula is true at the node and it is labeled at the node. If the operand formula is not true (in other words, it is missing at a node), then the resulting NOT formula is true and it is labeled at the node. In the second case, the operators are CTL temporal operators. We only need to consider EX$f$, E($f$U$g$), and EG($f$), because all other CTL operators can be expressed in these three.

To verify EX$f$, we must assume formula $f$ has been verified. In other words, all nodes at which $f$ holds are labeled with $f$. Then, every node with a *successor* node that is labeled with $f$ is a node satisfying EX$f$. Therefore, all these nodes are labeled with EX$f$. This is summarized as follows:

### Algorithm for Checking EX($f$)

input: a Kripke structure K and a CTL formula $EX(f)$.

output: labeling of the states where $EX(f)$ holds.

Verify_EX($f$): // check CTL formula $EX(f)$
for each state s of K, add label $EX(f)$ if $f$ is labeled at a successor of s.

To verify E($f$U$g$), we need to assume formulas $f$ and $g$ have been verified. Then, E($f$U$g$) is true at a node if there is a path from the node to a $g$-labeled node, and at every node along that partial path $f$ is labeled but $g$ is not. This process can be described by induction on its successors as the following: A node satisfies E($f$U$g$) if $g$ is labeled at the node or $f$ but not $g$ is labeled at the node and its successor is either labeled $g$ or E($f$U$g$). This is summarized as follows:

---

**Algorithm for Checking E(fUg)**

input: a Kripke structure K and a CTL formula E(fUg).

output: labeling of the states where E(fUg) holds.

Verify_EU(f, g): // check CTL formula E(fUg)

   1. M = empty.
   2. Add label E(fUg) to all states that have label g. Call this set of states L.
   3. For every state in L, if there is a predecessor, p, that is not in L, add label E(fUg) to p. Add p to set M. Set M consists of newly added nodes.
   4. Set L = M and M = empty.
   5. Repeat steps 3 and 4 until L is empty.

---

Finally, to verify EG(f), we again need to assume f has been processed. The semantics of EG(f)—there is a path along which f is true at every node—together with the fact that the path is infinite in a finite Kripke structure implies that EG(f) is true at a node if there is a path from the node to a loop of states such that f is true along the path from the start to an entry to the loop and f is true at all the nodes in the loop. This understanding gives rise to the following algorithm. First we remove all nodes from the Kripke structure that do not have label f. Therefore, every node remaining has label f. Second, we determine all SCCs in the resulting structure. Finally, EG(f) is true at a node if there is a path from the node to an SCC. This algorithm is as follows:

---

**Algorithm for Checking EG(f)**

input: a Kripke structure K and a CTL formula EG(f).

output: labeling of the states where EG(f) holds.

Verify_EG(f): // check CTL formula EG(f)

   1. Remove all states that do not have label f.
   2. Find all SCCs.
   3. Add label EG(f) to all states in the SCCs. Call this set L.
   4. For every state in L, if there is a predecessor, p, that is not in L, add label EG(f) to p. Add p to set M.
   5. Set L = M and M = empty.
   6. Repeat steps 4 and 5 until L is empty.

To check other CTL temporal operators, one can either express the operator in terms of the three base operators and apply the previous algorithms or check the operator directly on the Kripke structure.

---

**Example 9.9**

We can verify the CTL formula $F = E(\bar{v}U(EG(p+q)))$ on the Kripke structure shown in Figure 9.15A. There are five variables: $u$, $v$, $p$, $q$, and $w$. This formula can be broken into three levels. The first-level formulas are propositional: $f = \bar{v}$ and $g = p+q$. The second-level formula is $h = EGg$. The third-level formula is $F = E(fUh)$.

We first label the nodes at which the first-level formulas are true. Formula $f = \bar{v}$ is true at a node if the node does not have label $v$. Formula $g = p+q$ is true at a node if the node has either label $p$ or $q$. The result is shown in Figure 9.15B. To check formula $h = EGg$, we remove nodes without label $g$ and find all SCCs in the resulting graph, as shown in Figure 9.16A. Then all nodes that can reach an SCC are labeled with $h$. We annotate these $h$ labels to the Kripke structure, as shown in Figure 9.16. Finally, in checking formula $F = E(fUh)$, we label a node with $F$ if either it is labeled $h$ or it has label $f$, but not $h$, and it has a successor labeled $h$ or $F$. The states in which the formula holds are shaded in Figure 9.16B. Because the initial state is among these states, formula $E(\bar{v}U(EG(p+q)))$ holds for the machine represented by this Kripke structure.

---

If you are interested in checking CTL* and LTL formulas, please consult the bibliography. CTL model checking can also be done using the so-called *fix-point operator* and we study this method in a later section.

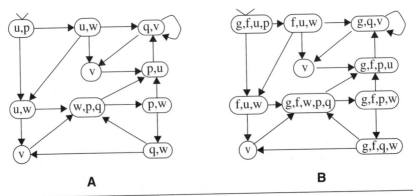

**Figure 9.15** Verifying CTL formula $E(\bar{v}U(EG(p+q)))$. (A) Original Kripke structure. (B) Kripke structure after $f$ and $g$ are verified.

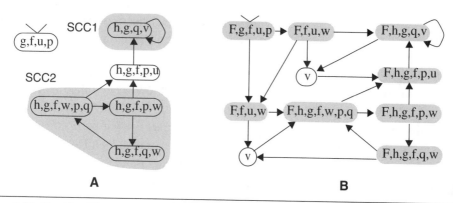

**Figure 9.16** Checking *EG(g)* and *E(fUh)*. (A) Kripke structure after *EG(g)* verified. (B) Final Kripke structure after *E(fUh)* is verified.

### 9.2.4  Checking with Fairness Constraints

Checking with fairness constraints means that only the paths satisfying the fairness constraints are considered. Such paths are called fair paths. A fairness constraint can be sets of states or CTL formulas, which we call *fair state sets* and *fair formulas* respectively. A fair path traverses each of the fair state sets infinitely often, or each of the fair formulas hold infinitely often along the path.

In the following discussion, let's assume fairness constraints are expressed as a set of states. The result can be extended to sets of states or a set of fair formulas. As we discussed in a previous section, CTL formulas with fairness constraints cannot be expressed as CTL formulas. Thus, we have to enhance our CTL checking algorithms to deal with fairness constraints.

Because we are dealing with an infinite path in a finite structure, a fair path is a path that emanates from a state and reaches an SCC that contains the fair states or satisfies fair formulas. Therefore, a fair path starting from a node can be identified in three steps:

1. Find all SCCs.
2. Remove the SCCs that do not contain the fair states.
3. Determine whether the state reaches one of the remaining SCCs.

If step 3 succeeds, then there is a fair path starting from the state. These three steps form an algorithm for determining all states that have a fair path. If we define a fairness variable *fair* such that it is true at a state if there is a fair path starting from the state, then this algorithm verifies formula *fair* (in other words, it labels all states in which *fair* is true).

Because all CTL operators can be expressed in terms of the three operators $EX(f)$, $E(fUg)$, and $EG(f)$, we just consider checking these three with fairness constraints. As usual, we need to assume $f$ and $g$ have been processed according to fair semantics.

For $EX(f)$ to be true at a state, there must be a fair path starting at the state and $f$ holds at the next state. We then make use of the fairness variable $fair$. Note that if $fair$ holds at state $s$, then $fair$ also holds at all states that can reach $s$. Then, $EX(f)$ is true with fairness constraints if and only if $f$ holds at the next state and there is a fair path starting from the next state. In other words, $EX(f)$ holds with fairness constraints if and only if $EX(f \cdot fair)$ holds in the ordinary CTL semantics without fairness. That is, we can use an ordinary CTL checker to verify $EX(f \cdot fair)$ with a passing that means $EX(f)$ holds with fairness constraints.

Formula $E(fUg)$ with fairness constraints means that there is a fair path along which $fUg$ holds. If a such fair path exists and $fUg$ holds along the path, then $g \cdot fair$ also holds. Therefore, $E(fUg)$ holds with fairness constraints if and only if $E(fU(g \cdot fair))$ holds in the ordinary CTL semantics without fairness constraints.

The procedure to check $EG(f)$ at state $s$ with fairness constraints is as follows: First, all states without label $f$ are removed. Second, all SCCs are determined. Third, all SCCs that do not contain the fair states are removed. Finally, we need to determine whether state $s$ reaches one of the remaining SCCs. If the last step succeeds, it implies that state $s$ reaches an SCC that contains all fair states. Then, the path starting at $s$, reaching this SCC, and traversing the fair states infinitely often is a fair path along which $f$ holds globally. Therefore, $EG(f)$ holds at $s$ subject to the fairness constraints.

## 9.3   Symbolic Computation and Model Checking

With the model-checking techniques discussed earlier, the Kripke structure is assumed to be explicitly represented by a graph: Each state is represented by a node and each transition is represented by an edge. This method of representation and the associated checking algorithms are called *explicit model checking*, and are infeasible for circuits with a large number of state elements. A circuit with 100 FFs amounts to $2^{100}$ states. If each node is represented with 4 bytes in a computer program, it would take $4 \times 2^{100}$ or approximately $4 \times 10^{30}$ bytes just to read in the circuit. Note that a terabyte is only $10^{12}$. Runtime-wise, if a node takes one nanosecond to process, it would take $10^{30} \times 10^{-9} = 10^{21}$ seconds or $3.17 \times 10^{13}$ years to read in the circuit. Today, it is common to have more than a million FFs in a circuit. Therefore, to be able to handle sizable circuits, implicit representation and algorithms must be used. With an *implicit* representation, graphs and their traversal are converted to Boolean functions and Boolean operations. Therefore, all model-checking operations are operations on Boolean functions. Using symbolic methods, properties on circuits with hundreds of state elements have been formally verified.

Let's first study the symbolic representation of finite-state machines and then symbolic state traversal. Then we will apply these implicit techniques to verification by language containment and CTL model checking.

### 9.3.1 Symbolic Finite-State Machine Representation and State Traversal

Recall that a finite-state machine is a quintuple, $(Q, \Sigma, \delta, q_0, F)$, where $Q$ is the set of states, $\Sigma$ is the set of input symbols, $\delta$ is the transition function with $\delta : Q \times \Sigma \to Q$, $q_0$ is an initial state, and $F$ is a set of states designated as acceptance or final states. The transition function $\delta$ is based on the current state and an input symbol, and it produces the next state. Another way to represent $\delta$ is with $q^{i+1} = \delta(q^i, a)$, where $q^i$ is the present state at cycle $i$, $q^{i+1}$ is the next state, and $a \in \Sigma$ is an input symbol. The key to model and traverse a finite-state machine symbolically lies in representing a set of states as well as the transition function as characteristic functions.

A characteristic function $Q(r)$ for a set of states $S$ is such that

$$Q(r) = \begin{cases} 1, r \in S \\ 0, r \notin S \end{cases}$$

If each state is encoded in bits, then the characteristic function is just a disjunction of the terms representing all the states in the set. For instance, if eight states are encoded with 3 bits—$a$, $b$, and $c$—the characteristic function $Q(a, b, c)$ representing the set consisting of state 101 and 110 is $a\bar{b}c + ab\bar{c}$, which is 1 if and only if $abc = 101$ or 110.

To represent the state transition function we would define the characteristic function as follows:

$$T(p, n, a) = \begin{cases} 1, n = \delta(p, a) \\ 0, otherwise \end{cases}$$

where $p$ is a present state, $n$ is a next state, and $a$ is an input symbol. That is, $T(p, n, a)$ is 1 if and only if the present state, next state, and the input symbol satisfy the transition function. This method of representation using characteristic functions is called *relational representation*. Thus, $T(p, n, a)$ is sometimes called a *transition relation*. Another method of representation is *functional representation*: The transition function is represented as $q^{i+1} = \delta(q^i, a)$. An important benefit gained through the use of characteristic functions is that all the arguments ($p$, $n$, and $a$) can be a set of states and a set of input symbols, as opposed to a single state. In contrast, a transition function representation cannot represent the situation in which more than one next state exists for a given present state and an input symbol.

To demonstrate symbolic traversal, we need to compute the set of all possible next states, $N(s)$, from a given set of present states, $P(s)$, under all input symbols. $N(s)$ is a characteristic function for the set of next states; thus, $s$ is a next state if there is a present state $p$ and an input symbol such that $T(p, s, a) = 1$. Because $p$ is a present state, $P(s) = 1$. The statement "there is a" translates into the existential quantifier. Putting these together symbolically, we have

$$N(s) = \begin{cases} 1, \exists(p, a)(T(p, s, a) \cdot P(p) = 1) \\ 0, otherwise \end{cases}$$

---

**Example 9.10**

Consider the circuit in Figure 9.17 and compute the following:

  **1.** Derive a relational representation for the transition function.
  **2.** Compute the next state if the present state is 00 and the input symbol is 10.
  **3.** Compute the set of all next states if the present state is either 00 or 11.

Based on the circuit diagram we can relate the next-state bits $s_1$ and $s_2$ to present-state bits $p_1$ and $p_2$, and input bits $a_1$ and $a_2$, as

$$s_1 = \bar{s}_2 + \bar{a}_1 a_2 \overline{(p_1 p_2)}$$

$$s_2 = p_1 + a_2$$

To make these equations into a characteristic function, or a relation, we form a function that becomes 1 if and only if the right-hand side of the previous equation is equal to the left-hand side. Therefore, the transition relation is

$$T(p_1, p_2, s_1, s_2, a_1, a_2) = \overline{(s_1 \oplus (\bar{s}_2 + \bar{a}_1 a_2 \overline{(p_1 p_2)}))} \overline{(s_2 \oplus (p_1 + a_2))}$$

**Example 9.10    (Continued)**

The next state, given the present state 00 and input 10, is $T(0, 0, s_1, s_2, 0, 1)$, which simplifies to $s_1 \cdot s_2$. Thus, the next state $s_2 s_1$ is 11, consistent with a calculation from the circuit diagram.

For step 3 we first must represent the set of present states, 00 and 11, with a characteristic function: $P(p_1, p_2) = \bar{p}_1 \bar{p}_2 + p_1 p_2$. The set of all next states for all possible inputs is

$$N(s_1, s_2) = \exists(a_1, a_2, p_1, p_2) T(p_1, p_2, s_1, s_2, a_1, a_2) P(p_1, p_2)$$

Recall that $\exists(x)f(x) = f_x + f_{\bar{x}}$. Thus, the previous expression becomes

$$
\begin{aligned}
&\exists(a_1, a_2, p_1, p_2) T(p_1, p_2, s_1, s_2, a_1, a_2) P(p_1, p_2) \\
&= (T(p_1, p_2, s_1, s_2, 0, 0) + T(p_1, p_2, s_1, s_2, 0, 1) \\
&+ T(p_1, p_2, s_1, s_2, 1, 0) + T(p_1, p_2, s_1, s_2, 1, 1)) P(p_1, p_2)
\end{aligned}
$$

After simplification, it becomes

$$N(s_1, s_2) = \exists(p_1, p_2)(s_1 \bar{p}_1 \bar{p}_2 + s_2 \bar{p}_1 \bar{p}_2 + s_2 p_1 p_2) = s_1 + s_2$$

Because $N(s_1, s_2)$ is a characteristic function for next states, any values of $(s_1, s_2)$ that make $N(s_1, s_2)$ evaluate to 1 are next states. Therefore, the next states $(s_1, s_2)$ are 10, 01, and 11. I leave you to confirm that the set of next states is correct. To appreciate the elegance of relational representation, it is an interesting exercise to repeat these computations using functional representation.

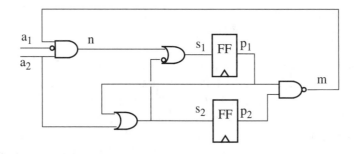

**Figure 9.17**  A circuit for symbolic traversal

The transition function of a finite-state machine takes in a set of present states $P(s)$ and produces a set of next states $N(s)$. $N(s)$ is called the *image* of $P(s)$ under the transition function. Conversely, $P(s)$ is called the *preimage* of $N(s)$. Starting from a set of initial states $I(s)$, we can iteratively compute the image of the initial states, then the image of the image just computed, and so on. When the image contains all reachable states, further computations will not yield a larger image. At this stage, the image is identical to the preimage. When this occurs, the computation stops. Image computation is also called *forward traversal*, because each iteration of computing next states is one cycle advance in time. The algorithm computing reachable states from a set of initial states is as follows:

---

**Forward Reachable States by Symbolic Computation**

---

Input: transition relation $T(p, s, a)$ and initial state $I(s)$

Output: a characteristic function $R(s)$ of all reachable states


ReachableState($T$, $I$):

  **1.** Set $S = I$.
  **2.** Compute $N(s) = \exists (p, a)(T(p, s, a) \cdot S(p))$.
  **3.** $R = S + N$.
  **4.** If $R \neq S$, set $S = R$ and repeat steps 2 and 3; otherwise, return $R$.

---

Every step in the previous image computation algorithm is executed as BDD operations, as opposed to graphical operations. First, the transition relation and state set $S$ are represented using BDDs. In step 2, the BDD of $S(p)$ is ANDed with that of $T(p, s, a)$, and the result is existentially quantified, again as a BDD operation. In step 3, checking whether $S = R$ is executed by comparing the BDDs of $S$ and $R$. The result is a BDD for the characteristic function of all reachable states.

To verify that a design satisfies a safety property (such as, an error should never occur), the property is expressed as a set of faulty states such that the property is violated if any of these faulty states is reached. This set of faulty states is represented by a characteristic function in the form of a BDD: $F(s)$. This property is satisfied if and only if none of the bad states is contained in the set of reachable states, $R(s)$. Symbolically, $R(s) \cdot F(s) = 0$. The AND of these two BDDs is identically equal to constant 0. At any iteration of computing $R(s)$, if $(R(s) \cdot F(s)) \neq 0$, then an error state is reachable and the property fails so that no further computation is necessary. Furthermore, if $R(s) \cdot F(s) = D(s) \neq 0$, then BDD $D(s)$ is the set of all reachable faulty states. This method of verification using forward traversal is visualized in Figure 9.18. The crescents represent the states reached at each iteration—$N(s)$. In the diagram, state $e$, a faulty state, is reachable from an initial state at iteration $n$, implying the design contains an error. This verification algorithm is summarized as follows:

---

**Forward Faulty State Reachability Analysis**

---

Input: transition relation $T(p, s, a)$, initial state $I(s)$, and a fault state $F(s)$

Output: a resolution on whether any faulty state is reachable

FaultyStateReachability($T$, $I$, $F$):

**1.** Set $S = I$.
**2.** If $(S \cdot F) \neq 0$, return YES.
**3.** Compute $N(s) = \exists(p, a)(T(p, s, a) \cdot S(p))$.
**4.** $R = S + N$.
**5.** If $R \neq S$, set $S = R$ and repeat steps 2 through 5; otherwise, return NO.

---

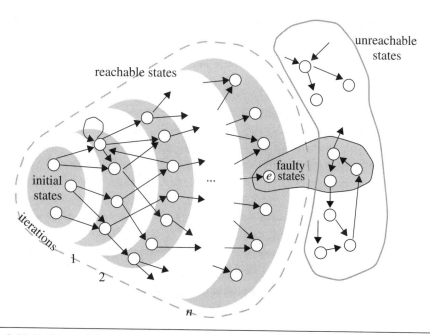

**Figure 9.18**  Verifying safety property using forward traversal

An alternative to using forward traversal is to traverse backward starting from a set of faulty states $F(s)$. Let $B(s)$ be the set of previous states of present state set $P(s)$. Then $B(s)$ is formulated as

$$B(s) = \begin{cases} 1, \exists(p, a)(T(s, p, a) \cdot P(p) = 1) \\ 0, otherwise \end{cases}$$

Compare $B(s)$ with $N(s)$ and note that the positions of the arguments in the transition function are reversed. Backward traversal is exactly the same as forward traversal if $N(s)$ is replaced by $B(s)$, and $I(s)$ by $F(s)$. At any iteration, if $BR(s)$, the backward reachable set, has a nonempty intersection with the initial state set, then it means that a faulty state is reachable from an initial state and, the design has an error. Backward traversal terminates when all backward reachable states have been reached. Namely, $BR(s)$ remains unchanged during an iteration. For completeness, a backward traversal algorithm and a backward faulty state reachability analysis algorithm are summarized as follows:

---

**Backward Reachable States by Symbolic Computation**

---

Input: transition relation $T(p, s, a)$ and a set of states $F(s)$

Output: a characteristic function $BR(s)$ of all states that reach $F(s)$

ReachableState($T$, $F$):

   **1.** Set $S = F$.
   **2.** Compute $B(s) = \exists(p, a)(T(s, p, a) \cdot S(p))$.
   **3.** $BR = S + B$.
   **4.** If $BR \neq S$, set $S = BR$ and repeat steps 2 and 3; otherwise, return $BR$.

---

**Backward Faulty State Reachability Analysis**

---

Input: transition relation $T(p, s, a)$, initial state $I(s)$, and a fault state $F(s)$

Output: a resolution on whether any faulty state is reachable

---

**Backward Faulty State Reachability Analysis (Continued)**

---

BackwardFaultyStateReachability($T$, $I$, $F$):

**1.** Set $S = F$.

**2.** If $(S \cdot I) \neq 0$, return YES.

**3.** Compute $B(s) = \exists (p, a)(T(s, p, a) \cdot S(p))$.

**4.** $BR = S + B$.

**5.** If $BR \neq S$, set $S = BR$ and repeat steps 2 through 5; otherwise, return NO.

---

Whether forward or backward traversal first reaches a conclusion depends on the design under verification and is difficult to predict in advance. However, one can combine both forward and backward traversal, called the *hybrid traversal method*. An error is detected whenever $R(s)$ and $F(s)$ or $BR(s)$ and $I(s)$ have a nonempty intersection. The hybrid traversal terminates when either $R(s)$ or $BR(s)$ becomes unchanged.

## 9.3.2 Counterexample Generation

When a verification algorithm declares that an error has been discovered, it should generate an input sequence and an initial state, called a *counterexample*, so that the designer can simulate the input sequence from the initial state to reproduce the error and debug it. Let's assume a forward traversal algorithm was used to detect an error, and $N_i$ is $N$ at iteration $i$. Suppose that a bad state is contained in the $i$th iteration of the traversal. Then, $N_i \cdot F \neq 0$. We then pick a state $e \in N_i \cdot F$ as an error state $E$ and, with a slight modification, apply the backward traversal algorithm to the initial state. As we back trace, we want to limit the preimage of backward traversal to be within the image of the forward traversal that found the error to guarantee that we will reach an initial state. Specifically, in the first iteration of backward traversal, $S = \{e\}$, we compute $BN_1$, the preimage of $S$. Before computing the preimage of $BN_1$ in iteration 2, instead of setting $S = BN_1$, we intersect $BN_1$ with the image at iteration $i - 1$ of the forward traversal, $N_{i-1}$. In other words, we set $S = BN_1 \cdot N_{i-1}$. The reason for performing this intersection is to ensure that we back trace from states that also lie in the forward-traversed image so that an initial state will be reached. Without this intersection, it is possible that we would never reach an initial state from the faulty state, as explained next. We then compute the preimage of $S$. The traversal stops when an initial state is contained in $BN$. Because it took $i$ iterations for the forward traversal to reach a faulty state, it must take at most $i$ iterations to reach an initial state from the faulty state. However, an initial state may be reached in fewer than $i$ iterations.

This algorithm of generating a counterexample is depicted in Figure 9.19. In the figure, the shaded ovals are forward reachable states, whereas the unfilled ovals are backward

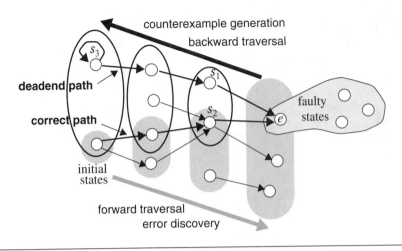

**Figure 9.19** Generating a counterexample for an error found in a forward traversal

reachable states. In three iterations, a faulty state, $e$, is reachable from the initial state. To generate a counterexample from this faulty state, we can back trace from $e$. In the first backward traversal, two states lie in the preimage of $e$, $s_1$, and $s_2$. $s_1$ does not lie in the forward traversal image, but $s_2$ does. We should pick $s_2$ in our next back-tracing iteration to ensure that we will eventually reach an initial state. If we pick $s_1$, we would end up in the deadend state $s_3$ and never reach an initial state, as the upper boldface path indicates. To guarantee that backward-traversed states lie in the forward image, such as $s_2$, it is necessary to intersect the backward preimage with the forward image, which in this case yields $s_2$. From $s_2$, we continue back tracing until it reaches the initial state, as indicated in the lower path.

### 9.3.3 Equivalence Checking

To determine whether two sequential circuits are functionally equivalent, we can just compare their next-state functions using BDDs if their states are in one-to-one correspondence. This was discussed in Section 8.3. However, when two circuits' states are not in one-to-one correspondence, model-checking technique must be used. The two circuits' outputs are pairwise XORed, and the XOR outputs are ORed together as output $D$, as shown in Figure 9.20. Therefore, the two circuits are equivalent if and only if output $D$ is identically 0.

Let transition functions and output functions of the two circuits be $T_1(\vec{s}_1, \vec{n}_1, \vec{x})$, $O_1(\vec{y}, \vec{s}_1, \vec{x})$, $T_2(\vec{s}_2, \vec{n}_2, \vec{x})$, and $O_2(\vec{z}, \vec{s}_2, \vec{x})$ respectively, where $y, z, s_1, s_2, n_1, n_2$, and $x$

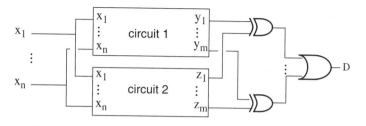

**Figure 9.20** Auxiliary circuit for determining sequential equivalence

are output, present state, next state, and input vectors. The output functions are characteristic functions, that evaluate to 1 if and only if output $y$ or $z$ is the output value at state $s$ and input $x$. Then, the transition function for the product machine of these two circuits is

$$T(\vec{s}_1, \vec{s}_2, \vec{n}_1, \vec{n}_2, \vec{x}) = T_1(\vec{s}_1, \vec{n}_1, \vec{x}) \cdot T_2(\vec{s}_2, \vec{n}_2, \vec{x}).$$

Output $D$ is

$$D(\vec{s}_1, \vec{s}_2, \vec{n}_1, \vec{n}_2, \vec{x}) = O_1(\vec{y}, \vec{s}_1, \vec{x})O_2(\vec{z}, \vec{s}_2, \vec{x})(\vec{y} \neq \vec{z})$$

where $(\vec{y} \neq \vec{z})$ is the characteristic function that evaluates to 1 whenever $\vec{y}$ is not equal to $\vec{z}$. For a single variable, $(\vec{y} \neq \vec{z})$ becomes $y \oplus z$. For vectors, $(\vec{y} \neq \vec{z})$ becomes $\sum y_i \oplus z_i$. Suppose $I_1$ and $I_2$ are initial states of the two circuits. Then the initial state of the product machine is $I = I_1 \cdot I_2$. We can apply the faulty state reachability analysis algorithm to this product machine with faulty state function $D(\vec{s}_1, \vec{s}_2, \vec{n}_1, \vec{n}_2, \vec{x})$. The algorithm evaluates $D(\vec{s}_1, \vec{s}_2, \vec{n}_1, \vec{n}_2, \vec{x})$ at each reachable state. If $D(\vec{s}_1, \vec{s}_2, \vec{n}_1, \vec{n}_2, \vec{x})$ is 1 at a reachable state, the outputs of the two circuits are not equal; therefore, the two circuits are equivalent if and only if the answer from the algorithm is NO.

**Example 9.11**

We can apply the previous algorithm to determine whether the circuits in Figure 9.21 are equivalent, assuming the initial states for both circuits are all zeros. Because the two circuits have different numbers of FFs, there is no one-to-one correspondence between the FFs. Thus, we cannot determine their equivalence by simply comparing their next-state functions. Let's call the circuit in Figure 9.21A circuit 1 and the other circuit 2.

*continues*

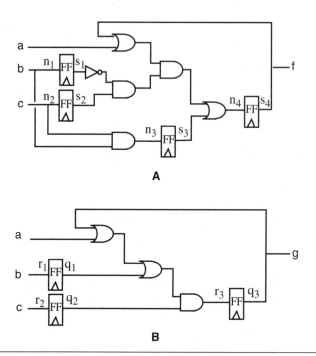

**Figure 9.21** (A, B) Example circuits for determining sequential equivalence

---

**Example 9.11    (Continued)**

The transition function and output function for circuit 1 are

$$T_1(s_1, s_2, s_3, s_4, n_1, n_2, n_3, n_4, a, b, c) = (\overline{n_1 \oplus b})(\overline{n_2 \oplus c})(\overline{n_3 \oplus bc})(\overline{n_4 \oplus (s_3 + \bar{s}_1 s_2(a + s_4))})$$

$$O_1(f, s_4) = \overline{f \oplus s_4}$$

Similarly, the transition function and output function for circuit 2 are

$$T_2(q_1, q_2, q_3, r_1, r_2, r_3, a, b, c) = (\overline{r_1 \oplus b})(\overline{r_2 \oplus c})(\overline{r_3 \oplus q_2(q_1 + q_3 + a)})$$

$$C_2(g, q_3) = \overline{g \oplus q_3}$$

**Example 9.11 (Continued)**

Thus, the transition function for the product machine is

$$T(s_1, s_2, s_3, s_4, q_1, q_2, q_3, n_1, n_2, n_3, n_4, r_1, r_2, r_3, a, b, c)$$
$$= T_1(s_1, s_2, s_3, s_4, n_1, n_2, n_3, n_4, a, b, c)T_2(q_1, q_2, q_3, r_1, r_2, r_3, a, b, c)$$

$$D(f, g, q_3, s_4) = O_1(f, s_4) \cdot C_2(g, q_3) \cdot (f \oplus g)$$
$$= (\overline{f \oplus s_4}) \cdot (\overline{g \oplus q_3}) \cdot (f \oplus g)$$

The initial state is

$$I(s_1, s_2, s_3, s_4, q_1, q_2, q_3) = \bar{s}_1 \bar{s}_2 \bar{s}_3 \bar{s}_4 \bar{q}_1 \bar{q}_2 \bar{q}_3$$

We can then apply the forward reachability verification algorithm. The set of reachable states in one iteration is calculated by enumerating over all present states and inputs, and is accomplished by quantifying over the present state variables and input variables:

$$N_1 = \exists(a, b, c, s_1, s_2, s_3, s_4, q_1, q_2, q_3)(T \cdot I)$$

$$= \exists(a, b, c)(\overline{n_1 \oplus b})(\overline{n_2 \oplus c})(\overline{n_3 \oplus bc})(\overline{r_1 \oplus b})(\overline{r_2 \oplus c})$$
$$\exists(s_1, s_2, s_3, s_4, q_1, q_2, q_3)(\overline{n_4 \oplus (s_3 + s_1 s_2 (a + s_4))})(\overline{r_3 \oplus q_2(q_1 + q_3 + a)})$$

$$= \exists(a, b, c)\bar{n}_4 \bar{r}_3((\overline{n_1 \oplus b})(\overline{n_2 \oplus c})(\overline{n_3 \oplus bc})(\overline{r_1 \oplus b})(\overline{r_2 \oplus c}))$$
$$= \bar{n}_4 \bar{r}_3(n_1 n_2 n_3 r_1 r_2 + \bar{n}_1 n_2 \bar{n}_3 r_1 \bar{r}_2 + \bar{n}_1 \bar{n}_2 \bar{n}_3 \bar{r}_1 \bar{r}_2 + n_1 \bar{n}_2 \bar{n}_3 r_1 \bar{r}_2)$$

For iteration 2, we first replace the next-state variables with present state variables—namely, replacing $n_i$ by $s_i$ and $r_i$ by $q_i$. Therefore, the reachable states after one iteration are

$$R_1 = \bar{s}_4 \bar{q}_3(s_1 s_2 s_3 q_1 q_2 + \bar{s}_1 s_2 \bar{s}_3 q_1 q_2 + \bar{s}_1 \bar{s}_2 \bar{s}_3 \bar{q}_1 \bar{q}_2 + s_1 \bar{s}_2 \bar{s}_3 q_1 \bar{q}_2)$$

Now, we check whether $D(f, g, q_3, s_4)$ has a nonempty intersection with $R_1$ by forming

$$D \cdot R_1 = (\overline{f \oplus s_4}) \cdot (\overline{g \oplus q_3}) \cdot (f \oplus g)$$
$$(\bar{s}_4 \bar{q}_3(s_1 s_2 s_3 q_1 q_2 + \bar{s}_1 s_2 \bar{s}_3 q_1 q_2 + \bar{s}_1 \bar{s}_2 \bar{s}_3 \bar{q}_1 \bar{q}_2 + s_1 \bar{s}_2 \bar{s}_3 q_1 \bar{q}_2))$$

which is identically equal to 0. A program can confirm this by building a BDD. However, we will observe that for the expression to become 1 it is necessary that $s_4 = q_3 = 0$, which implies $f = 0$ and $g = 0$. Thus $f \oplus g = 0$, making the expression 0.

*continues*

**Example 9.11    (Continued)**

For iteration 2, the set of states in two cycles is given by

$$N_2 = \exists(a, b, c, s_1, s_2, s_3, s_4, q_1, q_2, q_3)(T \cdot R_1)$$

After some calculations, we obtain

$$N_2 = (\bar{n}_4 \bar{r}_3 + n_4 r_3)(n_1 n_2 n_3 r_1 r_2 + \bar{n}_1 n_2 \bar{n}_3 r_1 r_2 + \bar{n}_1 \bar{n}_2 \bar{n}_3 \bar{r}_1 r_2 + n_1 \bar{n}_2 \bar{n}_3 r_1 \bar{r}_2)$$

Hence, the set of reachable state after two cycles is, after making the present state variable substitutions,

$$R_2 = R_1 + N_2 = (\bar{s}_4 \bar{q}_3 + s_4 q_3)(s_1 s_2 s_3 q_1 q_2 + \bar{s}_1 s_2 \bar{s}_3 q_1 q_2 + \bar{s}_1 \bar{s}_2 \bar{s}_3 \bar{q}_1 q_2 + s_1 \bar{s}_2 \bar{s}_3 q_1 \bar{q}_2)$$

Here we check whether $D(f, g, q_3, s_4)$ has a nonempty intersection with $R_2$. In this case,

$$D \cdot R_2 = (\overline{f \oplus s_4}) \cdot (\overline{g \oplus q_3}) \cdot (f \oplus g)$$
$$(\bar{s}_4 \bar{q}_3 + s_4 q_3)(s_1 s_2 s_3 q_1 q_2 + \bar{s}_1 s_2 \bar{s}_3 q_1 q_2 + \bar{s}_1 \bar{s}_2 \bar{s}_3 \bar{q}_1 q_2 + s_1 \bar{s}_2 \bar{s}_3 q_1 \bar{q}_2)$$

is identically equal to 0, because $(\bar{s}_4 \bar{q}_3 + s_4 q_3)$ implies $s_4 = q_3$, which forces $f = g$ and $f \oplus g = 0$. Therefore, $D \cdot R_2 = 0$.

For iteration 3, the set of states reachable in three cycles is given by

$$N_3 = \exists(a, b, c, s_1, s_2, s_3, s_4, q_1, q_2, q_3)(T \cdot R_2)$$

and it produces, again after state variable substitutions

$$N_3 = (\bar{s}_4 \bar{q}_3 + s_4 q_3)(s_1 s_2 s_3 q_1 q_2 + \bar{s}_1 s_2 \bar{s}_3 q_1 q_2 + \bar{s}_1 \bar{s}_2 \bar{s}_3 \bar{q}_1 q_2 + s_1 \bar{s}_2 \bar{s}_3 q_1 \bar{q}_2)$$

The states reachable after three cycles are

$$R_3 = R_2 + N_3 = (\bar{s}_4 \bar{q}_3 + s_4 q_3)(s_1 s_2 s_3 q_1 q_2 + \bar{s}_1 s_2 \bar{s}_3 q_1 q_2 + \bar{s}_1 \bar{s}_2 \bar{s}_3 \bar{q}_1 q_2 + s_1 \bar{s}_2 \bar{s}_3 q_1 \bar{q}_2)$$

Because $R_3 = R_2$, we have reached all states and the algorithm stops. Therefore, we can conclude that the two circuits are equivalent.

### 9.3.4   Language Containment and Fairness Constraints

As discussed earlier, one method of model checking is to prove that the language of the design is contained in that of the property. Language containment can be transformed into the problem of deciding whether an automaton has empty language. If an automaton has empty language, then none of the acceptance states is reachable, which is attainable by applying reachability algorithms.

Because most designs are continuously running and do not have final states on which the designs halt, input strings are infinite. To deal with infinite strings, a different kind of acceptance conditions is required. $\omega$-automata accept infinite strings and have the same components as those accepting only finite strings except for the acceptance structure. In a $\omega$-automaton, the set of states or edges that are traversed infinitely often are checked for acceptance. Different kinds of acceptance structures give rise to different kinds of $\omega$-automata. In a Buchi automaton, the acceptance structure is a set of state $A$, and an infinite string is accepted if some of the states in $A$ are visited infinitely often as the string is run. Let $inf(r)$ denote the states visited infinitely often when input string $r$, an infinite string, is run. Then $r$ is accepted if $inf(r) \cap A \neq \varnothing$. The set of all accepted strings is the language of the $\omega$-automaton. Other types of $\omega$-automata are Muller, Robin, and Street, and L-automaton.

Recall that fairness constraints require certain states to be traversed infinitely often. Therefore, fairness constraints can be cast into a form of acceptance conditions of $\omega$-automata. Thus, algorithms in this section apply equally for fairness constraints. Next we will look at the language containment of Buchi automata. Buchi automata are closed under complementation and intersection—meaning that for every Buchi automaton $M$, there exists a Buchi automaton that accepts the complement of the language of $M$ and a Buchi automaton that accepts intersection of the languages of two Buchi automata. Therefore, for Buchi automata, the problem of language containment becomes the problem of determining language emptiness. More precisely, if $M_1$ and $M_2$ are Buchi automata, then there exist Buchi automata $M_p$ and $\overline{M_2}$ such that $L(\overline{M_2}) = \overline{L(M_2)}$ and $L(M_p) = L(M_1) \cap L(\overline{M_2})$. Therefore, $L(M_1) \subseteq L(M_2)$ if and only if $L(M_p) = L(M_1 \cdot \overline{M_2}) = \varnothing$.

A Buchi automaton has empty language if no infinite string traverses any state in $A$, which implies that no initial state can reach an SCC that contains a state in $A$. To be able to reason about this symbolically, we first show how transitive closure is computed symbolically, and then we apply transitive closure to language containment.

Recall that the transitive closure of a state is the set of states reachable from the state. Let $C(s, q)$ be a characteristic function for the transitive closure of state $q$. In other words,

$C(s, q) = 1$ if state $s$ is reachable from $q$ in any number of transitions. Then $C(s, q)$ is calculated in a number of iterations. In the first iteration, $C_1(s, q) = \exists(a)T(q, s, a)$, which is the set of next states. In the second iteration, we compute the image of $C_1(s, q)$ and add to it $C_1(s, q)$—namely,

$$C_2(s, q) = \exists(a, p)T(p, s, a)C_1(p, q) + C_1(s, q)$$

The first term is the reachable state in one step from $C_1$, and, by adding $C_1$, $C_2$ is the set of states reachable in one or two transitions. In the $i$th iteration,

$$C_i(s, q) = \exists(a, p)T(p, s, a)C_{i-1}(p, q) + C_{i-1}(s, q)$$

Because the number of states is finite, there is an integer $n$ such that $C_{n+1}(s, q) = C_n(s, q)$. This happens when all reachable states from $q$ are included. Then, $C(s, q) = C_n(s, q)$.

Once we have computed the transitive closure of a state, the transitive closure of a set of states $Q(s)$ is $C(s, Q) = \exists(p)C(s, p)Q(p)$. The algorithm for computing transitive closure is summarized as follows:

---

**Transitive Closure by Symbolic Computation**

Input: transition relation $T(p, s, a)$ and a set of states $Q(s)$

Output: a characteristic function $C(s)$ for the transitive closure of $Q(s)$

TransitiveClosureStateSet($T$, $Q$): // transitive closure of a set of states

$C(s, q) = $ TransitiveClosureState($T$, $q$);

$C(s, Q) = \exists(p)C(s, p)Q(p)$.

TransitiveClosureState($T$, $q$): // transitive closure of state $q$.

1. Compute $C(s, q) = \exists(a)T(q, s, a)$.
2. Compute $N(s, q) = \exists(p, a)(T(p, s, a) \cdot C(p, q))$.
3. If $C(s, q)$ is *not* equal to $C(s, q) + N(s, q)$,
   then $C(s, q) = C(s, q) + N(s, q)$, and repeats step 2 and 3.
4. Otherwise, return $C(s, q)$.

**Example 9.12**

Compute symbolically the transitive closure of the graph in Figure 9.22. With the states encoding as shown, we can construct the transition relation $T(x, y, r, s)$, where $x$ and $y$ are present state bits, and $r$ and $s$ are next state bits. At state $a$, the next states are $b$ and $c$. Thus, the transition relation contains the term $\bar{x}\bar{y}(\bar{r}s + r\bar{s})$. At state $b$, the next state is $b$, giving $\bar{x}y\bar{r}s$. At state $c$, the next state is $d$, and hence the transition term is $x\bar{y}rs$. Finally, at state $d$, the next state is $b$ or $c$, giving $xy(\bar{r}s + r\bar{s})$. Adding all the terms gives a transition relation of

$$T(x, y, r, s) = \bar{x}\bar{y}(\bar{r}s + r\bar{s}) + \bar{x}y\bar{r}s + x\bar{y}rs + xy(\bar{r}s + r\bar{s}) = (\bar{x} + y)\bar{r}s + x\bar{y}rs + (xy + \bar{x}\bar{y})r\bar{s}$$

Now we need to follow the previous algorithm to compute the transitive closure of state $(xy)$, $C(m, n, x, y)$.

In step 1,

$$C(m, n, x, y) = \exists(a)\, T(x, y, m, n, a) = (\bar{x} + y)\bar{m}n + x\bar{y}mn + (xy + \bar{x}\bar{y})m\bar{n}$$

In step 2,

$$T(v, w, m, n) = (\bar{v} + w)\bar{m}n + v\bar{w}mn + (vw + \bar{v}\bar{w})m\bar{n}$$
$$N(m, n, x, y) = \exists(v, w)(T(v, w, m, n) \cdot C(v, w, x, y))$$
$$= (\bar{m}n + m\bar{n})(C(0, 0, x, y) + C(1, 1, x, y)) + \bar{m}nC(0, 1, x, y) + mnC(1, 0, x, y)$$

And

$$C(0, 0, x, y) = 0, \quad C(0, 1, x, y) = \bar{x} + y, \quad C(1, 0, x, y) = xy + \bar{x}\bar{y}, \quad C(1, 1, x, y) = x\bar{y}$$

It follows that

$$N(m, n, x, y) = (\bar{m}n + m\bar{n})(x\bar{y}) + \bar{m}n(\bar{x} + y) + mn(xy + \bar{x}\bar{y})$$
$$= \bar{m}n + m\bar{n}x\bar{y} + mn(xy + \bar{x}\bar{y})$$

Because $C(m, n, x, y) = (\bar{x} + y)\bar{m}n + x\bar{y}mn + (xy + \bar{x}\bar{y})m\bar{n}$, and $C(m, n, x, y)$ is not equal to $C(m, n, x, y) + N(m, n, x, y)$, we set

$$C(m, n, x, y) = C(m, n, x, y) + N(m, n, x, y)$$

*continues*

**Example 9.12    (Continued)**

Thus,

$$C(m, n, x, y) = ((\bar{x} + y)\overline{m}n + x\bar{y}mn + (xy + \bar{x}\bar{y})m\bar{n}) + (\overline{m}n + m\bar{n}x\bar{y} + mn(xy + \bar{x}\bar{y}))$$
$$= \overline{m}n + m(x + \bar{y})$$

Next we must repeat steps 2 and 3 of the transitive closure algorithm:

$$N(m, n, x, y) = (\overline{m}n + m\bar{n})(C(0, 0, x, y) + C(1, 1, x, y)) + \overline{m}\,\overline{n}C(0, 1, x, y) + m\overline{n}C(1, 0, x, y)$$

And

$$C(0, 0, x, y) = 0, \ \ C(0, 1, x, y) = 1, \ \ C(1, 0, x, y) = x + \bar{y}, \ \ C(1, 1, x, y) = x + \bar{y}$$

Therefore,

$$N(m, n, x, y) = (\overline{m}n + m\bar{n})(x + \bar{y}) + \overline{m}\,\overline{n} + m\bar{n}(x + \bar{y}) = \overline{m}n + m\bar{n}(x + \bar{y})$$

$$C(m, n, x, y) + N(m, n, x, y) = (\overline{m}n + m(x + \bar{y})) + (\overline{m}\,\overline{n} + m\bar{n}(x + \bar{y})) = \overline{m}n + m(x + \bar{y})$$

which is equal to $C(m, n, x, y)$. So we are done, and the transitive closure of state $(xy)$ is $C(m, n, x, y) = \overline{m}n + m(x + \bar{y})$.

As a check, if $xy = 00$ (in other words, state $a$), $C(m, n, 0, 0) = m + n$, which is 1 for states 10, 01, and 11, meaning states $b, c,$ and $d$ are in the transitive closure of state $a$, as expected. If $xy = 01$ (in other words, state $b$), then $C(m, n, 0, 0) = \overline{m}n$, meaning only state $b$ is in its transitive closure. Again, this is consistent with what the graph indicates. If $xy = 10$ (state $c$), then $C(m, n, 1, 0) = m + n$, meaning states $b, c,$ and $d$ are in the transitive closure of state $c$. Finally, If $xy = 11$ (state $d$), then $C(m, n, 1, 1) = m + n$, meaning states $b, c,$ and $d$ are in the transitive closure of state $d$.

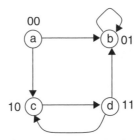

**Figure  9.22** Graph for symbolic computation of transitive closure

Now we can compute SCCs using transitive closure. Recall that an SCC is a set of states such that every state is reachable from every other state in the SCC. Let $SCC(r, s)$ be the characteristic function of the SCC of state $s$. Then $SCC(r, s) = C(r, s)C(s, r)$, where $C(r, s)$ is the transitive closure of state $s$. To understand this equality, refer to Figure 9.23 and consider state $p$ in the SCC of $s$. We want to show that $SCC(p, s) = 1$. Because $p$ is reachable from $s$ and vice versa, $C(p, s)$ and $C(s, p)$ are 1. Hence, $SCC(p, s)$ is 1. Conversely, we want to show that $SCC(p, s) = 1$ implies $p$ is in the same SCC of $s$. $SCC(p, s) = 1$ implies that $C(p, s) = 1$ and $C(s, p) = 1$, which in turn implies $p$ is reachable from $s$ and vice versa. Therefore, $p$ is in the SCC of $s$.

Now we can apply the previous SCC computing algorithm to check language emptiness. To determine whether a Buchi automaton with acceptance state $A(s)$ and initial state $I(s)$ has empty language, we check whether there is an initial state that reaches an SCC that contains a state in $A$. First, $\exists(r)(SCC(s, r)A(r))$ represents all states in an SCC containing an acceptance state. To understand, the formula is 1 if and only if there is a state $r$ that is an acceptance state and it belongs to an SCC. The formula is a characteristic function for all states in the SCC of $r$. Let $C(s)$ be the transitive closure of initial states $I(s)$. Then, similarly, the state of SCCs reachable from an initial state is given by $\exists(q)(SCC(s, q)C(q))$. Putting these two conditions together, the state of SCCs reachable from an initial state and containing an acceptance state is

$$\exists(v)SCC(s, v)\exists(r)(SCC(v, r)C(r))A(v)$$

This formula describes an SCC that contains two states, $r$ and $v$, such that $r$ is reachable from an initial state and $v$ is an acceptance state.

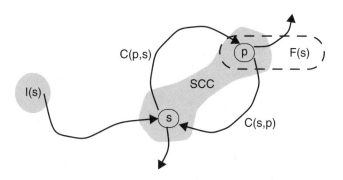

**Figure 9.23** An SCC illustrating the expression of *SCC(r,s)*

Therefore, a Buchi automaton has empty language if and only if the following expression is identically equal to 0:

$$\exists(v)SCC(s, v)\exists(r)(SCC(v, r)C(r))A(v)$$

Let $F(s)$ be a fairness constraint (a characteristic function of a set of fair states). We can determine SCCs that further satisfy $F(s)$. This is accomplished by further quantifying SCC over $F(s)$:

$$\exists(u)SCC(s, u)\exists(v)SCC(u, v)\exists(r)SCC(v, r)C(r)A(v)F(u)$$

The algorithm for language containment of Buchi automata is as follows:

---

**Algorithm for Determining Buchi Automaton Language Emptiness**

---

Input: $M$ is a Buchi automaton with transition relation $T(p, s, a)$, initial state $I(s)$, acceptance state $A(s)$, and fairness condition $F(s)$.

Output: decision of $L(M) = \emptyset$

1. Compute transitive closure $C(r, s)$.
2. Compute $SCC(r, s) = C(r, s)C(s, r)$.
3. $M$ has empty language if and only if

   $$\exists(u)SCC(s, u)\exists(v)SCC(u, v)\exists(r)SCC(v, r)C(r)A(v)F(u)$$

   is identically equal to 0.

---

**Example 9.13**

---

Let us compute all SCCs for the graph in Figure 9.22. The transitive closure is

$$C(m, n, x, y) = \overline{m}n + m(x + \overline{y})$$

$$SCC(m, n, x, y) = C(m, n, x, y)C(x, y, m, n)$$

$$= (\overline{m}n + m(x + \overline{y}))(\overline{x}y + x(m + \overline{n}))$$

$$= mx + \overline{m}n\overline{x}y$$

| Example 9.13 (Continued) |
|---|

As a check, the SCC of node $d$ consists of nodes $c$ and $d$. The SCC of $d$ is given by $SCC(m, n, 1, 1) = m$, meaning nodes 10 or 11 are in the SCC, as expected. The SCC of node $b$ should be itself. Indeed, $SCC(m, n, 0, 1) = \overline{m}n$, which is satisfied only at node $b$. The SCC of node $c$ contains $c$ and $d$. $SCC(m, n, 1, 0) = m$, giving nodes $c$ and $d$. Finally, node $a$ does not belong to any SCC. Indeed, $SCC(m, n, 0, 0) = 0$.

## 9.4 Symbolic CTL Model Checking

Earlier we studied checking CTL formulas using graph traversal algorithms. Building on the knowledge of traversing graphs symbolically—traversing graphs through Boolean function computations—we are in a position to derive symbolic algorithms to check CTL formulas. Keep in mind that the Boolean functions in symbolic algorithms are represented using BDDs. Therefore, all Boolean operations are BDD operations. We will first look at fix-point computation and relate it to CTL formulas. Once expressed using fix-point notation, the representation lends itself to symbolic computation.

### 9.4.1 Fix-point Computation

A function $y = f(x)$ has a *fix-point* if there exists $p$ such that $p = f(p)$, and $p$ is called a *fix-point* of the function. On one hand, not all functions have fix-points. Function $f(x) = x + 1$ in a real domain and range does not have a fix-point. On the other hand, some functions have multiple fix-points. For example, $f(x) = x^2 - 2$ in a real domain and range has two fix-points: $-1$ and $2$. One way of finding a fix-point is to select an initial value of $x$, say $x_0$, and compute iteratively the sequence $x_1 = f(x_0)$, $x_2 = f(x_1)$, ..., $x_{i+1} = f(x_i)$, until $x = f(x)$. This sequence of computations can be regarded as applying mapping $f$ to $x_0$ multiple times until it reaches a fix-point. The choice of $x_0$ plays an important role in determining whether a fix-point will be found and which fix-point will be found.

Let's consider functions with a domain and range that are subsets of states. In other words, functions that take as input a subset of states and produce another subset of states. Let $\tau(S)$ denote a such function and $\tau^i(S)$ be $i$ applications of $\tau$ to $S$. In other words, $\tau^i(S) = \tau(\tau(...\tau(S)))$ $i$ times. Furthermore, let's say that $\tau(S)$ is *monotonic* if $A \subseteq B$ implies $\tau(A) \subseteq \tau(B)$. Let $U$ be the set of all states. Theorem 9.1 provides conditions on $\tau$ so that it has fix-points.

---

**Theorem 9.1**

---

If $\tau$ is monotonic and $U$ is finite, then $\tau^i(\varnothing) \subseteq \tau^{i+1}(\varnothing)$ and $\tau^i(U) \supseteq \tau^{i+1}(U)$. Furthermore, there exist integers $a$ and $b$ such that $\tau^a(\varnothing) = \tau^{a+1}(\varnothing)$ and $\tau^b(U) = \tau^{b+1}(U)$.

---

We can call $\tau^a(\varnothing)$ and $\tau^b(U)$ the least and greatest fix-point of $\tau$, and denote them by $lfp(\tau)$ and $gfp(\tau)$ respectively. Now we can express the six CTL operators using fix-point notation. In Theorem 9.2, for brevity a CTL formula stands for the set of states at which the formula holds. That is, formula $f$ stands for the characteristic function $\lambda(s, f) = \{s | f \, holds \, at \, state \, s\}$. It is for notational brevity that we use this shorthand.

---

**Theorem 9.2**

---

The six CTL operators can be expressed as fix-point computations:

1. $AF(f) = lfp(\tau(S))$, where $\tau(S) = f + AX(S)$.
2. $EF(f) = lfp(\tau(S))$, where $\tau(S) = f + EX(S)$.
3. $AG(f) = gfp(\tau(S))$, where $\tau(S) = f \cdot AX(S)$.
4. $EG(f) = gfp(\tau(S))$, where $\tau(S) = f \cdot EX(S)$.
5. $A(fUg) = lfp(\tau(S))$, where $\tau(S) = g + f \cdot AX(S)$.
6. $E(fUg) = lfp(\tau(S))$, where $\tau(S) = g + f \cdot EX(S)$.

---

To understand the notation in Theorem 9.2, consider $AF(f) = lfp(\tau(S))$, where $\tau(S) = f + AX(S)$. This says that the set of states $AF(f)$ holds is the least fix-point of $\tau(S)$. $\tau(S)$ takes in a set of states $S$ and computes the set of states with next states that satisfy $S$. In other words, in $S$, this result is then ORed with the set of states at which formula $f$ holds. Using characteristic function notation, it is $\lambda(s, AF(f)) = lfp(\tau(S))$ and $\tau(S) = \lambda(s, f) + \lambda(s, AX(S))$. Operators $AX$ and $EX$ can be computed without using a fix-point operator, as

$$\lambda(s, AX(f)) = \forall(n) T(s, n)\lambda(n, f)$$

and

$$\lambda(s, EX(f)) = \exists(n)(T(s, n)\lambda(n, f))$$

where $T(s, n)$ is the transition relation of the underlying Kripke structure. The universal and existential quantifiers are computed as conjunction and disjunction as follows:

$$\lambda(s, AX(f)) = \prod_n T(s, n)\lambda(n, f)$$

and

$$\lambda(s, EX(f)) = \sum_n T(s, n)\lambda(n, f).$$

Function $\tau(S)$ transforms predicates $f$ and $g$ into a function and is called a *predicate transformer*.

To have an intuitive understanding of these identities, we need to examine $AF(f)$. $AF(f)$ holds at a state if all paths from the state will eventually lead to states, including the state itself, at which $f$ holds. The least fix-point computation $AF(f) = lfp(f + AX(S))$ consists of a sequence of operations, as shown here.

In iteration 1, $\tau^1 = f + AX(\varnothing)$. The states included in $\tau^1$ are the states at which $f$ holds and the states at which $AX(\varnothing)$ holds. $AX(\varnothing)$ contains the states with next states that satisfy $\varnothing$, which is not satisfiable. Thus, $AX(\varnothing)$ gives $\varnothing$. Therefore, the states returned by $\tau^1$ are the set of states at which $f$ holds—$\lambda(s, f)$. Thus, states in $\tau^1$ satisfy $AX(f)$.

In iteration 2, $\tau^2 = f + AX(f)$. The states included are those that satisfy $f$ and all the next states that satisfy $f$. Thus, $\tau^2$ satisfies $AX(f)$.

In the $i$th iteration,

$$\tau^i = f + AX(f) + AX(AX(f)) + \dots = f + AX(f) + AX^2(f) + \dots + AX^{i-1}(f)$$

Thus, a state is in $\tau^i$ if it satisfies $f$ or all paths from it reach states that satisfy $f$ in $i$ or fewer transitions. Therefore, states in $\tau^i$ satisfy $AX(f)$. Hence, the states in the least fix-point of $\tau(S) = f + AX(S)$ satisfy $AX(f)$. Figure 9.24 shows a Kripke structure and illustrates the progressive inclusion of states through iterations in fix-point computation. The solid circles represent included states, and states labeled $f$ satisfy $f$.

Conversely, consider state $v$ that all its paths eventually lead to states satisfying $f$. Let $n$ be the earliest time when all paths from $v$ have reached states satisfying $f$. Then $v$ must be included in $\tau^n$, which is a subset of the least fix-point of $\tau(S)$. Therefore, $v$ must be in $lfp(\tau(S))$.

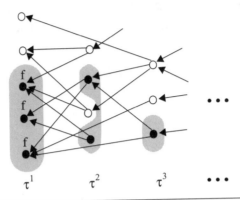

**Figure 9.24** Successive inclusion of states satisfying *AF*(*f*) in fix-point computation

Similarly, $AG(f) = gfp(\tau(S))$, where $\tau(S) = f \cdot AX(S)$ expands to

$$f \cdot AX(f) \cdot AX^2(f) \cdot AX^3(f)\ldots$$

which implies that a state $v$ satisfies $AG(f)$ if it satisfies $f$, $AX(f)$, $AX^2(f)$, and so on, which is equivalent to saying that $v$ satisfies $f$, and all its next states satisfy $f$, and all its next next states satisfy $f$, and so on. This is just what $AG(f)$ is. Other identities can be interpreted similarly.

To complete our treatment of symbolic checking of CTL formulas, we need to examine the case when the formula $f$ is a Boolean formula without CTL operators. We can compute $\lambda(s, f)$ from the bottom up using the following identities:

$$\lambda(s, \neg f) = \overline{\lambda(s, f)}$$

and

$$\lambda(s, f + g) = \lambda(s, f) + \lambda(s, g)$$

Note that characteristic functions are represented using BDDs. Therefore, fix-point computation, complementation, and ORing of characteristic functions are executed as efficiently as BDD operations.

### Example 9.14

Let's take a look at $E(aU(b + \bar{c}))$ in the Kripke structure in Figure 9.25. We use 3 bits—$q_2$, $q_1$, and $q_0$—to encode the eight states. For simplicity, the code of a state is the binary of the state number. For example, $s_6$ has code 110. The transition relation is

$$T(p, n) = (p = s_0)(n = s_2 \vee s_7) + (p = s_1)(n = s_0 \vee s_2) + (p = s_2)(n = s_3) +$$
$$(p = s_3)(n = s_4 \vee s_6) + (p = s_4)(n = s_5) + (p = s_5)(n = s_1) +$$
$$(p = s_6)(n = s_5 \vee s_7) + (p = s_7)(n = s_5)$$

Plugging in the state bits, we obtain

$$T(p_0, p_1, p_2, n_0, n_1, n_2) = \bar{p}_0\bar{p}_1\bar{p}_2(\bar{n}_0 n_1 \bar{n}_2 + n_0 n_1 n_2) + \bar{p}_0\bar{p}_1 p_2(\bar{n}_0 \bar{n}_1 \bar{n}_2 + \bar{n}_0 n_1 \bar{n}_2)$$
$$+ \bar{p}_0 p_1 \bar{p}_2(\bar{n}_0 n_1 n_2) + \bar{p}_0 p_1 p_2(n_0 \bar{n}_1 \bar{n}_2 + n_0 n_1 \bar{n}_2) + p_0 \bar{p}_1 \bar{p}_2(n_0 \bar{n}_1 n_2) + p_0 \bar{p}_1 p_2(\bar{n}_0 \bar{n}_1 n_2)$$
$$+ p_0 p_1 \bar{p}_2(n_0 \bar{n}_1 n_2 + n_0 n_1 n_2) + p_0 p_1 p_2(n_0 \bar{n}_1 n_2)$$

Notice the complexity of the transition function. It is a major source of computational cost. This transition function is used in computing $EX(f)$, as in

$$\lambda(s, EX(f)) = \sum_n T(s, n)\lambda(n, f)$$

To compute $E(aU(b + \bar{c}))$, we first compute $f = a$ and $g = b + \bar{c}$, and then $E(fUg)$. The set of the states satisfying atomic formulas, $a$, consists of $s_2$, $s_3$, $s_6$, and $s_7$. So, the characteristic function $\lambda(q_0, q_1, q_2, a)$ is

$$\bar{q}_2 q_1 \bar{q}_0 + \bar{q}_2 q_1 q_0 + q_2 q_1 \bar{q}_0 + q_2 q_1 q_0 = q_1$$

Similarly, $\lambda(q_0, q_1, q_2, b) = \bar{q}_2 \bar{q}_1 \bar{q}_0 + q_2 \bar{q}_1 \bar{q}_0$ and

$$\lambda(q_0, q_1, q_2, c) = q_2 q_1 \bar{q}_0 + q_2 \bar{q}_1 q_0 + q_2 \bar{q}_1 \bar{q}_0 + \bar{q}_2 q_1 q_0 + \bar{q}_2 q_1 \bar{q}_0 + \bar{q}_2 \bar{q}_1 q_0$$
$$= q_1 \bar{q}_0 + q_2 \bar{q}_1 + \bar{q}_2 q_0$$

From these atomic propositions, we can compute the following:

$$\lambda(q_0, q_1, q_2, \bar{c}) = \overline{q_1 \bar{q}_0 + q_2 \bar{q}_1 + \bar{q}_2 q_0}$$
$$= \bar{q}_2 \bar{q}_1 \bar{q}_0 + q_2 q_1 q_0$$

$$\lambda(q_0, q_1, q_2, b + \bar{c}) = \lambda(q_0, q_1, q_2, b) + \lambda(q_0, q_1, q_2, \bar{c})$$
$$= q_2 q_1 q_0 + \bar{q}_1 \bar{q}_0$$

*continues*

**Example 9.14    (Continued)**

Now, we have $\lambda(q_0, q_1, q_2, f) = q_1$ and $\lambda(q_0, q_1, q_2, g) = q_2 q_1 q_0 + \overline{q_1}\overline{q_0}$. As a check, we label the states with $f$ and $g$, as in Figure 9.26A. It can be seen that both $\lambda(q_0, q_1, q_2, f)$ and $\lambda(q_0, q_1, q_2, g)$ give the correct sets of states.

Next, we use the fix-point formulation to compute $E(fUg)$. That is, $E(fUg) = lfp(\tau(S))$, where $\tau(S) = g + f \cdot EX(S)$.

In iteration 1, we have $\tau^1(S) = g + f \cdot EX(\varnothing) = \lambda(q_0, q_1, q_2, g) = q_2 q_1 q_0 + \overline{q_1}\overline{q_0}$. The states included in this iterations are 000, 100, and 111 (or, $s_0$, $s_4$, and $s_7$). They are shaded in Figure 9.26A. This iteration finds the states only satisfy $g$.

In iteration 2,

$$\tau^2(S) = g + f \cdot EX(\tau^1) = \lambda(q_0, q_1, q_2, g) + \lambda(q_0, q_1, q_2, f) EX(g)$$

where

$$EX(g) = \exists(t_0, t_1, t_2) T(q_0, q_1, q_2, t_2, t_1, t_0)(\lambda(t_0, t_1, t_2, g))$$
$$= q_2 q_1 \overline{q_0} + \overline{q_2} q_0$$

After simplification, $\tau^2(S) = q_2 q_1 + \overline{q_1}\overline{q_0} + q_1 q_0$. The states are 000, 011, 100, 110, and 111. They are shaded in Figure 9.26B. This iteration includes states that are labeled $g$ or $f$, and those with a next state labeled $g$. The former is a result of iteration 1, and the latter is a result of the current iteration. Because $\tau^2(S)$ is not equal to $\tau^1(S)$, we continue.

In iteration 3,

$$\tau^3(S) = g + f \cdot EX(\tau^2) = \lambda(q_0, q_1, q_2, g) + \lambda(q_0, q_1, q_2, f) EX(\tau^2)$$

$$EX(\tau^2) = \exists(t_0, t_1, t_2) T(q_0, q_1, q_2, t_2, t_1, t_0)(t_2 t_1 + \overline{t_2}\overline{t_0} + t_1 t_0)$$
$$= \overline{q_2} + q_1 \overline{q_0}$$

Substituting the expressions, we get $\tau^3(S) = q_1 + \overline{q_0}$, which is not equal to $\tau^2(S)$. The states are 000, 010, 111, 100, 110, and 111 (or, $s_0$, $s_2$, $s_3$, $s_4$, $s_6$, and $s_7$). They are shaded in Figure 9.27. The newly added state $s_2$ is two cycles from a state labeled $g$.

In iteration 4,

$$\tau^4(S) = g + f \cdot EX(\tau^3) = \lambda(q_0, q_1, q_2, g) + \lambda(q_0, q_1, q_2, f) EX(\tau^3)$$

**Example 9.14    (Continued)**

$$EX(\tau^3) = \exists (t_0, t_1, t_2) T(q_0, q_1, q_2, t_2, t_1, t_0)(t_1 + \bar{t_0})$$
$$= \bar{q_2} + q_1 \bar{q_0}$$

Substituting the expressions, we get $\tau^4(S) = q_1 + \bar{q_0}$, which is equal to $\tau^2(S)$.

Therefore, we have reached a fix-point. The characteristic function for the states satisfying $E(aU(b + \bar{c}))$ is $q_1 + \bar{q_0}$. The states are $s_0$, $s_2$, $s_3$, $s_4$, $s_6$, and $s_7$.

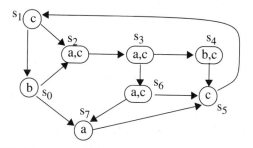

**Figure 9.25** A Kripke structure for symbolically checking $E(aU(b + \bar{c}))$

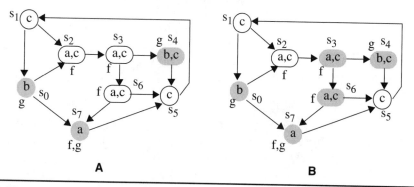

A                                    B

**Figure 9.26** Progressive inclusion of states satisfying $E(aU(b + \bar{c}))$. (A) Kripke structure after $f$ and $g$ are verified. (B) Kripke structure after two iteration of $\tau(s) = g + f \cdot EX(S)$.

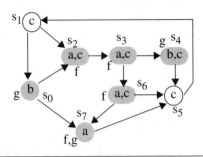

**Figure 9.27** States attained in iteration 3 in computing $E(aU(b + \bar{c}))$

## 9.4.2   CTL Checking with Fairness Constraints

A fairness constraint $\phi = \{\phi_1, ..., \phi_n\}$ mandates that all its formulas, $\phi_1, ..., \phi_n$, be satisfied infinitely often along a path. A path satisfying $\phi$ is called a *fair path*. Checking a CTL formula $f$ with a fairness constraint $\phi$ means that only fair paths are considered and we find all states satisfying $f$ along the fair paths. Because all CTL operators can be expressed in three base operators ($EX$, $EU$, and $EG$), we will study checking of these three operators with fairness constraints. To distinguish CTL checking from CTL checking with fairness constraints, we denote checking with fairness with a subscript $\phi$, as in $EX_\phi(f)$, $E_\phi(fUg)$, and $EG_\phi(f)$. We first compute $EG_\phi$, then $EX_\phi$ and $EU_\phi$, which makes use of the algorithm of $EG_\phi$.

$EG_\phi(f)$ means that there exists a path along which $f$ holds at every state, and every formula of $\phi$ holds infinitely often. In Figure 9.28, $EG_\phi(f)$ with $\phi = \{\phi_1, \phi_2\}$ holds at state

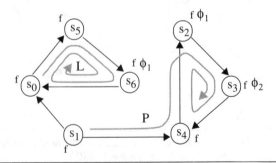

**Figure 9.28** A Kripke structure having states satisfying $EG_\phi(f)$

$s_1$, because path $P$ satisfies $f$ at every state, and it satisfies both $\phi_1$ and $\phi_2$. Similarly, $EG_\phi(f)$ holds at $s_2$, $s_3$, and $s_4$. But $EG_\phi(f)$ fails at $s_0$, $s_5$, and $s_6$, because path $L$, the only path, does not satisfy *both* fairness constraints. $EG_\phi(f)$ can be expressed using fix-point notation and ordinary CTL operators

$$EG_\phi(f) = gfp(\tau(S))$$

where

$$\tau(S) = f \cdot \prod_{i=1}^{n} (EX(E(fU(S \cdot \phi_i))))$$

When we compare $EG_\phi(f)$ with $EG(f) = gfp(f \cdot EX(S))$, we see that an $EU$ operator is nested inside the $EX$ operator. To get an intuitive understanding of the identity, we must assume there is only one fairness formula—$\phi = \{\phi\}$—and unroll it a couple times.

In iteration 1, $\tau^1 = f \cdot EX(E(fU(\phi)))$, which holds at state $s$ if $f$ holds at $s$ and its next state satisfies $fU\phi$. Such a state, $s_0$, is illustrated in Figure 9.29A.

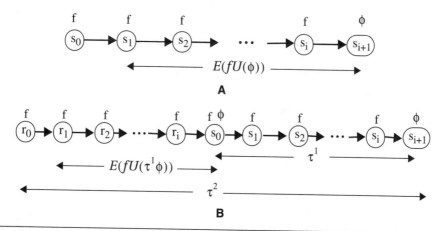

**Figure 9.29** Unrolling the fix-point computation of $EG_\phi(f)$. (A) A sequence of states after one iteration expansion of $EG_\phi(f)$. (B) A sequence of states after two iterations of expanding $EG_\phi(f)$.

In iteration 2, $\tau^2 = f \cdot EX(E(fU(\tau^1 \cdot \phi)))$, which is the same description as $\tau^1$ except for how the path ends—it must end at a state in $\tau^1$, which contains $s_0$. Concatenating the path described by $\tau^2$ with the path for $s_0$ from $\tau^1$, we have the path shown in Figure 9.29B. This path satisfies $f$ at every state and the fairness constraint is satisfied twice. Further iterations simply append paths from the previous iterations. Each iteration only adds states that satisfy $f$ and a state that satisfies $\phi$. At the $n$th iteration, the resulting path satisfies $\phi$ $n$ times. Therefore, the result is a path, or a set of paths, that satisfies $f$ at every state and satisfies $\phi$ infinitely often, which is simply $EG_\phi(f)$.

To compute $EX_\phi(f)$, we use the same technique we introduced in explicit checking of $EX_\phi(f)$—namely, creating variable $fair$, which evaluates true at a state if there is a fair path emanating from the state. Then $EX_\phi(f) = EX(f \cdot fair)$. To obtain the characteristic function for a set of states satisfying $fair$, $\lambda(s, fair)$, we use the identity $\lambda(s, fair) = EG_\phi(1)$. Therefore, the set of states satisfying $EX_\phi(f)$ is the set of states with next states that are in both $\lambda(s, f)$ and $\lambda(s, fair)$. In other words, $EX_\phi(f) = EX(f) \cdot EG_\phi(1)$. Similarly, $E_\phi(fUg) = E(fUg \cdot fair)$, which mandates the ending states also be in $\lambda(s, fair)$.

---

### Example 9.15

If we take a look at Figure 9.30, we use 2 bits $(s_1 s_0)$ to encode the states, and the encoding of a state matches the state's subscript. With fairness constraint $\phi = \{s_1\}$, which indicates that states $q_2$ and $q_3$ are fair states, we can compute $EG_\phi(f)$.

First we calculate the set of states at which $f$ holds. It consists of states $q_0$, $q_1$, and $q_3$, giving $\lambda(s, f) = s_0 + \bar{s}_1$. The state set satisfying the fairness constraint is $s_1$.

In $EG_\phi(f) = gfp(f \cdot EX(E(fU(S \cdot \phi))))$, there are two fix-point computations here. The inner fix-point computation comes from $E(fU(S \cdot \phi))$.

In iteration 1 of the outer fix-point computation, $\tau^1$, we need to compute $f \cdot EX(E(fU\phi))$. The inner fix-point computation is $E(fU\phi) = lfp(\phi + fEX(S))$. This fix-point computation unfolds as follows:

$$\kappa^1 = \phi$$

$$\kappa^2 = \phi + fEX(\kappa^1) = s_1 + (s_0 + \bar{s}_1)EX(s_1) = \bar{s}_0 + s_1$$

$$\kappa^3 = \phi + fEX(\kappa^2) = s_1 + (s_0 + \bar{s}_1)EX(\bar{s}_0 + s_1) = \bar{s}_0 + s_1 = \kappa^2 \text{ (a fix-point)}$$

---

**Example 9.15     (Continued)**

Thus, $E(fU\phi) = \bar{s}_0 + s_1$, giving

$$\tau^1 = f \cdot EX(E(fU\phi)) = (s_0 + \bar{s}_1)EX(\bar{s}_0 + s_1) = \bar{s}_1\bar{s}_0 + s_1s_0$$

In iteration 2 of the outer fix-point computation, we compute $\tau^2 = f \cdot EX(E(fU(\tau^1 \cdot \phi)))$. $\tau^1 \cdot \phi = (\bar{s}_1\bar{s}_0 + s_1s_0) \cdot (s_1) = s_1s_0$. So $\tau^2 = f \cdot EX(E(fU(s_1s_0)))$.

The inner fix-point computation is $E(fU(s_1s_0)) = lfp(s_1s_0 + fEX(S))$.

$$\kappa^1 = s_1s_0$$

$$\kappa^2 = s_1s_0 + fEX(\kappa^1) = s_1s_0 + (s_0 + \bar{s}_1)EX(s_1s_0) = \bar{s}_1\bar{s}_0 + s_1s_0$$

$$\kappa^3 = s_1s_0 + fEX(\kappa^2) = s_1s_0 + (s_0 + \bar{s}_1)EX(\bar{s}_1\bar{s}_0 + s_1s_0) = \bar{s}_1\bar{s}_0 + s_1s_0 = \kappa^2 \text{ (a fix-point)}$$

So, $E(fU(\bar{s}_1\bar{s}_0)) = \bar{s}_1\bar{s}_0 + s_1s_0$.

$\tau^2 = f \cdot EX(E(fU(s_1s_0))) = (s_0 + \bar{s}_1) \cdot EX(\bar{s}_1\bar{s}_0 + s_1s_0) = \bar{s}_1\bar{s}_0 + s_1s_0 = \tau^1$, a fix-point for the outer fix-point computation. Therefore, the entire computation terminates, giving

$$EG_\phi(f) = \bar{s}_1\bar{s}_0 + s_1s_0$$

Based on this result, only states $q_0$ and $q_3$ satisfy $EG_\phi(f)$. Examining Figure 9.30, we can confirm that these two states traverse the self-loop at state $q_3$ satisfy $\phi$ infinitely often and every state along the path also satisfies $f$. State $q_1$ is excluded because it does not satisfy $\phi$. State $q_2$ is omitted because it does not satisfy $f$.

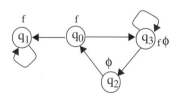

**Figure 9.30** Checking fair CTL formula $EG_\phi(x)$ with fairness constraint $\phi = \{s_1\}$

## 9.5 Computational Improvements

What distinguishes formal verification from simulation-based verification is the efficiency of implicit enumeration of the entire state space. A cornerstone in providing the efficiency is BDDs. Because BDDs are sensitive to variable ordering, almost all commercial formal verification tools at the time of this writing can deal only with moderate-size circuits (for example, circuits at the block level). Accomplishing this level of capacity requires the use of computational efficiency-improving techniques. Getting an understanding of the principles behind these techniques helps the user to take full advantage of formal verification software (for example, by tailoring the circuit to suit the tool's temperament). A formal verification tool developer will also benefit from understanding these techniques.

The key in improving computational efficiency lies in minimizing the BDD size. Whether in language containment or CTL model checking, a crucial component in their symbolic manipulation is computing the image of a set of states—namely,

$$N(\vec{n}) = \exists(\vec{p}, \vec{a})(T(\vec{p}, \vec{n}, \vec{a}) \cdot P(\vec{p}))$$

which produces the set of next states from the set of present states. In language containment, this next-state image computation is used in obtaining transitive closure and state traversal. In CTL model checking, all fix-point computations are merely iterations of either *AX* or *EX*, which is a direct image computation. To appreciate the complexity of transition function, consider the transition function in Example 9.11, as shown here:

$$T(s_1, s_2, s_3, s_4, q_1, q_2, q_3, n_1, n_2, n_3, n_4, r_1, r_2, r_3, a, b, c)$$
$$= T_1(s_1, s_2, s_3, s_4, n_1, n_2, n_3, n_4, a, b, c) T_2(q_1, q_2, q_3, r_1, r_2, r_3, a, b, c)$$

where

$$T_1(s_1, s_2, s_3, s_4, n_1, n_2, n_3, n_4, a, b, c) = (\overline{n_1 \oplus b})(\overline{n_2 \oplus c})(\overline{n_3 \oplus bc})(\overline{n_4 \oplus (s_3 + s_1 s_2(a + s_4))})$$

$$T_2(q_1, q_2, q_3, r_1, r_2, r_3, a, b, c) = (\overline{r_1 \oplus b})(\overline{r_2 \oplus c})(\overline{r_3 \oplus q_2(q_1 + q_3 + a)})$$

The techniques discussed in this section all focus on making this image computation more efficient.

### 9.5.1 Early Quantification

In $N(n) = \exists(p, a)(T(p, n, a) \cdot P(p))$, a straightforward approach is to compute the conjunction of $T(p, n, a)$ and $P(p)$ first, and then quantify the result over present state $p$ and input $a$. The result from the conjunction before quantification can be so massively complex that it exceeds allotted memory capacity. Because quantification eliminates variables, the sooner a function is quantified, the smaller the function becomes, and the more efficient the rest of computation involving the function gets. One may observe that $P(p)$ does not involve input variable $a$ and hence it can be taken out from the quantification of variable $a$. That is, $N(n) = \exists(p, a)(T(p, n, a) \cdot P(p))$ becomes $N(n) = \exists p(\exists a T(p, n, a)) \cdot P(p))$. This is a consequence of the following identity:

$$\exists x(f(x, y) \cdot g(y)) = f(1, y)g(y) + f(0, y)g(y) = (f(1, y) + f(0, y))g(y) = (\exists x f(x, y)) \cdot g(y)$$

This early quantification of input $a$ on $T(p, n, a)$ is performed once before image computation begins and its result is reused thereafter for all image computations. Therefore, with early quantification, image computation takes two steps: (1) compute $\tau(p, n) = \exists a T(p, n, a))$ and (2) $N(n) = \exists p(\tau(p, n) \cdot P(p))$. Because $\tau(p, n)$ has fewer variables than $T(p, n, a)$, the BDD size of $\tau(p, n)$ is smaller than that of $T(p, n, a)$. $\tau(p, n)$ is the characteristic function for the set of all legal transitions from present state $p$ to next state $n$.

---

**Example 9.16**

We can compare the sizes of the following transition relation with and without early quantification, $T(p, q, r, s, a, b) = (\overline{r \oplus a + p})(\overline{s \oplus bpq})$, where $a$ and $b$ are inputs.

Without early quantification, it is

$$\begin{aligned}
T &= (r(a + p) + \bar{r} \cdot (\overline{a + p}))(sbpq + s(\overline{bpq})) \\
&= abrspq + brspq + a\bar{b}rs + \bar{b}rsp + \bar{a}\bar{b}rsp + ars\bar{p} + \bar{a}rs\bar{p} + ars\bar{q} + rsp\bar{q} + \bar{a}rs\bar{p}\bar{g}
\end{aligned}$$

With early quantification over input variables $a$ and $b$, it is

$$\begin{aligned}
\tau &= \exists(a, b)T(p, q, r, s, a, b) \\
&= \exists(b)(r + \bar{p})(s \oplus bpq) \\
&= (r + \bar{p})(s + \underline{pq}) \\
&= rs + rpq + \bar{p}s
\end{aligned}$$

The ratio of the numbers of terms between the two is 10:3.

A further improvement of this algorithm comes from the observation that the transition relation can be partitioned into a conjunction of transition relations of the state bits. That is, to build a transition relation for a circuit having $n$ state bits, one can first construct transition relations for each of the state bits, $t_i(n_i, \vec{p}, \vec{a})$, and AND these bit transition relations together to obtain the overall transition relation:

$$T(\vec{p}, \vec{n}, \vec{a}) = \prod_{i=1}^{n} t_i(\vec{p}, n_i, \vec{a})$$

The early quantification algorithm just described applies to the overall transition relation after the bit transition relations have been ANDed together. Instead of beginning quantification only after the overall transition relation is available, an improvement is to apply early quantification on the bit transition relations as soon as possible. Early quantification on bit transition relations is possible if not all bit transition relations have the same support (in other words, some variables are missing in some bit transition relations). Then, early quantification can be applied to these missing variables. As an example, consider two bit transition relations $t_1(p_1, n_1, a_1, a_2)$ and $t_2(p_1, n_1, a_2)$. Because variable $a_1$ is missing in $t_2(p_1, n_1, a_2)$, we quantify $a_1$ on $t_1(p_1, n_1, a_1, a_2)$ early, as shown here:

$$\begin{aligned}N(n_1) &= \exists(p_1, a_1, a_2)T(p_1, n_1, a_1, a_2)P(p_1) \\ &= \exists(p_1, a_1, a_2)(t_1(p_1, n_1, a_1, a_2)t_2(p_1, n_1, a_2)P(p_1)) \\ &= \exists(p_1, a_2)((\exists(a_1)t_1(p_1, n_1, a_1, a_2))t_2(p_1, n_1, a_2)P(p_1)) \\ &= \exists(p_1, a_2)(r(p_1, n_1, a_2)t_2(p_1, n_1, a_2))P(p_1)\end{aligned}$$

where $r(p_1, n_1, a_2) = \exists(a_1)t_1(p_1, n_1, a_1, a_2)$

It should be clear that the correctness of the equations is again guaranteed by the identity

$$\exists x(f(x, y) \cdot g(y)) = (\exists xf(x, y)) \cdot g(y)$$

If the missing variable is a state variable, $p_i$, the present state set $P(p)$ also has to be included in early quantification. Suppose the 2-bit transition relations are $t_1(p_1, p_2, n_1, a_1)$ and $t_2(p_2, n_1, a_1)$, and the present state function is $P(p_1, p_2)$. Then, early quantification on $p_1$ has to include $P(p_1, p_2)$:

$$\begin{aligned}N(n_1) &= \exists(p_1, p_2, a_1)T(p_1, p_2, n_1, a_1)P(p_1, p_2) \\ &= \exists(p_1, p_2, a_1)(t_1(p_1, p_2, n_1, a_1)(t_2(p_2, n_1, a_1))P(p_1, p_2)) \\ &= \exists(p_2, a_1)((\exists(p_1)t_1(p_1, p_2, n_1, a_1)P(p_1, p_2))t_2(p_2, n_1, a_1)) \\ &= \exists(p_2, a_1)(r(p_2, n_1, a_1)t_2(p_2, n_1, a_1))\end{aligned}$$

where $r(p_2, n_1, a_1) = \exists(p_1)t_1(p_1, p_2, n_1, a_1)P(p_1, p_2)$

Of course, if $p_1$ is also missing in the present state function (in other words, $P(p_2)$), $r(n_1, p_2, a_1)$ will not include $P(p_2)$. This technique of partitioning the transition relation into bit transition relations and applying quantification on the bit functions is sometimes called *transition relation partitioning*.

Extending this idea to a general situation, we have the following algorithm. The idea is to find successively a variable $v$ that partitions the set of functions into one group with supports that contain $v$ and another group with supports that do not. $F^v$ denotes the group of functions $f_1, ..., f_n$ with supports that contain variable $v$, and $F^{-v}$ is the other group that does not contain $v$ in its support. Then, early quantification is applied to functions in $F^v$. In the early phase of the algorithm, there may be many variables satisfying the partitioning requirement. Because it is better to quantify fewer functions during the early stages, a tie-breaking rule is to choose the variable that minimizes the number of functions in $F^v$.

---

**Early Quantification of Conjunction of Functions**

Inputs: a set of functions $f_1, ..., f_n$, and a set of variables $V = \{v_1, ..., v_m\}$

Output: $\exists(V)F$, where $F = f_1 \cdot ... \cdot f_n$

While a $v \in V$ exists that partitions $F$ into $F^v$ and $F^{-v}$

  $V = V - \{v\}$

  $R = \exists v F^v$ and $S = F^{-v}$

  $F = R \cdot S;$ // end of while

$return(\exists(V)F)$

---

## 9.5.2   Generalized Cofactor

In $N(\vec{n}) = \exists(\vec{p}, \vec{a})(T(\vec{p}, \vec{n}, \vec{a}) \cdot P(\vec{p}))$, it was observed that, because of the conjunction of $T(\vec{p}, \vec{n}, \vec{a})$ with $P(\vec{p})$, the entire expression $\exists(\vec{p}, \vec{a})(T(\vec{p}, \vec{n}, \vec{a}) \cdot P(\vec{p}))$ becomes zero when $\vec{p}$ is not in $P(\vec{p})$. Therefore, at the values of $\vec{p}$ that make $P(\vec{p}) = 0$, $T(\vec{p}, \vec{n}, \vec{a})$ can take on any value without affecting the result of $N(\vec{n})$. That is, these values of $\vec{p}$ are don't-cares for $T(\vec{p}, \vec{n}, \vec{a})$. Hence, by optimizing over these don't-care values, a simpler but equivalent expression for $T(\vec{p}, \vec{n}, \vec{a})$ can be obtained. Using this simpler expression for $N(\vec{n})$ improves computing efficiency. This is the essence of the idea behind the generalized cofactor method.

Recall that the generalized cofactor of $f$ with respect to $g$ is defined as

$$f_g = \begin{cases} don't\text{-}care, & if\ g = 0 \\ f, & if\ g = 1 \end{cases}$$

It can be interpreted as the image of $f$ restricted to $g$, or the image of $g$ through $f$. Using this interpretation, the state image computation can be related to the generalized cofactor.

With early quantification, $N(n) = \exists p(\tau(p, n) \cdot P(p))$, where $\tau(p, n) = \exists a T(p, n, a))$. The expression $\exists p(\tau(p, n) \cdot P(p))$ is precisely the image of $P(p)$ under $\tau(p, n)$. Therefore, $N(\vec{n}) = \tau(p, n)_P$.

---

**Example 9.17**

---

We can compute the following image using a generalized cofactor:

$$\tau(p_1, p_2, n_1, n_2) = \bar{n}_1 \bar{n}_2 \bar{p}_1 + \bar{n}_2 p_1 \bar{p}_2 + n_1 p_1 \bar{p}_2$$

and

$$P(p_1, p_2) = \bar{p}_1 + \bar{p}_2$$

We optimize $\tau(p_1, p_2, n_1, n_2)$ with don't-care set $\bar{P}(p_1, p_2) = p_1 p_2$. For this simple function, we use the Karnaugh map as shown in Figure 9.31. The simplified function is

$$\tau^*(p_1, p_2, n_1, n_2) = \bar{n}_1 \bar{n}_2 + n_1 p_1$$

To check, we compare $N(\vec{n}) = \exists \vec{p}(\tau(\vec{p}, \vec{n}) \cdot P(\vec{p}))$ with $N^*(\vec{n}) = \exists \vec{p}(\tau^*(\vec{p}, \vec{n}) \cdot P(\vec{p}))$:

$$N(\vec{n}) = \exists \vec{p}(\tau(\vec{p}, \vec{n}) \cdot P(\vec{p}))$$
$$= \exists \vec{p}((\bar{n}_1 \bar{n}_2 \bar{p}_1 + \bar{n}_2 p_1 \bar{p}_2 + n_1 p_1 \bar{p}_2) \cdot (\bar{p}_1 + \bar{p}_2))$$
$$= \exists p_2(\bar{n}_1 \bar{n}_2 + \bar{n}_2 \bar{p}_2 + n_1 \bar{p}_2)$$
$$= n_1 + \bar{n}_2$$

**Example 9.17    (Continued)**

And

$$N^*(\vec{n}) = \exists \vec{p}(\tau^*(\vec{p}, \vec{n}) \cdot P(\vec{p}))$$
$$= \exists \vec{p}((\bar{n}_1 \bar{n}_2 + n_1 p_1) \cdot (\bar{p}_1 + \bar{p}_2))$$
$$= \exists p_2(\bar{n}_1 \bar{n}_2 + n_1 \bar{p}_2)$$
$$= n_1 + \bar{n}_2$$

Surely, they are identical and $\tau^*(\vec{n})$ is noticeably simpler than $\tau(\vec{n})$.


## 9.6   Using Model-Checking Tools

In this section we will look at practical issues encountered when using model checkers. There are three stages in the cycle of verifying using a model checker: property writing, input constraining, and property debugging.

Before writing properties for a design, the verification engineer has to have a firm grasp of the specifications of the design. This is usually accomplished by reading the specification documentation and communicating with the architects.

Because of memory capacity and runtime limitations, most industrial designs cannot be accepted as a whole by a model checker. Instead, blocks or modules relevant to the

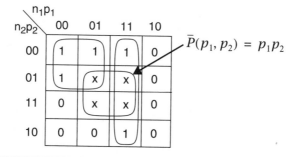

**Figure 9.31** Compute the state image using a generalized cofactor

properties under verification are carved out and run on a model checker. As a result, the user often will have to constrain the inputs of the block to be consistent with those values in the embedded environment. Without constraining to legal inputs, false failures may occur. Because the block is carved out from the design, some of the primary inputs to the block are not primary inputs to the design, but are intermediate nodes. The legal values of these intermediate-turn-primary inputs may not be obvious to the verification engineer, and in most situations they must be determined through debugging false failures. There-fore, the constraining process demands a major engineering effort. Constraining inputs, in essence, builds a model of the environment in which the block resides in the design. If the block is in a feedback loop of the design, constraining amounts to recreating the feedback loop around the block. This use of constraints is illustrated in Figure 9.32.

Once a property fails, the tool will generate a counterexample for the user to simulate to determine why the property failed. By understanding the counterexample, the user determines whether the failure was a design fault or whether it was caused by an illegal input. In the former case, the design is reworked; in the latter case, more input con-straints are added to eliminate the illegal values. Overconstraining, meaning some legal values are prevented from feeding the inputs, can cause false successes, which are far more damaging than false failures. These steps are repeated until all properties have been proved correct.

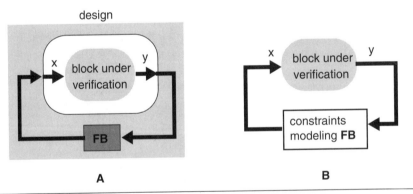

**Figure 9.32** (A) The block under verification is embedded in the design. (B) The block is verified as a stand-alone. The constraints are added to ensure that only legal inputs are given to the block.

## 9.7  Summary

In this chapter we first studied ways to specify a property and discussed property specification using finite-state automata and temporal logic. Along the way, we introduced the computational tree, the Kripke structure, and the fairness condition.

Then we examined graph-based algorithms for checking properties. The cases we considered were checking safety property, checking property using language containment, and checking properties specified in CTL formulas. We also examined the situations when fairness constraints were incorporated.

Once the property-checking algorithms had been introduced we studied symbolic model-checking algorithms using BDDs. First we showed how state traversal and image computation could be accomplished using BDDs. Then we applied these techniques to checking the functional equivalence of sequential circuits, language containment, and CTL formulas, with and without fairness constraints.

Finally, we discussed commonly used techniques for improving symbolic computation: early quantification and generalized cofactor. We concluded by addressing issues encountered in verifying industrial designs using model-checking tools.

## 9.8  Problems

**1.** Derive a Kripke structure for the circuit shown in Figure 9.33.

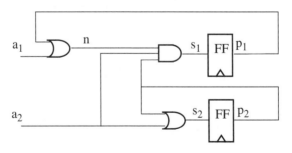

**Figure 9.33** Circuit for creating a Kripke structure

2. Describe the following design properties using CTL:

   a. When the brake fluid level is detected as low in a car, the brake light will eventually turn on.
   b. If the gas pedal is released, the gear may end up in a lower gear.
   c. As long as a driver is going through the sequence—press gas pedal, turn ignition key—the engine will always start.

3. Consider the Kripke structure in Figure 9.34.

   a. Find a path that is fair with respect to fairness condition $\{a\bar{b}, cd\}$.
   b. Write a CTL formula with $AF$ operator that is true at state $S$.
   c. How about using the $EU$ and $AG$ operators?
   d. Write an LTL formula that is true at state $S$ that is not a CTL formula.

4. Construct a finite-state automaton that describes the following properties:

   a. Whenever state $A$ is entered, state B will be reached, at most, five cycles later.
   b. Once state $A$ is visited twice, $B$ is eventually visited twice before state $A$ is visited the third time.

5. Does the automaton in Figure 9.35 satisfy the property that state $X$ will be eventually reached from state $Y$? Which state will eventually be reached from $Y$?

6. Consider the Buchi automaton in Figure 9.36, which has the acceptance condition $A = \{s_3, s_5\}$, and the initial state is $s_6$. Which of the following strings are in its language?

   a. $0(101)^{\text{inf}}$, where $(x)^{\text{inf}}$ means string x is repeated infinitely many times
   b. $(0)^{\text{inf}}$
   c. $(101001)^{\text{inf}}$

7. Construct a Kripke structure such that there is a state at which CTL formula $EG(a \rightarrow A(\bar{b}U(\bar{a}c)))$ holds.

8. Find all states satisfying the following CTL formulas in the Kripke structure of Figure 9.37.

   a. $E(aU(b + \bar{c}))$
   b. $AF(a \rightarrow EGc)$

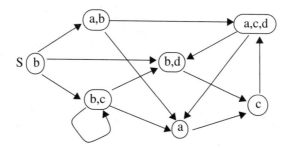

**Figure 9.34** A Kripke structure for illustrating CTL formulas

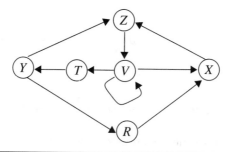

**Figure 9.35** An automaton illustrating eventuality

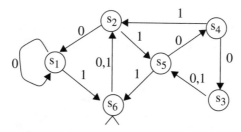

**Figure 9.36** A Buchi automaton for complementation

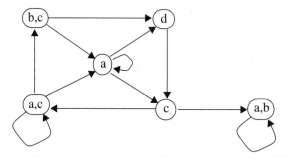

**Figure 9.37** Kripke structure for verifying CTL formulas

9. As introduced in the chapter, variable $fair$ is true at a state if there is a fair path emanating from the state.

    a. Given fairness condition $\{a + d\}$, label all fair states for the Kripke structure in Figure 9.38.

    b. Verify $EG_f(a + \bar{c})$.

10. For the circuit in Figure 9.39, assume the initial state is 00.

    a. Determine the transition relation $T(\vec{p}, \vec{n}, \vec{a})$.

    b. Find all reachable states. What is the longest distance of a state from an initial state? The distance is the number of cycles needed to reach the state from an initial state.

11. Suppose state $F = 10$ is a faulty state. Can the design reach it? If so, derive a counterexample.

12. Determine whether the circuits in Figure 9.40 are equivalent. If not, produce a counterexample. The initial states of FFs in both circuits are zero.

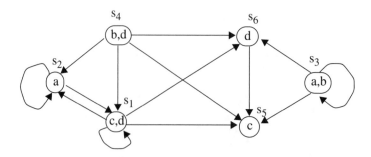

**Figure 9.38** A Kripke structure for evaluating fairness constraints

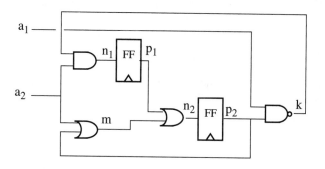

**Figure 9.39** A circuit for reachability analysis

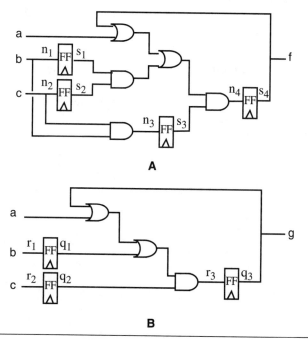

**Figure 9.40** (A, B) Circuits for equivalence checking

**13.** Consider Figure 9.41. Symbolically compute the following:

    a. Determine its transitive closure.

    b. Determine its SCCs.

**14.** Suppose Figure 9.41 is a state diagram of a Buchi automaton (with input symbols omitted) with acceptance state $c$ and initial state $a$. Determine symbolically the following propositions.

    a. Does the automaton have empty language?

    b. Given the fairness constraint consisting of state $b$, represent the fairness constraint using a characteristic function.

    c. Is the language of the automaton empty under the fairness condition?

**15.** Find a function that has neither a least nor a greatest fix-point.

**16.** Use fix-point computation to check CTL formula $E(aU(b+\bar{c}))$ for the Kripke structure in Figure 9.37.

**17.** Repeat problem 16 for $AF(a \rightarrow EGc)$.

**18.** Refer to the Kripke structure in Figure 9.42. Compute symbolically the following:

    a. With fairness constraint $\phi$ as labeled, determine all states at which variable *fair* holds.

    b. Verify $E_\phi(fUg)$.

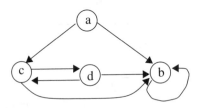

**Figure 9.41** A $\omega$-automaton for symbolic computation and determining language emptiness

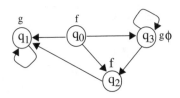

**Figure 9.42** Checking fair CTL formula $EG_\phi(x)$ with fairness constraint $\phi = \{s_1\}$

**19.** Consider the circuit in Figure 9.43.

    a. Derive its transition relation.

    b. Compare the complexity of the transition relation with and without early
       quantification. The measure for complexity is the number of cubes.

**20.** Given transition relation $T(\vec{p}, \vec{n}) = p_1 n_1 n_2 + \bar{p}_1 \bar{p}_2 \bar{n}_2 + p_1 n_1 + \bar{p}_1 \bar{n}_1$ and present
state $P(\vec{p}) = p_1 p_2 + n_1 n_2 + \bar{p}_1 \bar{p}_2 + p_1 n_1,$ use the idea of generalized cofactor to
simplify the transition function.

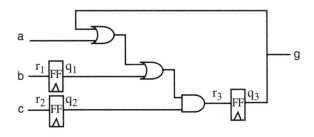

**Figure 9.43** Circuit for early quantification

# Bibliography

1. Abramovici, M., M. A. Breuer, and A. D. Friedman. *Digital Systems Testing and Testable Design*. New York: IEEE Computer Society Press, 1994.
2. Abts, D. "Integrating Code Coverage Analysis into a Large Scale ASIC Design Verification Flow." In *Proceedings of the International HDL Conference*, 1999.
3. Akers, S. B. "Binary Decision Diagrams." *IEEE Transactions on Computers* 37, no. 6 (June 1978): 40–45.
4. Aloul F. and K. Sakallah. "Satometer: How Much Have We Searched?" In *Proceedings of the 39th Design Automation Conference*, 2002.
5. Aloul, F., A. Ramani, I. Markov, and K. Sakallah. "PBS: A Backtrack Search Pseudo-Boolean Solver." In *International Symposium on the Theory and Applications of Satisfiability Testing*, Cincinnati, OH, 2002.
6. Aloul, F., M. Mneimneh, and K. Sakallah. "Search-Based SAT Using Zero-Suppressed BDDs." In *Proceedings of the IEEE/ACM Design, Automation, and Test in Europe*, 2002.
7. Alur, R., S. Kannan, and M. Yannakakis. "Communicating Hierarchical State Machines." In *Automata, Languages and Programming*. 26th International Colloquium. New York: Springer-Verlag, 1999, 169–178.
8. Amdahl, G. "Validity of the Single Processor Approach to Achieving Large Scale Computing Capabilities." In *Proceedings of the Spring Joint Computer Conference*, Washington, D.C., 1967.
9. Anderson, R., P. Beame, S. Burns, W. Chan, F. Modugno, D. Notkin, and J. Reese. "Model Checking Large Software Specifications." *Symposium on the Foundations of Software Engineering*, San Francisco, 1996.
10. Andersson, G., P. Bjesse, B. Cook, and Z. Hanna. "A Proof Engine Approach to Solving Combinational Design Automation Problems." In *Proceedings of the 39th Design Automation Conference*, 2002.
11. Arnold, M., N. J. Sample, and J. D. Schuler. "Guidelines for Safe Simulation and Synthesis of Implicit Style Verilog." In *Proceedings of the International Verilog HDL Conference,*1998.
12. Ashar, P. and S. Malik. "Fast Functional Simulation Using Branching Programs." In *Proceedings of the International Conference on Computer-Aided Design*, 1995.

**13.** Bacchus, F. "Exploring the Computational Tradeoff of More Reasoning and Less Searching." In *International Symposium on Theory and Applications of Satisfiability Testing*, Cincinnati, OH, 2002.

**14.** Bacchus, F. "Enhancing Davis-Putnam with Extended Binary Clause Reasoning." In *Proceedings of the National Conference on Artificial Intelligence*, 2002.

**15.** Bahar, R., E. Frohm, C. Gaona, G. Hachtel, E. Macii, A. Pardo, and F. Somenzi. "Algebraic Decision Diagrams and Their Applications." In *Proceedings of the International Conference on Computer-Aided Design*, 1993.

**16.** Ball, T., R. Majumdar, T. Millstein, and S. Rajamani, "Automatic Predicate Abstraction of C Programs." *ACM SIGPLAN Notices* 36, no. 5 (May 2001): 203–213.

**17.** Ball, T. and S. Rajamani. "Bebop: A Symbolic Model Checker for Boolean Programs." In *Proceedings of SPIN 2000 Workshop on Model Checking of Software*. Lecture Notes in Computer Science, 1885. New York: Springer-Verlag, 2000.

**18.** Barbacci M. and D. P. Siewiorek. "Automated Exploration of the Design Space for Register Transfer (RT) Systems." In *Proceedings of the First Annual Symposium on Computer Architecture*, 1973.

**19.** Barnes, P. and M. Warren. "A Fast and Safe Verification Methodology Using VCS." Synopsys User's Group (SNUG99), San Jose, CA, 1999.

**20.** Barzilai, Z., D. K. Beece, L. M. Huisman, V. S. Iyengar, and G. M. Silberman. "SLS-A Fast Switch-level Simulator." *IEEE Transactions on Computer-Aided Design* 7, no. 8 (August 1988): 838–849.

**21.** Barzilai, Z., J. L. Carter, B. K. Rosen, and J. D. Rutledge. "HSS-A High-Speed Simulator." *IEEE Transactions on Computer- Aided Design* 6, no. 4 (July 1987): 601–616.

**22.** Bayardo, R., Jr., and J. D. Pehoushek. "Counting Models Using Connected Components." In *Proceedings of the National Conference on Artificial Intelligence*. Menlo Park, CA: AAAI Press, 2000.

**23.** Bayardo, R., Jr., and R. Schrag. "Using CSP Look-Back Techniques to Solve Real-World SAT Instances." In *Proceedings of the National Conference on Artificial Intelligence*, 1997.

**24.** Becker, B. and R. Drechsler. "How Many Decomposition Types Do We Need?" In *Proceedings of the European Design and Test Conference*, 1995.

**25.** Becker, B., R. Drechsler, and R. Enders. "On the Computational Power of Bit-level and Word-Level Decision Diagrams." In *Proceedings of the Asia Pacific Design Automation Conference*, 1997.

**26.** Becker, B., R. Drechsler, and R. Werchner. "On the Relation Between BDDs and PDDs." *Information and Computation*, 123, no. 2 (1995): 185–197.

**27.** Beizer, B. *Software Testing Techniques*. 2d ed. New York: Van Nostrand Rheinhold, 1990.

**28.** Bell, J.R. "Threaded Code." *Journal of the ACM*, 16, no. 6 (June 1973): 777–785.

**29.** Ben-Ari, M., Z. Manna, and A. Pnueli. "The Temporal Logic of Branching Time." In *Proceedings of 8th Annual Symposium on Principles of Programming Languages,* 1981.

**30.** Benedikt, M., P. Godefroid, and T. Reps. "Model Checking of Unrestricted Hierarchical State Machines." In *Automata, Languages and Programming, 28th International Colloquium.* Lecture Notes in Computer Science 2076. New York: Springer-Verlag, 2001.

**31.** Bening, L. "A Two-State Methodology for RTL Logic Simulation." In *Proceedings of the 36th Conference on Design Automation,* 1999.

**32.** Bening, L. "An RTL Design Verification Linting Methodology." In *Proceedings of the International HDL Conference,* 1999.

**33.** Bening, L., B. Hornung, and R. Pflederer. "Hardware Description Language-Embedded Regular Expression Support for Module Iteration and Interconnection." In *Proceedings of the International HDL Conference,* 2001.

**34.** Bening, L., and H. Forster. *Principles of Verifiable RTL Design—A Functional Coding Style Supporting Verification Processes.* 2d ed. New York: Kluwer Academic Publishers, 2001.

**35.** Bening, L., T. A. Lane, C. R. Alexander, and J. E. Smith. "Developments in Logic Network Path Delay Analysis." In *Proceedings of the ACM/IEEE Design Automation Conference,*1982.

**36.** Bening, L., T. Brewer, H. D. Foster, J. S. Quigley, R. A. Sussman, P. F. Vogel, and A. W. Wells. "Physical Design of 0.35m Gate Arrays for Symmetric Multiprocessing Servers." *Hewlett-Packard Journal* (April 1997).

**37.** Berge', J., O. Levia, and J. Rouillard. "High-level System Modeling Specification Languages." Current Issues in Electronic Modeling, 3. New York: Kluwer Academic Publishers, 1995.

**38.** Bergeron, J. *Writing Testbenches: Functional Verification of HDL Models.* 2d ed. New York: Kluwer Academic Publishers, 2003.

**39.** Bergmann, J. and M. Horowitz. "Improving Coverage Analysis and Test Generation for Large Designs." In *Proceedings of the International Conference on Computer-Aided Design,* 1999.

**40.** Berman, C. and L. Trevillyan. "Functional Comparison of Logic Designs for VLSI Circuits." In *Proceedings of the International Conference on Computer-Aided Design,* 1989.

**41.** Berman, C. "Circuit Width, Register Allocation, and Ordered Binary Decision Diagrams." *IEEE Transactions on Computer-Aided Design,* 10 (1991).

**42.** Bern, J., J. Gergov, C. Meinel, and A. Slobodova. "Boolean Manipulation with Free BDDs. First Experimental Results." In *Proceedings of the European Design and Test Conference,* 1994.

**43.** Bertacco, V. and M. Damiani. "The Disjunctive Decomposition of Logic Functions." In *Proceedings of the International Conference on Computer-Aided Design*, 1997.

**44.** Bertacco, V., M. Damiani, and S. Quer. "Cycle-based Symbolic Simulation of Gate-level Synchronous Circuits." In *Proceedings of the 36th Conference on Design Automation*, 1999.

**45.** Bianco, A. and L. de Alfaro. "Model Checking of Probabilistic and Nondeterministic Systems." In *Foundations of Software Technology and Theoretical Computer Science*. Lecture Notes in Computer Science 1026. New York: Springer-Verlag, 1995.

**46.** Biere, A., A. Cimatti, E. Clarke, M. Fujita, and Y. Zhu. "Symbolic Model Checking Using SAT Procedures Instead of BDDs." In *Proceedings of the 36th Conference on Design Automation*, 1999.

**47.** Biere, A., A. Cimatti, E. Clarke, and Y. Zhu. "Symbolic Model Checking Without BDDs." In *Proceedings of the 5th International Conference on Tools and Algorithms for the Construction and Analysis of Systems*. Lecture Notes in Computer Science 1579. New York: Springer-Verlag, 1999.

**48.** Birtwistle G. and A. Davis, editors. *Asynchronous Digital Circuit Design*. Workshops in Computing. New York: Springer-Verlag, 1995.

**49.** Bjesse, P., T. Leonard, and A. Mokkedem. "Finding Bugs in an Alpha Microprocessor Using Satisfiability Solvers." In *Proceedings of 13th Conference on Computer-Aided Verification*, 2001.

**50.** Blank, T. "A Survey of Hardware Accelerators Used in Computer-aided Design." *IEEE Design and Test* (August 1984): 21–39 .

**51.** Bochmann, G. "Hardware Specification with Temporal Logic: An Example." *IEEE Transactions on Computer*s 31, no. 3 (March 1982): 223–231.

**52.** Bollig, R., et. al. "Improving the Variable Ordering of ROBDDs is NP Complete." *IEEE Transactions on Computers* 45, no. 9 (1996): 993–1002.

**53.** Bouajjani, A., B. Jonsson, M. Nilsson, and T. Touili. "Regular Model Checking." In *Proceedings of the 12th International Conference on Computer-Aided Verification*. Lecture Notes in Computer Science, 1855. New York: Springer-Verlag, 2000.

**54.** Boujjani, A., J. Esparza, and O. Maler. "Reachability Analysis of Pushdown Automata: Applications to Model Checking." In *Concurrency Theory, Eighth International Conference*. Lecture Notes in Computer Science, 1243. New York: Springer-Verlag, 1997.

**55.** Boyer R. and J. Moore. *A Computational Logic Handbook*. New York: Academic Press, 1988.

**56.** Brace, K., R. Rudell, and R. Bryant. "Efficient Implementation of a BDD Package." In *Proceedings of the 27th ACM/IEEE Design Automation Conference*, 1990.

**57.** Brand, D. "Verification of Large Synthesized Designs." In *Proceedings of the International Conference on Computer-Aided Design*, 1993.

**58.** Brayton, R., et al. "VIS: A System for Verification and Synthesis." In *Proceedings of the Eighth International Conference on Computer-Aided Verification*. Lecture Notes on Computer Science, 1102. New York: Springer-Verlag, 1996.

**59.** Brayton, R., G. Hachtel, C. McMullen, and A. Sangiovanni-Vincentelli. *Logic Minimization Algorithms for VLSI Synthesis*. The Kluwer International Series in Engineering and Computer Science. New York: Kluwer Academic Publishers, 1986.

**60.** Breuer, M. "A Note on Three-valued Logic Simulation." *IEEE Transactions on Computers* C-21( April 1972): 399–402.

**61.** Breuer, M. and A. D. Friedman. *Diagnosis and Reliable Design of Digital Systems*. Woodland Hills, CA: Computer Science Press, 1976.

**62.** Bronstein, A. and C. Talcott. "Formal Verification of Synchronous Circuits Based on String-functional Semantics: The 7 Paillet Circuits in Boyer-Moore." In *Proceedings of the International Workshop on Automatic Verification Methods for Finite State Systems*. Lecture Notes in Computer Science, 407. New York: Springer-Verlag, 1989.

**63.** Bruni, R. and A. Sassano. "Restoring Satisfiability or Maintaining Unsatisfiability by Finding Small Unsatisfiable Subformulae." In *International Symposium on the Theory and Applications of Satisfiability Testing*, 2001.

**64.** Bryant R. and Y. A. Chen. "Verification or Arithmetic Circuits with Binary Moment Diagrams." In *Proceedings of the 32nd ACM/IEEE Design Automation Conference*, 1995.

**65.** Bryant, R. "On the Complexity of VLSI Implementations and Graph Representations of Boolean Functions with Applications to Integer Multiplication." *IEEE Transactions on Computers* 40, no. 2 (February 1991).

**66.** Bryant, R.E., D. Beatty, K. Brace, K. Cho, and T. Sheffler. "COMOS: A Compiled Simulator for MOS Circuits." In *Proceedings of the 24th Design Automation Conference*, 1987.

**67.** Bryant, R. "Graph-based Algorithms for Boolean-function Manipulation." *IEEE Transactions on Computers*, C-35, no. 8, 1986.

**68.** Bryant, R. "On the Complexity of VLSI Implementations and Graph Representations of Boolean Functions with Application to Integer Multiplication." *IEEE Transactions on Computers* 40, no. 2 (February 1991): 205–213.

**69.** Buchnik, E., and S. Ur. "Compacting Regression-suites On-the-Fly." In *Proceedings of the 4th Asia Pacific Software Engineering Conference*, 1997.

**70.** Burch, J. and D. Dill. "Automatic Verification of Pipelined Microprocessor Control." In *Proceedings of the International Conference on Computer-Aided Verification*, 1994.

71. Burch, J. and D. Long. "Efficient Boolean Function Matching." In *Proceedings of the International Conference on Computer-Aided Design*, 1992.

72. Burch, J. and V. Singhal. "Tight Integration of Combinational Verification Methods." In *Proceedings of the International Conference on Computer-Aided Design*, 1998.

73. Burch, J., E. Clarke, D. Dill, L. Hwang, and K. McMillan. "Symbolic Model Checking: States and Beyond." *Information and Computation* 98: no. 2 (June 1992).

74. Burch, J., E. Clarke, D. Long, K. MacMillan, and D. Dill. "Symbolic Model Checking for Sequential Circuit Verification. *IEEE Transactions on Computer-Aided Design of Integrated Circuits and Systems*, 1994.

75. Burch, J., E. Clarke, K. McMillan, D. Dill, and L. Hwang. "Symbolic Model Checking: $10^{200}$ States and Beyond." In *Proceedings of the Fifth Annual IEEE Symposium on Logic in Computer Science*, 1990.

76. Burch, J, and E. Clarke, and D. Long. "Representing Circuits More Efficiently in Symbolic Model Checking." In *Proceedings of the 28th ACM/IEEE Automation Conference*, 1991.

77. Butler, K., D. Ross, R. Kapur, and M. Mercer. "Heuristics to Compute Variable Orderings for Efficient Manipulation of Ordered Binary Decision Diagrams." In *Proceedings of the 28th ACM/IEEE Design Automation Conference*, 1991.

78. McGeer, P.C., K. L. McMillan, A. Saldanha, A. L. Sangiovanni-Vincentelli, and P. Scaglia. "Fast Discrete Function Evaluation Using Decision Diagrams," In *Proceedings of the International Conference on Computer-Aided Design*, 1995.

79. Cabodi, G., P. Camurati, F. Corno, P. Prinetto, and M. S. Reorda. "A New Model for Improving Symbolic Product Machine Traversal." In *Proceedings of the 29th ACM/IEEE Design Automation Conference*, 1992.

80. Cabodi, G., P. Camurati, L. Lavagno, and S. Quer. "Disjunctive Partitioning and Partial Iterative Squaring: An Effective Approach for Symbolic Traversal of Large Circuits." In *Proceedings of the 34th Conference on Design Automation*, 1997.

81. Cabodi, G., P. Camurati, and S. Quer. "Efficient State Space Pruning in Symbolic Backward Traversal." In *Proceedings of the International Conference on Computer Design*, 1994.

82. Cabodi, G., P. Camurati, and S. Quer. "Improved Reachability Analysis of Large Finite State Machine." In *Proceedings of the International Conference on Computer-Aided Design*, 1996.

83. Cabodi, G. R. Camurati, and S. Quer. "Symbolic Exploration of Large Circuits with Enhanced Forward/Backward Traversals." In *Proceedings of the European Design Automation Conference*, 1994.

**84.** Cadambi, S., C. Mulpuri, and P. N. Ashar. "A Fast, Inexpensive and Scalable Hardware Acceleration Technique for Functional Simulation." In *Proceedings of the 39th Conference on Design Automation*, 2002.

**85.** Camiller, A., C. Harkness, M. Heap, W. Lam, D. Prekeges, and J. Wickeraad. "Introduction to Formal Hardware Verification: A Tutorial." Presented at the Design Technology Conference, Hewlett-Packard Co., Santa Clara, CA., May 1994.

**86.** Camilleri, A., M.J.C. Gordon, and T.F. Melham. "Hardware Verification Using Higher Order Logic." From *HDL Descriptions to Guaranteed Correct Circuit Design*s. D. Borrione, ed. Amsterdam: North-Holland, 1986.

**87.** Camurati, P., T. Margaria, and R Prinetto. "Resolution-Based Correctness Proofs of Synchronous Circuits." In *Proceedings of the European Design Automation Conference*, 1991.

**88.** Carter, W.C. "A Time for Reflection." In *Proceedings of the 12th IEEE International Symposium on Fault Tolerant Computing*, 1982.

**89.** Cerny E. and C. Mauras. "Tautology Checking Using Cross-controllability and Cross-Observability Relations." In *Proceedings of the International Conference on Computer-Aided Design*, 1990.

**90.** Cerny, E., B. Berkane, P. Girodias, and K. Khordoc. *Hierarchical Annotated Action Diagram.* New York: Kluwer Academic Publishers, 1998.

**91.** Chakrabarti, A., L. de Alfaro, T. Henzinger, M. Jurdzinski, and F. Mang. "Interface Compatibility Checking for Software Modules." In *Proceedings of the 14th International Conference on Computer-Aided Verification.* Lecture Notes in Computer Science, 2404. New York: Springer-Verlag, 2002.

**92.** Chakradhar, S. and V. Agrawal. "A Transitive Closure-based Algorithm for Test Generation." In *Proceedings of the 28th Design Automation Conference*, 1991.

**93.** Chappell, S. and S. Yau. "A Three-Value Design Verification System." In *Proceedings of the Fall Joint Computer Conference*, 1971.

**94.** Chappell, B. "The Fine Art of IC Design." *Computer* 32: no. 7 (July 1999).

**95.** Chatalic, P. and L. Simon. "ZRes: The Old Davis-Putnam Procedure Meets ZBDDs." In *Proceedings of the 17th International Conference on Automated Deduction.*

**96.** Cheng, K. and A. Krishnakumar. "Automatic Functional Test Generation Using the Extended Finite State Machine Model." In *Proceedings of the 30th Design Automation Conference*, 1993.

**97.** Cho, H., G.D. Hachtel, E. Macii, B. Plessier, and F. Somenzi. "Algorithms for Approximate FSM Traversal." In *Proceedings of the 30th Design Automation Conference*, 1993.

**98.** Cho, H., G.D. Hachtel, E. Macii, M. Poncino, and F. Somenzi. "A State Space Decomposition Algorithm for Approximate FSM Traversal." In *Proceedings of the European Design and Test Conference*, 1994.

**99.** Cho, H., G.D. Hachtel, and F. Somenzi. "Redundancy Identification/Removal and Test Generation for Sequential Circuits Using Implicit State Enumeration." *IEEE Transactions on Computer-Aided Design of Integrated Circuits and Systems.*

**100.** Cho, H., S.W. Jeong, F. Somenzi, and C. Pixley. "Multiple Observation Time Single Reference Test Generation Using Synchronizing Sequences." In *European Conference on Design Automation with European Event in ASIC Design*, 1993.

**101.** Clarke, E. and E. Emerson. "Design and Synthesis of Synchronization Skeletons Using Branching Time Temporal Logic." In *Proceedings of the Workshop on Logic of Programs*. Lecture Notes on Computer Science. New York: SpringerVerlag, 1981.

**102.** Clarke, E. and R. Kurshan. "Computer-Aided Verification." *IEEE Spectrum* (June 1996): 61–67.

**103.** Clarke, E., O. Grumberg, K. McMillan, and X. Zhao. "Efficient Generation of Counterexamples and Witnesses in Symbolic Model Checking." In *Proceedings of the 32nd Design Automation Conference*, 1995.

**104.** Clarke, E., O. Grumberg, and D. Long. *Verification Tools for Finite-State Concurrent Systems. A Decade of Concurrency—Reflections and Perspectives.* Lecture Notes in Computer Science, 803. New York: Springer-Verlag, 1993.

**105.** Clarke, E., O. Grumberg, and K. Hamaguchi. "Another Look at LTL Model Checking." In *Proceedings of the 6th International Conference on Computer-Aided Verification*. Lecture Notes in Computer Science, 818. New York: Springer-Verlag, 1994.

**106.** Clarke, E., and J. Wing. "Formal Methods: State of the Art and Future Directions." *CMU Computer Science Technical Report* CMU-CS-96-178, August 1996.

**107.** Clarke, E., K.L. McMillian, X. Zhao, M. Fujita, and J. Yang. "Spectral Transforms for Large Boolean Functions with Application to Technology Mapping." In *Proceedings of the 30th ACM/IEEE Design Automation Conference*, 1993.

**108.** Clarke, E., O. Grumberg, and D. Peled. *Model Checking*. Cambridge, MA: MIT Press, 2000.

**109.** Cook, S. and D. Mitchell. "Finding Hard Instances of the Satisfiability Problem: A Survey." In *Satisfiability Problem: Theory and Applications*. Du, Gu, and Pardalos, editors. Dimacs Series in Discrete Mathematics and Theoretical Computer Science 35. American Mathematical Society, 1997.

**110.** Cook, S. "The Complexity of Theorem-Proving Procedures." In *Proceedings of the Third IEEE Symposium on the Foundations of Computer Science*, 1971.

**111.** Copty, F., L. Fix, E. Giunchiglia, G. Kamhi, A. Tacchella, and M. Vardi. "Benefits of Bounded Model Checking at an Industrial Setting." In *Proceedings of the 13th Conference on Computer-Aided Verification*, 2001.

112. Coudert, O., C. Berthet, and J.C. Madre. "Verification of Sequential Machines Using Boolean Functional Vectors." In *Proceedings of the IFIP International Workshop Applied Formal Methods for Correct VLSI Design*. L.J.M. Claesen, ed. Leuven, Belgium: North-Holland, 1989.

113. Coudert, O., C. Berthet, and J.C. Madre. "Verification of Synchronous Sequential Machines Based on Symbolic Execution." In *Workshop on Automatic Verification Methods for Finite State Systems*, 1989.

114. Coudert, O., C. Berthet, and J. Madre. "Verification of Synchronous Sequential Machines Based on Symbolic Execution." In *Automatic Verification Methods for Finite State Systems, International Workshop*. Lecture Notes in Computer Science 407. New York: Springer-Verlag, 1989.

115. Courcoubetis, C., M. Vardi, P. Wolper, and M. Yannakakis. "Memory Efficient Algorithms for the Verification of Temporal Properties." In *Formal Methods in System Design*. New York: Kluwer Academic Publishers, 1992.

116. Cyrluk, D., P. Lincoln, and N. Shankar. "On Shostak's Decision Procedure for Combinations of Theories." In *Conference on Automated Deduction*. LNAI 1104. New York: Springer-Verlag, 1996.

117. DeVane, C. "Efficient Circuit Partitioning to Extend Cycle Simulation Beyond Synchronous Circuits." In *Proceedings of the International Conference on Computer-Aided Design*, 1997.

118. Devadas, S., A. Ghosh, K. Keutzer. "An Observability-Based Code Coverage Metric for Functional Simulation." In *Proceedings of the International Conference on Computer-Aided Design*, 1996.

119. Devadas, S., H.-K.T. Ma, and A.R. Newton. "On the Verification of Sequential Machines at Differing Levels of Abstraction." *IEEE Transactions on Computer-Aided Design of Integrated Circuits and Systems* 7, no. 6 (June 1988).

120. Dewey, A., ed. "Three Decades of HDLs Part 1: CDL Through TI-HDL." *IEEE Design and Test* 9, no. 2 (June 1992).

121. Dewey, A, ed. "Three Decades of HDLs Part 2: Conlan Through Verilog." *IEEE Design and Test* 9, no. 3 (September 1992).

122. Dill, D., A. Drexler, A. Hu, and H. Yang. "Protocol Verification as a Hardware Design Aid." Presented at the International Conference on Computer Design, Boston, MA, 1992.

123. Dill, D. and S. Tasiran. "Simulation Meets Formal Verification." In *Proceedings of the International Conference on Computer-Aided Design*, 1999.

124. Drechsler, R., N. Drechsler, and W. Gunther. "Fast Exact Minimization of BDDs." In *Proceedings of the Design Automation Conference*, 1998.

125. Drusinsky, D. and D. Harel. "On the Power of Bounded Concurrency: Finite Automata." *Journal of the ACM* 41, no. 3 (1994).

**126.** Dubois, O. and G. Dequen. "A Backbone-Search Heuristic for Efficient Solving of Hard 3-SAT Formulae." In *Proceedings of the Seventeenth International Joint Conference on Artificial Intelligence*, 2001.

**127.** Eiriksson, A. "Integrating Formal Verification Methods with A Conventional Project Design Flow." In *Proceedings of the 33rd Conference on Design Automation*, 1996.

**128.** Ellsberger, J., D. Hogrefe, and A. Sarma. *SDL: Formal Object-oriented Language for Communicating Systems*. Upper Saddle River, NJ: Prentice Hall PTR, 1997.

**129.** Emerson, E. and A. Sistia. "Symmetry and Model Checking." In *Proceedings of the 5th International Conference on Computer-Aided Verification*, 1993. Lecture Notes in Computer Science, 697. C. Courcoubetis, ed. New York: Springer-Verlag, 1993.

**130.** Emerson, E.A. and C.-L. Lei. "Modalities for Model Checking: Branching Time Strikes Back." In *Proceedings of the Twelfth Annual ACM Symposium on Principles of Programming Languages*. New York: ACM Press, 1985.

**131.** Emerson, E.A. and C.S. Jutia. "The Complexity of Free Automata and Logics of Programs." In *The 29th Annual IEEE-CS Symposium in Foundations of Computer Science*, 1988.

**132.** Emerson, E.A. and E.M. Clarke. "Characterizing Correctness Properties of Parallel Programs as Fixpoints." In *Proceedings of the Seventh International Colloquium on Automata, Languages and Programming*. Lecture Notes in Computer Science, 85. New York: Springer-Verlag, 1981.

**133.** Emerson, E.A. "Temporal and Modal Logic." *Handbook of Theoretical Computer Science, Volume B*. J. van Leeuwen, editor. Amsterdam: Elsevier Science Publishers, 1990.

**134.** Esparza, J., D. Hansel, P. Rossmanith, and S. Schwoon. "Efficient Algorithms for Model Checking Pushdown Systems." In *Computer Aided Verification, 12th International Conference*. Lecture Notes in Computer Science, 1855. New York: Springer-Verlag, 2000.

**135.** Fallah, F., S. Devadas, and K. Keutzer. "OCCOM: Efficient Computation of Observability-Based Code Coverage Metrics for Functional Verification." In *Proceedings of the 35th Conference on Design Automation*, 1998.

**136.** Feng, X. and A. Hu. "Automatic Formal Verification for Scheduled VLIW Code." *ACM SIGPLAN Joint Conference: Languages, Compilers, and Tools for Embedded Systems, and Software and Compilers for Embedded Systems*. ACM Press, 2002.

**137.** Fitting, M. *First-Order Logic and Automated Theorem Proving*. Texts and Monographs in Computer Science. New York: Springer-Verlag, 1990.

**138.** Foster, H. "Formal Verification of the Hewlett-Packard V-Class Servers." In *Proceedings of DesignCon99 On-Chip Design Conference*, 1999.

**139.** Foster, H. and C. Coelho. "Assertions Targeting a Diverse Set of Verification Tools." In *Proceedings of the International HDL Conference*, 2001.

**140.** Freeman, J. "Improvements to Propositional Satisfiability Search Algorithms." Ph.D. thesis, University of Pennsylvania, 1995.

**141.** Friedman, S. and K. Supowit. "Finding the Optimal Variable Ordering for Binary Decision Diagrams." *IEEE Transactions on Computers* 39 (1990).

**142.** Fujii, H., G. Ootomo, and C. Hori. "Interleaving Based Variable Ordering Methods for Ordered Binary Decision Diagrams." In *IEEE/ACMInternational Conference on Computer-Aided Design*, 1993.

**143.** Fujita, M., H. Fujisawa, and N. Kawato. "Evaluation and Improvements of Boolean Comparison Method Based on Binary Decision Diagrams." In *Proceedings of the International Conference on Computer-Aided Design*, 1988.

**144.** Fujita, M., Y. Matsunaga, and T. Kakuda. "On Variable Ordering of Binary Decision Diagrams for the Application of Multi-Level Logic Synthesis." In *Proceedings of the European Design Automation Conference*, 1991.

**145.** Fujuwara, H. and S. Toida. "The Complexity of Fault Detection Problems for Combinational Logic Circuits." *IEEE Transactions on Computers* 31, no. 6 (June 1982).

**146.** Gai, S., F. Somenzi, and M. Spalla. "Fast and Coherent Simulation with Zero Delay Elements." *IEEE Transactions on Computer-Aided Design* 6, no. 1 (January 1987).

**147.** Garey, M. and D. Johnson. *Computers and Intractability: A Guide to the Theory of NP-Completeness.* W. H. Freeman Co., 1979.

**148.** Goldberg, E. and Y. Novikov. "Verification of Proofs of Unsatisfiability for CNF Formulas." In *Proceedings of the IEEE/ACM Design, Automation, and Test in Europe*, 2003.

**149.** Goldberg, E. and Y. Novikov. "BerkMin: A Fast and Robust SAT-solver." In *Proceedings of the IEEE/ACM Design, Automation, and Test in Europe*.

**150.** Gordon, M. "HOL: A Proof Generating System for Higher-Order Logic." *VLSI Specification, Verification and Synthesis*. G.M. Birtwistle and P.A. Subrahmanyam, editors. Boston: Kluwer Academic Publishers, 1988.

**151.** Grinwald, R., E. Harel, M. Orgad, S. Ur, and A. Ziv. "User Defined Coverage—A tool Supported Methodology for Design Verification." In *Proceedings of the 35th Conference on Design Automation*, 1998.

**152.** Gu, J., P. Purdom, J. Franco, and B. Wah. "Algorithms for the Satisfiability (SAT) Problem: A Survey." *Satisfiability Problem: Theory and Applications*. DIMACS Series in Discrete Mathematics and Theoretical Computer Science. D. Du, J. Gu, and P. Pardalos, editors. American Mathematical Society, 1997.

**153.** Gu, J. "Local Search for Satisfiability SAT Problem." *IEEE Transactions on Systems and Cybernetics* 23, no. 3 (1993).

**154.** Gupta, A., S. Malik, and P. Ashar. "Toward Formalizing a Validation Methodology Using Simulation Coverage." In Proceedings of the 34th Conference on Design Automation, 1997.

**155.** Gupta, A., Z. Yang, P. Ashar, and A. Gupta. "SAT-based Image Computation with Application in Reachability Analysis." In *Proceedings of Third International Conference Formal Methods in Computer-Aided Design*, 2000.

**156.** Gupta, A. "Formal Hardware Verification Methods: A Survey." *Journal of Formal Methods in System Design* 1 (1992).

**157.** Hefferan, P.H., R. J. Smith, V. Burdick, and D. L. Nelson. "The STE-264 Accelerated Electronic CAD System." In *Proceedings of the ACM/IEEE Design Automation Conference*, IEEE Computer Society Press, 1985.

**158.** Habet, D., C.M. Li, L. Devendeville, and M. Vasquez. "A Hybrid Approach for SAT." In *Proceedings of the 8th International Conference on Principles and Practice of Constraint Programming*, 2002.

**159.** Hachtel, G. and F. Somenzi. *Logic Synthesis and Verification Algorithms.* Boston: Kluwer Academic Publishers, 1996.

**160.** Hall, A. "Seven Myths of Formal Methods." *IEEE Software* 7, no. 5 (September 1990): 11–19.

**161.** Hansen, C. "Hardware Logic Simulation by Compilation." In *Proceedings of the 25th ACM/IEEE Design Automation Conferenc*e, 1988.

**162.** Haque, F., K. Khan, and J. Michelson. *The Art of Verification with Vera.* Verification Central, 2001.

**163.** Hardy, G. H. "A Mathematician's Apology." Cambridge University Press, Reprint edition, 1992.

**164.** Henzinger, T., R. Jhala, R. Majumdar, G. Necula, G. Sutre, and W. Weimer. "Temporal-safety Proofs for Systems Code." In *Proceedings of 14th Conference on Computer-Aided Verification.* Lecture Notes in Computer Science, 2404. New York: Springer-Verlag, 2002.

**165.** Hirsch, E. and A. Kojevnikov. "Solving Boolean Satisfiability Using Local Search Guided by Unit Clause Elimination." In *Proceedings of 7th International Conference on Principles and Practice of Constraint Programming*, 2001.

**166.** Ho, P., T. Shiple, K. Harer, J. Kukula, R. Damiano, V. Bertacco, J. Taylor, and J. Long. "Smart Simulation Using Collaborative Formal and Simulation Engines." In *Proceedings of the International Conference on Computer-Aided Design*, 2000.

**167.** Hollander, Y., M. Morley, and A. Noy. "The e Language: A Fresh Separation of Concerns." *Technology of Object-Oriented Languages and Systems* 38 (March 2001).

**168.** Holzmann, G. *Design and Validation of Computer Protocols.* Upper Saddle River, NJ: Prentice Hall, 1991.

169. Holzmann, G. "The Model Checker SPIN." *IEEE Transactions on Software Engineering* 23, no. 5 (1997): 279–295.

170. Hooker, J. and V. Vinay. "Branching Rules for Satisfiability." *Journal of Automated Reasoning* 15 (1995): 359–383.

171. Hoos, H. "On the Run-time Behaviour of Stochastic Local Search Algorithms for SAT." In *Proceedings of the Sixteenth National Conference on Artificial Intelligence,* 1999.

172. Hopcroft, J. and J. Ullman. *Introduction to Automata Theory, Languages, and Computation.* Boston: Addison-Wesley, 1979.

173. Horeth, S. and R. Drechsler. "Dynamic Minimization of Word-Level Decision Diagrams." In *Proceedings of the Design Automation and Test in Europe,* 1998.

174. Horgan, J., S. London, and M. Lyu. "Achieving Software Quality with Testing Coverage Measures." *Computer* 27, no. 9 (September 1994): 60–69.

175. Howe, H. "Pre- and Postsynthesis Simulation Mismatches." In *Proceedings of the 6th International Verilog HDL Conference,* 1997.

176. Hu, A. and D. Dill. "Reducing BDD Size by Exploiting Functional Dependencies." In *Proceedings of the 30th Design Automation Conference,* 1993.

177. Hu, A. "Formal Hardware Verification with BDDs: An Introduction." *IEEE Pacific Rim Conference on Communications, Computers, and Signal Processing,* 1997.

178. Huang, S. and K. Cheng. *Formal Equivalence Checking and Design Debugging.* Boston: Kluwer Academic Publishers, 1998.

179. Goldberg, E.I., M. R. Prasad, and R. K. Brayton. "Using SAT for Combinational Equivalence Checking." In *Proceedings of the IEEE/ACM Conference on Design, Automation and Test in Europe,* 2001.

180. Ip, C. and D. Dill. "Better Verification Through Symmetry." In *IFIP Conference on Computer Hardware Description Languages and Their Applications,* 1993.

181. Ip, C. and D. Dill. "Better Verification through Symmetry." *Formal Methods in System Design* 9, nos. 1/2 (August 1996): 41–75.

182. Ishura, N., H. Sawada, and S. Yajima. "Minimization of Binary Decision Diagrams Based on Exchange of Variables." In *International Conference on Computer-Aided Design,* 1991.

183. Jacobi, R., N. Calazans, and C. Trullemans. "Incremental Reduction of Binary Decision Diagrams." In *International Symposium on Circuits and Systems,* 1991.

184. Jain, P. and G. Gopalakrishnan. Efficient Symbolic Simulation-based Verification Using the Parametric Form of Boolean Expressions." *IEEE Transactions on Computer-Aided Design of Integrated Circuits and Systems* 13, no. 8 (August 1994).

185. Jain, J., M. Abadir, J. Bitner, D.S. Fusell, and J.A. Abraham. "IBDD's: An Efficient Functional Representation for Digital Circuits." In *Proceedings of the European Design Automation Conference,* 1992.

**186.** Johnson, D. and M. A. Trick, editors. *Second DIMACS Implementation Challenge: Cliques, Coloring and Satisfiability.* DIMACS Series in Discrete Mathematics and Theoretical Computer Science, 26. Providence, RI: American Mathematical Society, 1996.

**187.** Joyce, J. and C. Seger. "The HOL-Voss System: Model-Checking Inside a General-Purpose Theorem-Prover." In *International Workshop on Higher Order Logic Theorem Proving and Its Applications.* Lecture Notes in Computer Science, 780. New York: Springer-Verlag, 1994.

**188.** Kang, S. and S. Szygenda. "Modeling and Simulation of Design Errors." In *Proceedings of the International Conference on Computer Design: VLSI in Computers and Processors*, 1992.

**189.** Kantrowitz, M. and L. Noack. "I'm Done Simulating: Now What? Verification Coverage Analysis and Correctness Checking of the DECchip 21164 Alpha Microprocessor." In *Proceedings of 33rd Conference on Design Automation*, 1996.

**190.** Karplus, K. "Representing Boolean Functions with If-Then-Else Dags." Technical Report UCSC-CRL-88-28, Baskin Center for Computer Engineering and Information Sciences, 1988.

**191.** Kaufmann, M., A. Martin, and C. Pixley. "Design Constraints in Symbolic Model Checking." In *Proceedings of the 10th International Conference on Computer-Aided Verification.* Lecture Notes in Computer Science, 1427. New York: Springer-Verlag, 1998

**192.** Kautz, H. and B. Selman. "Unifying SAT-based and Graph-based Planning." In *Proceedings of the Sixteenth International Joint Conference on Artificial Intelligence*, 1999.

**193.** Keating, M. and P. Bricaud. *Reuse Methodology Manual.* Boston: Kluwer Academic Publishers, 1999.

**194.** Kebschull, U., E. Schubert, and W. Rosenstiel. "Multilevel Logic Synthesis Based on Functional Decision Diagrams." In *Proceedings of the 29th ACM/IEEE Design Automation Conference*, 1992.

**195.** Kohavi, Z. *Switching and Finite Automata Theory.* New York: McGraw-Hill, 1970.

**196.** Kripke, S. "Semantic Considerations on Modal Logic." In *Proceedings of a Colloquium: Modal and Many Valued Logics.* Acta Philosophica Fennica, 16 (August 1993): 83–94.

**197.** Kropf, T. *Introduction to Formal Hardware Verification.* New York: Springer-Verlag, 2000.

**198.** Kuehlmann, A. and F. Krohm. "Equivalence Checking Using Cuts and Heaps." In *Proceedings of the 34th Conference on Design Automation*, 1997.

**199.** Kuehlmann, A., M. Ganai, and V. Paruthi. "Circuit-based Boolean Reasoning." In *Proceedings of the 38th Design Automation Conference*, 2001.

**200.** Kumar, R., C. Blumenrohr, D. Eisenbiegler, and D. Schmid. "Formal Synthesis in Circuit Design—A Classification and Survey." In *Proceedings of the International Conference on Formal Methods in Computer-Aided Design*. Lecture Notes in Computer Science, 1166. New York: Springer-Verlag, 1996.

**201.** Kunz, W. and D. Pradhan. "Recursive Learning: A New Implication Technique for Efficient Solutions to CAD Problems: Test, Verification and Optimization." *IEEE Transactions on Computer-Aided Design* 13, no. 9 (September 1994):1143–1158.

**202.** Kunz, W. "HANNIBAL: An Efficient Tool for Logic Verification Based on Recursive Learning." In *Proceedings of the International Conference on Computer-Aided Design*, 1993.

**203.** Kupferman, O. M. Vardi, and P. Wolper. "An Automata-Theoretic Approach to Branching-Time Model Checking." *Journal of the ACM* 47, no. 2 (2000): 312–360.

**204.** Kurshan, R. *Computer-Aided Verification of Coordinating Processes: The Automata-Theoretic Approach*. Princeton, NJ: Princeton University Press, 1994.

**205.** Lam, W. and R. Brayton. "Alternating RQ Timed Automata." In *International Conference on Computer-Aided Verification*. Lecture Notes in Computer Science, Costa Courcoubetis, ed. New York: Springer-Verlag, 1993.

**206.** Lam, W. and R. Brayton. *Timed Boolean Functions—A Unified Formalism for Exact Timing Analysis*. Boston: Kluwer Academic Publishers, 1994.

**207.** Lam, W. Circuit "Partitioning and Scheduling for Massively Parallel Simulation." Sun Microsystems Laboratories Report SML 2000-0137 (October 1998).

**208.** Lam, W. "Method and Apparatus to Facilitate Generating Simulation Modules for Testing System Designs." U.S. Patent No. 6,715,134.

**209.** Lam, W. "Partitioning and Communication Bandwidth for Parallel Simulation of a Microprocessor: A Case Study." Sun Microsystems Laboratories Report SML 2000-0138 (October 1999).

**210.** Lam, W. "Boosting Simulation Performance By Dynamically Customizing Segmented Object Codes Based On Stimulus Coverage." U.S. Patent No. 6,775,810.

**211.** Lam, W. "Race and Deadlock Free Performance Optimization in Cycle Based Simulation." Sun Microsystems Laboratories Report SML 2000-0135 (March 1999).

**212.** Lewis, D.M. "Hierarchical Compiled Event-Driven Logic Simulation." In *Proceedings of the International Conference on Computer-Aided Design*, 1989.

**213.** Li, C. and S. Grard. "On the Limit of Branching Rules for Hard Random Unsatisfiable 3-SAT." In *Proceedings of the 14th European Conference on Artificial Intelligence*, 2000.

**214.** Liaw, H. and C. Lin. "On the OBDD Representation of General Boolean Functions." *IEEE Transactions on Computers* 41 (1992): 661–664.

**215.** Lichtenstein, O. and A. Pnueli. "Checking that Finite State Concurrent Programs Satisfy their Linear Specification." In *Proceedings of the Twelfth Annual ACM Symposium on Principles of Programming Languages*, 1985.

**216.** Long, D. *Model Checking, Abstraction, and Compositional Verification*. Ph.D. thesis, Carnegie Mellon University, 1993.

**217.** Madre, J. and J.P. Billon. "Proving Circuit Correctness Using Formal Comparison Between Expected and Extracted Behavior." In *Proceedings of the 25th ACM/IEEE Design Automation Conference*, 1988. IEEE Computer Society Press.

**218.** Malik, S., A. Wang, R. Brayton, and A. Sangiovanni-Vincentelli. "Logic Verification Using Binary Decision Diagrams in a Logic Synthesis Environment. In *Proceedings of the International Conference on Computer-Aided Design*, 1988.

**219.** Malka Y. and Avi Ziv. "Design Reliability—Estimation Through Statistical Analysis of Bug Discovery Data." In *Proceedings of the 35th Conference on Design Automation*, 1998.

**220.** Manna, Z. and A. Pnueli. *The Temporal Logic of Reactive and Concurrent Systems: Specification*. New York: Springer-Verlag, 1991.

**221.** Manna, Z. and A. Pnueli. "Verification of Concurrent Programs: Temporal Proof Principles." In *Proceedings of the Workshop on Logics of Programs*. Lecture Notes in Computer Science, 131. New York: Springer-Verlag, 1981.

**222.** Manquinho, V. and J. Marques-Silva. "On Using Satisfiability-Based Pruning Techniques in Covering Algorithms." In *Proceedings of the IEEE/ACM Design, Automation and Test in Europe Conference*, 2000.

**223.** Marques-Silva, J. and K. Sakallah. "GRASP—A New Search Algorithm for Satisfiability." In *Proceedings of IEEE/ACM International Conference on Computer-Aided Design*, 1996.

**224.** Marques-Silva, J. and K. Sakallah. "Efficient and Robust Test Generation-Based Timing Analysis." In *Proceedings of the International Symposium on Circuits and Systems*, 1994.

**225.** Marques-Silva, J. and K. Sakallah. "GRASP—A Search Algorithm for Propositional Satisfiability." *IEEE Transactions in Computers* 48, no. 5 (May 1999):506–521.

**226.** Marques-Silva, J. and T. Glass. "Combinational Equivalence Checking Using Satisfiability and Recursive Learning." In *Proceedings of the IEEE/ACM Design, Automation and Test in Europe*, 1999.

**227.** Matsunaga, Y. "An Efficient Equivalence Checker for Combinatorial Circuits." In *Proceedings of the Design Automation Conference*, 1996.

**228.** Maurer, P., Z. Wang, and C. D. Morency. "Techniques for Multi-Level Compiled Simulation." Technical Report CSE-89-04. University of South Florida, Department of Computer Science and Engineering.

**229.** McCluskey, E. *Logic Design Principles*. Upper Saddle River, NJ: Prentice Hall, 1986.

230. McMillan, K. "Applying SAT Methods in Unbounded Symbolic Model Checking." In *Proceedings of 14th Conference on Computer-Aided Verification*. New York: Springer-Verlag, 2002.

231. McMillan, K. *Symbolic Model Checking: An Approach to the State Explosion Problem*. Boston: Kluwer Academic Publishers, 1993.

232. Melham, T. "Abstraction Mechanisms for Hardware Verification." *VLSI Specification, Verification, and Synthesis*. Boston: Kluwer Academic Publishers, 1988.

233. Mills, D. and C. Cummings. "RTL Coding Styles That Yield Simulation and Synthesis Mismatches." Presented at Synopsys Users Group, San Jose, CA, 1999.

234. Minato, S., N. Ishiura, and S. Yajima. "Shared Binary Decision Diagram with Attributed Edges for Efficient Boolean Function Manipulation." In *Proceedings of the 27th ACM/IEEE Design Automation Conference*, 1990.

235. Minato, S. "Zero-Suppressed BDDs for Set Manipulation in Combinatorial Problems." In *Proceedings of the 30th ACM/IEEE Design Automation Conference*, 1993.

236. Mitchell, D., B. Selman, and H. Levesque. "Hard and Easy Distribution of SAT Problems." In *Proceedings of the Tenth National Conference on Artificial Intelligence*, 1992.

237. Mittra, S. *Principles of VERILOG PLI*. Boston: Kluwer Academic Publishers, 1999.

238. Moeller, O., and H. Ruess. "Solving Bit-Vector Equations." In *Proceedings of the International Conference on Formal Methods in Computer-Aided Design, 1998*. Lecture Notes in Computer Science, 1522. New York: Springer-Verlag, 1998.

239. Moon, I., J. Kukula, K. Ravi, and F. Somenzi. "To Split or to Conjoin: The Question in Image Computation." In *Proceedings of the 37th Conference on Design Automation*, 2000.

240. Moondanos, J., C. Seger, Z. Hanna, and D. Kaiss. "CLEVER: Divide and Conquer Combinational Logic Equivalence Verification with False Negative Elimination?" In *Computer-Aided Verification, 13th International Conference*, 2001.

241. Moskewicz, M., C. Madigan, Y. Zhao, L. Zhang, and S. Malik. "Chaff: Engineering an Efficient SAT Solver?" In *Proceedings of the 39th Design Automation Conference*, 2001.

242. Murata, T. "Petri Nets: Properties, Analysis and Applications." *Proceedings of the IEEE* 77, no. 1 (April 1989): 541–580.

243. Namjoshi, K. "Certifying Model Checkers." In *Proceedings of the 13th International Conference on Computer-Aided Verification*, 2001.

244. Necula, G. and P. Lee. "Safe Kernel Extensions Without Run-Time Checking." Presented at the Symposium on Operating System Design and Implementation, 1996.

**245.** Norris Ip, C. and David L. Dill. "Better Verification Through Symmetry." Presented at the International Conference on Computer Hardware Description Languages, 1993.

**246.** O'Leary, J., X. Zhao, R. Gerth, and C. Seger. "Formally Verifying IEEE Compliance of Floating-Point Hardware." *Intel Technology Journal* Q1 (1999).

**247.** Owre, S., J.M. Rushby, N. Shankar, and M.K. Srivas. "A Tutorial on Using PVS for Hardware Verification." In *Proceedings of the 2nd International Conference on Theorem Provers in Circuit Design*. Lecture Notes in Computer Science, 901. New York: Springer-Verlag,1995.

**248.** Papadimitriou, C. "On Selecting a Satisfying Truth Assignment." In *Proceedings of the 32nd Annual IEEE Symposium on Foundations of Computer Science*, 1991.

**249.** Pilarski, S. and G. Hu. "SAT With Partial Clauses and Back-Leaps." In *Proceedings of the 39th Design Automation Conference*, 2002.

**250.** Pnueli, A. "The Temporal Logic of Programs." In *Proceedings of the 18th IEEE Symposium on Foundations of Computer Science*.

**251.** Prasad, M., P. Chong, and K. Keutzer. "Why is ATPG Easy?" In *Proceedings of the 33rd Design Automation Conference*, 1996.

**252.** Prosser, P. "Hybrid Algorithms for the Constraint Satisfaction Problem." *Computational Intelligence* 9 (1993): 268–299.

**253.** Ravi, K. and F. Somenzi. "High Density Reachability Analysis." In *Proceedings of the International Conference on Computer-Aided Design*, 1995.

**254.** Rowson, J. and A. Sangiovanni-Vincentelli. "Interface-Based Design." In *Proceedings of the 34th Design Automation Conference*, 1997.

**255.** Rudell, R. "Dynamic Variable Ordering for Ordered Binary Decision Diagrams." In *Proceedings of the IEEE/ACM International Conference on Computer-Aided Design*, 1993.

**256.** Sangiovanni-Vincentelli, A., P. McGeer, and A. Saldanh. "Verification of Electronic-Systems." In *Proceedings of the 33rd Design Automation Conference*, 1996.

**257.** Seger, C. and R. Bryant. "Formal Verification by Symbolic Evaluation of Partially-Ordered Trajectories." *Formal Methods in System Design* 6, no. 2 (March 1995): 147–190.

**258.** Selman, B., H. Levesque, and D. Mitchell. "A New Method for Solving Hard Satisfiability Problems." In *Proceedings of the Tenth National Conference on Artificial Intelligence*, 1992.

**259.** Sheeran, M. and G. Stalmarck. "A Tutorial on Stalmarck's Proof Procedure for Prepositional Logic." In *Proceedings of the International Conference on Formal Methods in Computer Aided-Design*. Lecture Notes in Computer Science, 1522. New York: Springer-Verlag, 1998.

**260.** Sheeran, M., S. Singh, and G. Stalmark. "Checking Safety Properties Using Induction and a SAT-Solver." In *Proceedings of the Third International Conference on Formal Methods in Computer-Aided Design,* 2000.

**261.** Sieling, D. and L. Wegener. "Graph Driven BDDs—A New Data Structure for Boolean Functions." *Theoretical Computer Science* 141 (1995): 283–310.

**262.** Sieling, D. and L. Wegener. "Reduction of BDDs in Linear Time." *Information Processing Letters* 48, no. 3 (November 1993): 139–144.

**263.** Silburt, A., A. Evans, G. Vrckovmc, M. Dufresne, and T. Brown. "Functional Verification of ASICs in Silicon Intensive Systems." Presented at the IFIP WG 10.5 Advanced Research Working Conference on Correct Hardware Design and Verification Methods (CHARME'97), Montreal, Canada, October 1997.

**264.** Smith, S., M. R. Mercer, and B. Brock. "Demand Driven Simulation: BACKSIM." In *Proceedings of the 24th Design Automation Conference,* 1987.

**265.** Somenzi, F. "Algebraic Decision Diagrams and Their Applications." In *IEEE/ACM International Conference on Computer-Aided Design,* 1993. ACM/IEEE, IEEE Computer Society Press.

**266.** Song, D., S. Berezin, and A. Perric. "Athena, a Novel Approach to Efficient Automatic Security Protocol Analysis." *Journal of Computer Security* 9, nos. 1/2 (2001): 47–74.

**267.** Spears, W. "A NN Algorithm for Boolean Satisfiability Problems." In *Proceedings of the 1996 International Conference on Neural Networks,* 1996.

**268.** Spears, W. "Simulated Annealing for Hard Satisfiability Problems." *Second DIMACS Implementation Challenge: Cliques, Coloring and Satisfiability.* D.S. Johnson and M.A. Trick, editors. DIMACS Series in Discrete Mathematics and Theoretical Computer Science. Providence, R.I.: American Mathematical Society, 1993.

**269.** Stalmarck, G. "A System for Determining Propositional Logic Theorems by Applying Values and Rules to Triplets That Are Generated From a Formula." Technical report. European Patent 0403 454 (1995). U.S. Patent 5 276 897. Swedish Patent 67 076 (1989).

**270.** IEEE Standard 1364-1995 IEEE Standard Hardware Description Language Based on the Verilog Hardware Description Language. New York: IEEE, 1996.

**271.** Stephan, P., R. Brayton, and A. Sangiovanni-Vincentelli. "Combinational Test Generation Using Satisfiability." *IEEE Transactions on Computer-Aided Design* 9 (September 1996): 1167–1176.

**272.** Stern, U. and D. Dill. "Improved Probabilistic Verification by Hash Compaction. Correct Hardware Design and Verification Methods." Presented at the IFIP WG10.5 Advanced Research Working Conference Proceedings, 1995.

273. Stern, U. and D. Dill. "A New Scheme for Memory-Efficient Probabilistic Verification." Presented at the Joint International Conference on Formal Description Techniques for Distributed Systems and Communication Protocols, and Protocol Specification, Testing, and Verification, 1996.

274. Strichman, O. "Tuning SAT Checkers for Bounded Model-Checking." In *Proceedings of the Conference on Computer-Aided Verification*, 2000.

275. Stump, A. and D. Dill. "Faster Proof Checking in the Edinburgh Logical Framework." In *Proceedings of 18th International Conference on Automated Deduction*, 2002.

276. Sutherland, S. *The Verilog PLI Handbook: A User's Guide and Comprehensive Reference on the Verilog Programming Language Interface*. Boston: Kluwer Academic Publishers, 1999.

277. International Symposium on the Theory of Switching, Part I 29, 1957.

278. Thomas, D. and P. Moorby. *The Verilog Hardware Description Language*. 4th ed. Boston: Kluwer Academic Publishers, 1998.

279. Thomas, W. *Automata on Infinite Objects*. Handbook of Theoretical Computer Science, Volume B. Amsterdam: Elsevier Science Publishers, 1990.

280. H.J. Touati, H. Savoj, B. Lin, R.S. Brayton, and A. Sangiovanni-Vincentelli. "Implicit State Enumeration of Finite State Machines Using BDDs." In *IEEE/ACM International Conference on Computer-Aided Design*, 1990.

281. Turing, A. "Checking a Large Routine." Paper presented for the EDSAC Inaugural Conference, June 1949.

282. Ulrich, E. "Concurrent Simulation at the Switch, Gate, and Register Levels." In *Proceedings of the 1985 International Test Conference*, 1985.

283. Ulrich, E. "Event Manipulation for Discrete Simulations Requiring Large Numbers of Events." *Journal of the ACM* 21, no. 9 (September 1978): 777–785.

284. Ulrich, E. and D. Herbert. "Speed and Accuracy in Digital Network Simulation Based on Structural Modeling." In *Proceedings of the 19th Design Automation Conference*, 1982.

285. Vardi, M. and P. Wolper. "Reasoning About Infinite Computations." *Information and Computation*, 1994.

286. Velev, M. and R. Bryant. "Effective Use of Boolean Satisfiability Procedures in the Formal Verification of Superscalar and VLIW Microprocessors." In *Proceedings of the Design Automation Conference*, June 2001.

287. Walukiewicz, I. "Pushdown Processes: Games and Model-Checking." *Information and Computation* 164, no. 2 (2001): 234–263.

288. Wang, L., N. Hoover, E. Porter, and J. Zasio. "SSIM: A Software Levelized Compiled-Code Simulator." In *Proceedings of Design Automation Conference*, 1987.

289. Willems, B. and P. Wolper. "Partial-Order Methods for Model Checking: From Linear Time to Branching Time." In *Symposium on Logic in Computer Science (LICS)*, 1996.

290. Wilson, C. and D. Dill. "Reliable Verification Using Symbolic Simulation with Scalar Values." In *Proceedings of the Design Automation Conference*, 2000.

291. Wolper, P. "Expressing Interesting Properties of Programs in Prepositional Temporal Logic." In *Proceedings of the Thirteenth Annual ACM Symposium on Principles of Programming Languages*, 1986.

292. Yuan, J., K. Schultz, C. Pixley, H. Miller, and A. Aziz. "Modeling Design Constraints and Biasing Using BDDs in Simulation." In *Proceedings of the International Conference on Computer-Aided Design*, 1999.

293. Zhang, H. and M. E. Stickel. "An Efficient Algorithm for Unit Propagation." In *Proceedings of the Fourth International Symposium on Artificial Intelligence and Mathematics (AI-MATH'96)*, 1996

294. Zhang, L. and S. Malik. "The Quest for Efficient Boolean Satisfiability Solvers." In *Proceedings of the 14th Conference on Computer-Aided Verification (CAV2002)*. Lecture Notes in Computer Science 2404. New York: Springer-Verlag, 2002.

295. Zhang, L. and S. Malik. "Conflict Driven Learning in a Quantified Boolean Satisfiability Solver." In *Proceedings of the IEEE/ACM International Conference on Computer-Aided Design (ICCAD)*, 2002.

296. Zhang, L. and S. Malik. "Validating SAT Solvers Using an Independent Resolution-Based Checker: Practical Implementations and Other Applications." In *Proceedings of the IEEE/ACM Design, Automation, and Test in Europe (DATE)*, 2003.

297. Zhang, L., C. Madigan, M. Moskewicz, and S. Malik. "Efficient Conflict Driven Learning in a Boolean Satisfiability Solver." In *Proceedings of the IEEE/ACM International Conference on Computer-Aided Design (ICCAD)*, 2001.

# Index

## A

A operators, 481
A quantifiers, 476
Absolute transition delays, 149–150
Abstraction
  levels of, 44–46
  for properties, 466
acc_next_cell routine, 146
acc_next_driver routine, 115
acc_next_port routine, 146
acc_set_value routine, 146
Acceptance states and strings, 359,
  362–363
Actions in assertions, 234
Active events, 22, 82
Address maps, CAM for, 190–194
Address space for pseudorandom test
  generators, 229
ADDs (algebraic decision diagrams),
  421–422
AF operator in CTL, 480
Affirmative assertions, 232–233
AG operator in CTL, 480
Alert on detection method, 121
Alert on output transition method, 121
$alert task, 186
Alerts, 185–186
  levels of, 186
  self-checking codes for, 169
Algebraic decision diagrams (ADDs),
  421–422
Algorithmic abstraction level, 44
ALU (Arithmetic Logic Unit), modules
  for, 35
always blocks
  at signs in, 47
  profiling statistics for, 123
Ancestor vertices in depth-first searches,
  375

AND gates and operators
  for Boolean functions, 343–344
  in clock gating, 41–42
  in RTL coding, 36
  for sequences, 254, 484
  update events with, 84–85
Architectural function and specifications
  for pseudorandom test generators, 228
  in test plans, 217–222
Arithmetic Logic Unit (ALU), modules
  for, 35
Arrays
  instantiating, 45, 51
  macro files for, 54
  mapping to, 188–189
  as recognizable components, 47
  tracing, 58–59, 308–311
Asserting cuts, 441
Assertions, 232
  built-in, 247–248
  components of, 233–234
  concurrent, 249–250
    multiple clocks for, 258
    sequence connectives for, 253–258
    sequence constructors for,
      250–253
    sequences for, 259
    system functions for, 258
  container, 245–247
  defining, 232–233
  immediate, 248–249
  interval constraints for, 242–243
  one-hot and one-cold signals in,
    237–238
  parity in, 236
  problems, 280–286
  sequential, 238–241
  signal ranges as, 235

# SEMICONDUCTOR CLUSTER

◆ **ENGINEERING THE COMPLEX SOC:**
**Fast, Flexible Design with Configurable Processors**
**Chris Rowen / Steve Leibson** • ©2004, 0-13-145537-0
**A methodological breakthrough gives SoC designers and engineers the techniques they need to finish larger projects in less time.**

- Engineers can keep pace with the demand for increased productivity by employing new SoC techniques that traditional processor design methods would not allow for.
- The first book to provide a unified hardware/software view of SoC design using the multi-processor system-on-chip or "sea-of processors" approach.
- Written by the CEO of Tensilica- sharing his proven and successful design methodologies.

◆ **HANDBOOK OF DIGITAL TECHNIQUES FOR HIGH-SPEED DESIGN:**
**Design Examples Signaling and Memory Technologies Fiber Optics**
**Modeling and Simulation to Ensure Signal Integrity**
**Tom Granberg** • ©2004, 0-13-142291-X

- Broadest discussion of actual high-speed devices and real-world technologies in use today for designing multi-Gigahertz boards and systems.
- Includes design and simulation examples. Discusses SerDes, WarpLink, Bus LVDS, HSTL, SSTL, GigaPro, GigaComm, CML, and fiber optic X-modules.
- Includes designing with DDR, QDR, XDR, ODR, GDDR3, SigmaRAM, FCRAM, and RLDRAM
- Presents jitter masks, eye diagrams, IBIS modeling, constraint management, design flows and EDA tools, mesh and fabric point-to-point backplane architectures, differential and mixed-mode S-parameters, TDR/TDT, RapidIO, PCI-Express, EOCB.

◆ **FPGA-BASED SYSTEM DESIGN**
**Wayne Wolf** • ©2004, 0-13-142461-0

- Learn the VLSI characteristics of FPGAs, and the "whys and hows" of FPGA-based logic design.
- Up-to-date information and comparison of different modern FPGA devices.
- Makes use of modern HDL design techniques in Verilog and VHDL; and describes platform-based FPGA systems and multi-FPGA systems.

◆ **HIGH-SPEED SIGNAL PROPAGATION: ADVANCED BLACK MAGIC**
**Howard Johnson / Martin Graham** • ©2003, 0-13-084408-X

- The long-awaited companion to *High Speed Digital Design: A Handbook of Black Magic*, Johnson and Graham's previous universally acclaimed bestseller.
- This new destined-bestseller presents state-of-the-art techniques for building digital interconnections that transmit faster, farther, and more efficiently than ever before.
- Offers a complete and unified theory of signal propagation for all baseband digital media, from pcb traces to cables to chips.

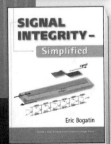

◆ **SIGNAL INTEGRITY — SIMPLIFIED**
**Eric Bogatin** • ©2004, Cloth, 0-13-066946-6

- Provides a thorough review of the fundamental principles associated with signal integrity — starting at the simplest level possible — does not hide behind mathematical derivations.
- Covers principles that can be applied to all interconnects: printed circuit boards, IC packages, connectors, wire and cable, on chip interconnects.
- Introduces the four important practical tools used to solve signal integrity problems: rules of thumb, approximations, numerical simulations, and measurements with TDR and VNA.

PRENTICE
HALL
PTR

# SEMICONDUCTOR CLUSTER

◆ **THE ESSENTIAL ELECTRONIC DESIGN AUTOMATION (EDA)**
   Mark D. Birnbaum • ©2004, 0-13-182829-0

• A highly accessible introduction to EDA business and technology, for anyone without a deep chip design background: sales/marketing/PR/legal/financial personnel, students and new entrants to the industry.
• Mark Birnbaum reviews the design problems EDA is intended to solve, each category of EDA tools, and the trends, standards, and emerging technologies that make EDA increasingly important.
• Includes easy primers on electricity, semiconductor manufacturing, computing, and common terms, plus extensive reference sources and a complete glossary.

◆ **THE ESSENTIAL GUIDE TO SEMICONDUCTORS**
   Jim Turley • ©2003, 0-13-046404-X

• Semiconductor technology, applications, and markets — explained in English.
• A complete professional's guide to the business and technology of semiconductor design and manufacturing.
• Leading semiconductor industry analyst, editor and lecturer illuminates every facet of the industry, explaining its fast-changing technologies, markets, and business models simply, clearly, and without unnecessary jargon.

◆ **VERILOG HDL: A Guide to Digital Design and Synthesis, Second Edition**
   Samir Palnitkar • ©2003, With CD-ROM, 0-13-044911-3

• Incorporates the many changes that have occurred since publication of the best selling first edition.
• Includes the latest information on Verilog — incorporating all enhancements described in IEEE 1364-2001 standard.
• Now includes the latest advances in verification techniques that are now an integral part of Verilog-based methodologies.
• Provides a working knowledge of a broad variety of Verilog-based topics for a global understanding of Verilog HDL-based design verification.

◆ **SIGNAL INTEGRITY ISSUES AND PRINTED CIRCUIT BOARD DESIGN**
   Doug Brooks • ©2004, 0-13-141884-X

• Finally, a book that covers the increasingly complex area of Signal Integrity issues in PCB design thoroughly, but without a need for great technical depth of understanding.
• Unique — only book that covers ALL signal integrity issues under one cover.
• Thorough — starts with basic engineering principles (Ohm' Law) and builds from there.
• Uses real-world relevant examples throughout that allow reader to visualize how high-end software simulators see various types of SI problems and their solutions.

◆ **FROM ASICS TO SOCS: A Practical Approach**
   Farzad Nekoogar / Faranak Nekoogar • ©2003, 0-13-033857-5

• This book deals with everyday real-world issues that ASIC/SOC designers face on the job with practical examples.
• Emphasizes principles and techniques as opposed to specific tools.
• Includes a section on FPGA to ASIC conversion.
• Modern physical design techniques are covered — providing tips and guidelines for designing front-end and back-end designs.

◆ **DESIGN VERIFICATION WITH *e***
   Samir Palnitkar • ©2003, 0-13-141309-0

• Functional verification has become the main bottleneck in the digital IC design process.
• No verification engineer can afford to ignore popular high-level verification languages (HVLs) such as e.
• Provides a working knowledge of a broad variety of e-based topics for a global understanding of e-based design verification.

# SEMICONDUCTOR CLUSTER

◆ **HIGH-SPEED DIGITAL DESIGN: A Handbook of Black Magic**
**Howard Johnson & Martin Graham • ©1993, 0-13-395724-1**

This best-selling book provides a practical approach to high-speed digital design and yet it presents the theory in a useful and concise manner to support it. It takes the black magic out of why high-speed systems work the way they do.

◆ **REAL WORLD FPGA DESIGN WITH VERILOG**
**Ken Coffman • ©2000, With CD-ROM, 0-13-099851-6**

Essential digital design strategies: recognizing the underlying analog building blocks used to create digital primitives; implementing logic with LUTs; clocking strategies, logic minimization, and more

◆ **MODERN VLSI DESIGN: System-on-a-Chip, Third Edition**
**Wayne Wolf • ©2002, 0-13-061970-1**

A "bottom-up" guide to the entire VLSI design process, focusing on state-of-the-art SoC techniques that maximize performance, reduce power usage, and minimize time to market. New coverage includes: the latest copper and advanced interconnect models, practical IP-based design guidance, Verilog and VHDL overviews.

◆ **THE ASIC HANDBOOK**
**Nigel Horspool / Peter Gorman • ©2001, 0-13-091558-0**

The practical, step-by-step guide to building ASIC's for every project leader, manager, and design engineer.

◆ **DIGITAL SIGNAL INTEGRITY:**
**Modeling and Simulation with Interconnects and Packages**
**Brian Young • ©2001, 0-13-028904-3**

An engineer's guide to modeling and simulating high-speed digital systems interconnects between components in order to achieve target system performance levels. The "next step" for engineers from Johnson's *High Speed Digital Design*.

◆ **DESIGN-FOR-TEST FOR DIGITAL IC'S AND EMBEDDED CORE SYSTEMS**
**Alfred Crouch • ©2000, 0-13-084827-1**

The first practical DFT guide from an industry insider. Skip the high-brow theories and mathematical formulas — get down to the business of digital design and testing as it's done in the real world.

◆ **VERILOG DESIGNER'S LIBRARY**
**Bob Zeidman • ©1999, 0-13-081154-8**

For Verilog users familiar with the basic structure of the language and want to develop real applications. Brings together an extensive library of Verilog routines, each designed to simplify and streamline a key task in integrated circuit design. Fully documented, well organized, and provided royalty-free on CD-ROM for your personal use, these routines offer the potential to dramatically reduce your development time – and your time to market.

◆ **A VHDL PRIMER, THIRD EDITION**
**Jayaram Bhasker • ©1999, 0-13-096575-8**

Want to leverage VHDL's remarkable power without bogging down in its notorious complexity? Get *A VHDL Primer, Third Edition*. This up-to-the-minute introduction to VHDL focuses on the features you need to get results — with extensive practical examples so you can start writing VHDL models immediately.

◆ **VHDL DESIGN REPRESENTATION AND SYNTHESIS, SECOND EDITION**
**James R. Armstrong / F. Gail Gray • ©2000, With CD-ROM, 0-13-021670-4**

Incorporates several design tools including, editors, simulators, checkers, analyzers, optimizers, and synthesizers. Also covers major VHDL topics including, major constructs, lexical description, source files, data types, data objects, and more.

PRENTICE
HALL
PTR